INDUSTRIAL ROBOTICS

GORDON M. MAIR

Department of Design, Manufacture and Engineering Management, University of Strathclyde

PRENTICE HALL

NEW YORK LONDON TORONTO SYDNEY TOKYO

First published 1988 by
Prentice Hall International (UK) Ltd,
66 Wood Lane End, Hemel Hempstead,
Hertfordshire, HP2 4RG.
A division of
Simon & Schuster International Group

Printed and bound in Great Britain at the University Press, Cambridge.

Library of Congress Cataloging-in-Publication Data

Mair, Gordon M., 1949–
Industrial robotics / Gordon M. Mair.
p. cm.
Bibliography: p.
Includes index.
ISBN 0-13-463217-6
1. Robots, Industrial. I. Title.
TS191.8.M32 1988
670.42′7—dc19 87-30554

British Library Cataloguing in Publication Data

British Library Cataloguing in Publication Data are available

1 2 3 4 5 92 91 90 89 88

ISBN 0-13-463217-6

ISBN 0-13-463209-5 PBK

For
my wife Linda
and children, Heather, Isla and Gordon

CONTENTS

PART II ROBOT DESIGN AND CONTROL

PREFACE

The purpose of this book is to provide a broad, comprehensive introduction to industrial robotics. It is intended to be suitable for three types of reader. Firstly, for those considering entering robotics, or any other discipline requiring a knowledge of advanced manufacturing technology, as a career. Secondly, for those requiring an appreciation of industrial robotics to complement other studies being undertaken in, for example, business management, or other commercial or social sciences courses. And finally, for engineers and managers in industry wishing to gain an overview and some guidance in the subject.

Since this book was first conceived there has been a proliferation of robotics literature in the form of magazines, journals, conference proceedings and books for both the serious student and the casual reader. However, it is thought that this text remains unique in its approach in that it attempts to take the reader, who is only assumed to possess a good general education, to a level of *understanding* of robotics which will be practically useful in the industrial context. Specific examples of readers could therefore be as follows: (a) an undergraduate at college or university; (b) a mature student undertaking a robotics module having had a background in another discipline. Many colleges and universities are adopting modular concepts in their course structures. This means that students from quite different backgrounds sometimes find modules offered by a particular department useful or interesting, an industrial robotics module being a good example; (c) an engineer or manager in a factory who requires a practical text to introduce himself to the subject before embarking on a robotics project.

The validity of this approach is based on the author's experience of lecturing the subject at various levels and close association with industry and further education institutions. The material introduced in the text has been used in various undergraduate and postgraduate courses, and in night school classes and day courses for representatives from industry and further education. It has also been used as a basis for introductory courses run in-house for industry. In each of the above cases it has been found that the knowledge base of people approaching the subject varies greatly in both depth and width. It is thus hoped that the text will be able to form a common platform to which a variety of readers can come to supplement their existing knowledge or build on towards other goals.

This text is not of a numerate or theoretical nature for three reasons. Primarily, once numerical examples of even moderate complexity are introduced a presupposition has to be made that a certain amount of mathematical knowledge has already been assimilated by the reader; this would contradict the essential concept of the book. Also, due to the amount of information that has been included, any mathematical examples and explanations would have made the book overly long. Lastly, there are already many books dealing with the theoretical aspects of the technology in varying degrees of intensity; examples of these may be found in Appendix 2.

The strategy adopted to achieve the aims mentioned above has been as follows. Firstly a general overview of robotics is given to put the subject into perspective. A closer examination of the industrial robot follows, looking in depth at the design of the components, how they operate, and their interrelationship. The use of robots is then discussed before once again taking a wider view to appreciate the total environment within which the robot exists.

Tactically the book is divided into three main parts, the Overview, Robot Design and Control, and Applying Robots. Any more detailed explanation of these would be redundant as the reader can grasp the content of the individual chapters by looking in the Contents.

It is important to remember that the industrial robot is only one element in a modern manufacturing system; this will be evident from the chapters towards the end of the book. However, it has emerged in recent years as a particularly suitable subject for study as it is a developing technology, incorporating the latest components and techniques in computing and mechanical, electrical, and electronic engineering amongst others. In the broader context, the study of industrial robotics gives the student an awareness of many of the aims, problems and concepts involved in today's industry. This text is intended as a vehicle by which the reader can embark on this study by acquiring an appreciation of the design, application and implications of industrial robots.

G.M.M.
1987

I

AN OVERVIEW

1

INTRODUCTION

DEFINITIONS

The definition, the forming of the concept, of an industrial robot is an essential first step in the study of industrial robots and robotics. Some formal definitions are therefore now given for consideration.

One common definition of a robot to be found in a dictionary, and popular in science fiction, is 'a mechanical man'. Now although a mechanical device in the form of a man could be made, and could also be capable of receiving and acting upon instructions, it is not a practical commercial proposition for industry at the moment. The machine would be of relatively limited intelligence and would be particularly expensive if it was to be capable of doing any useful work. A more rigid and appropriate definition is therefore required. This has been attempted by various national bodies throughout the world, and four of these are now considered.

1. *Definition adopted by the JIRA* (Japan Industrial Robot Association) (JIS B 0134-1986)

This definition is interesting as it is has the broadest scope of all the definitions in use. It includes all manipulating devices and may be considered as beginning with the definition of a 'manipulator' which it gives as follows.

> Manipulator. A machine, the mechanism of which usually consisting of a series of segments jointed or sliding relative to one another, for the purpose of grasping and moving objects usually in several degrees of freedom. It may be controlled by an operator, a programmable electronic controller, or any logic system (e.g. cam device, wired, etc.)

The definition then states that the term 'robot' will mean a manipulator for the manufacturing environment. It then further classifies robots into a number of subcategories which will be examined in Chapter 3.

This can give problems when attempting to make statistical comparisons of the number of robots in use worldwide. This is because the definition is so broad that some machines could be classified as robots in Japan but would be excluded from totals in other countries because of more stringent criteria.

Thus using this Japanese definition a remotely controlled device, guided by a human via radio signals, or signals transmitted through an umbilical cable, would be classed as a robot. So also would a piece of cam-operated automation equipment, in use on an assembly line and performing a fixed sequence of operations. Both of these would not be described as robots in this book. The first would be classed as a 'tele-operator' needing constant human supervision, and the second as a piece of 'hard automation' dedicated to performing one particular task. Therefore, for our purposes the Japanese definition is too general.

2. *Definition adopted by the AFRI* (Association Française de Robotique Industrielle)

In a manner similar to the Japanese all manipulators are included initially, i.e. manual handling devices, automatic handling devices, and robots. Essentially in this definition robots are separated from the automatic handling devices by stating that they must be servo controlled, (this term will be described later). They are then further subdivided into three categories, (1) programmable servoed robots, (2) second generation robots, and (3) third generation robots. Each of these categories signifies increasing degrees of sophistication, and again these terms will be considered fully at a more appropriate stage.

3. *Definition adopted by the RIA* (USA Robotic Industries Association)

> A robot is a reprogrammable multifunctional manipulator designed to move material, parts, tools, and specialized devices through variable programmed motions for the performance of a variety of tasks.

4. *Definition adopted by the BRA* (British Robot Association)

> An industrial robot is a reprogrammable device designed to both manipulate and transport parts, tools, or specialized manufacturing implements through variable programmed motions for the performance of specific manufacturing tasks.

The final two definitions are very similar and state quite precisely that robots are reprogrammable manipulators. Thus the robot as well as being able to operate on its own, according to previously given instructions, must also be able to have those instructions changed easily. This implies that a facility for storing instructions that can be changed at a later date must be provided. This facility can range in complexity from the basic mechanical stops, used on the most primitive non-servo-controlled robots, to advanced electronic memories used in computer controlled robots. In summary therefore an industrial robot can be considered as a *reprogrammable* device used for the *manipulation* of components, materials or tools. The methods used to program them and the ways in which manipulation is achieved vary considerably between robots, but the concept of *reprogrammable manipulation* remains constant.

THE ELEMENTS OF AN INDUSTRIAL ROBOT

To provide a clearer, more tangible, image the physical structure and composition of one type of industrial robot is now briefly considered. Later, the construction and method of operation will be examined more thoroughly.

The robot shown in Figure 1.1 is the ASEA IRB 2000 industrial robot and is but one example of the various robot designs available commercially. This type of robot can be applied to a wide range of tasks and it illustrates most of the basic elements.

The arm

This is the active manipulator that carries out the physical work, it is shown on the right-hand side of the figure. In this configuration the arm comprises a fixed base, at the top of which is a joint, or waist, connecting to the body of the robot, this allows the robot to rotate about a vertical axis. At the top of the body is another joint connecting the lower arm. This horizontal shoulder axis provides a second degree of freedom to the arm allowing it to sweep out a hemispherical shell. At the outer end of the lower arm is another horizontal joint, or elbow, attached to the forearm. This provides a further degree of freedom allowing the arm to move within the hemispherical work envelope. This robot has another three axes which constitute the 'wrist' movements of the arm. One joint has its axis lying along the center line of the forearm, the joint itself lying approximately midway along the arm. A further two axes lie at the extremity of the arm at right angles to each other. This robot, by utilizing all six axes, has the freedom to position the gripper or tool, frequently called the end effector, at any point in space and in any orientation.

To power the robot a drive system is required. The ASEA IRB 2000 robot uses electrical power with AC brushless servo motors being the prime movers. The two other commonly used drive systems are hydraulic and pneumatic, both of which are fluid power systems. In the case of electric robots the torque from the motors is transmitted to the required point through a system of gears, rods, screws, pulleys or belts. A few robots have their motors mounted coaxially with the joints themselves thus removing the necessity for linkages.

Unless the robot is a very simple type further components, which will be described in detail later, will be necessary to provide feedback information on the location, speed, and direction of movement, of the end effector at any point in time. These will take the form of encoders, resolvers, tachometers, etc. Had the robot been a fluid power machine there would also have been a need for control devices such as servo valves to be included.

All of these components for drive, transmission, control, and feedback, add mass to the arm and thus make it imperative that their weight is kept to a minimum. The arm framework that supports them must also be as light as possible and yet contain enough rigidity for accurate control.

The controls

Considering now the items on the left-hand side of the figure. There are two main elements shown; a keyboard teach pendant and the control cabinet. These items are related to the control function. The cabinet contains the computer and the means for interfacing the information therein to the robot arm and the human operator. The teach pendant is hand-held by the operator and used to instruct the robot in a step by step manner how to carry out a particular task. Also commonly used is a computer terminal with a visual display unit which allows two-way communication between the operator and the robot, it is also useful for preparing complex programs and is necessary for off-line programming.

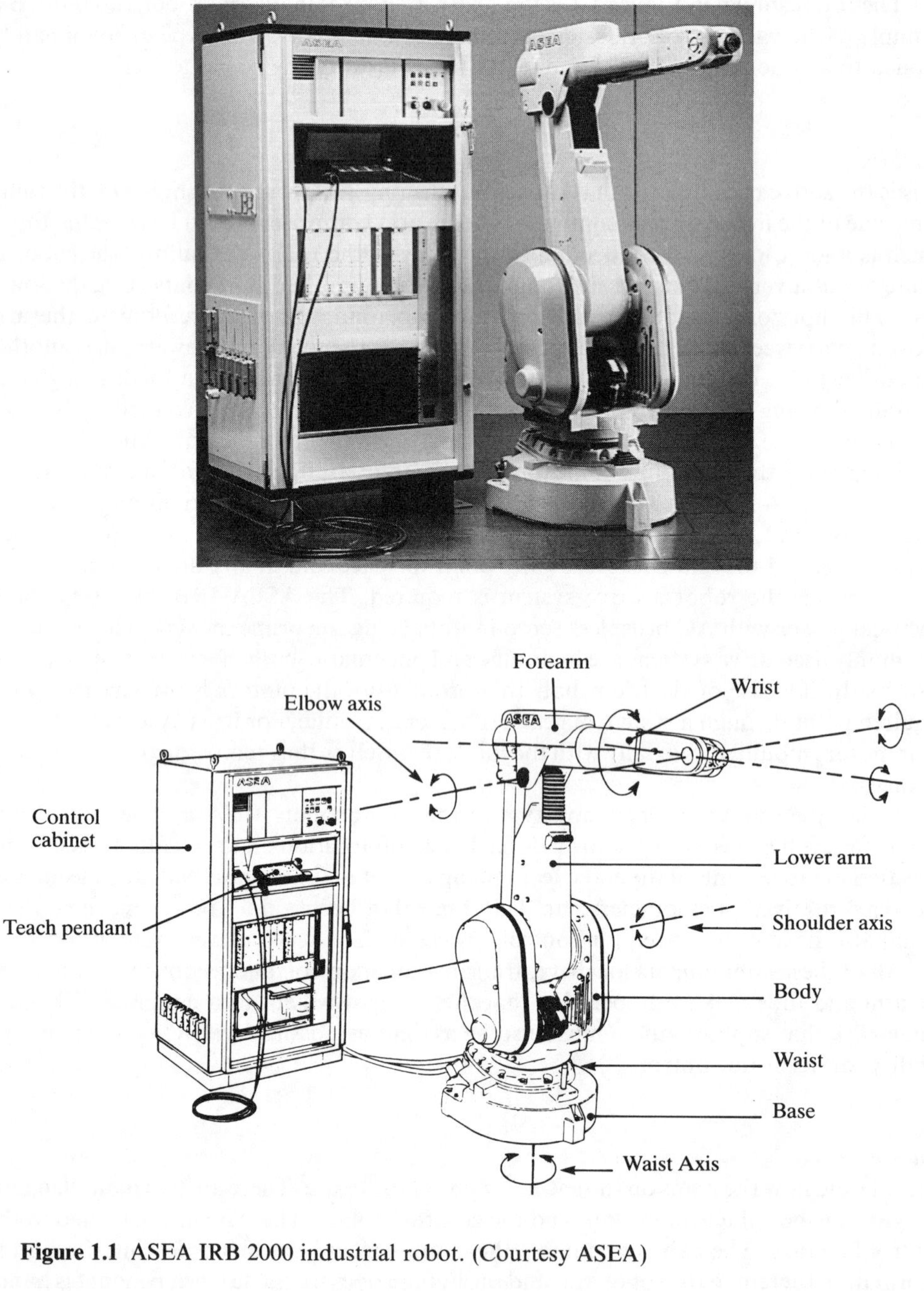

Figure 1.1 ASEA IRB 2000 industrial robot. (Courtesy ASEA)

The above two sections therefore serve to crystallize an image of an industrial robot. It is worthwhile considering now some devices that are similar in many ways to robots, but according to our concept cannot be classified as such.

DEVICES EXHIBITING SOME ROBOTIC CHARACTERISTICS

A computer numerical control (CNC) machine tool as shown in Figure 1.2 could not be called a robot as it does not manipulate and all it can ever do is cut material. Its function is to machine a component by removing metal using a cutting tool. It is, however, a reprogrammable device as it contains its own computer dedicated to controlling the

Figure 1.2 CNC machining center. (Courtesy Beaver Machine Tools Ltd)

machine functions. The machine can be taught by loading instructions into the computer via a control panel. The machine thus programmed will perform its task repeatedly, until a new job arrives and a new program is required. Each set of instructions, or program, can be retained on a storage medium such as paper tape and reloaded into the machine whenever necessary. Shown in the figure is a CNC machining center capable of performing a variety of cutting operations. This is only one example of the many types of CNC machine available, such as lathes and milling machines. There are also other reprogrammable machines not used for metal cutting, that embody the same principles, e.g. programmable knitting machines and drafting equipment.

The other two devices shown (Figures 1.3 and 1.4) are both 'tele-operators', i.e. devices which act as an extension of a human being and respond only to continously fed instructions from the operator. Figure 1.3 shows the 'Spider' diving suit, designed to allow a diver to work at low depths without decompression problems. The end effectors at the extremities of the suit arms are powered by hydraulics, these are operated directly by the diver and no automatic action occurs. This is an exoskeletal structure in that the operator

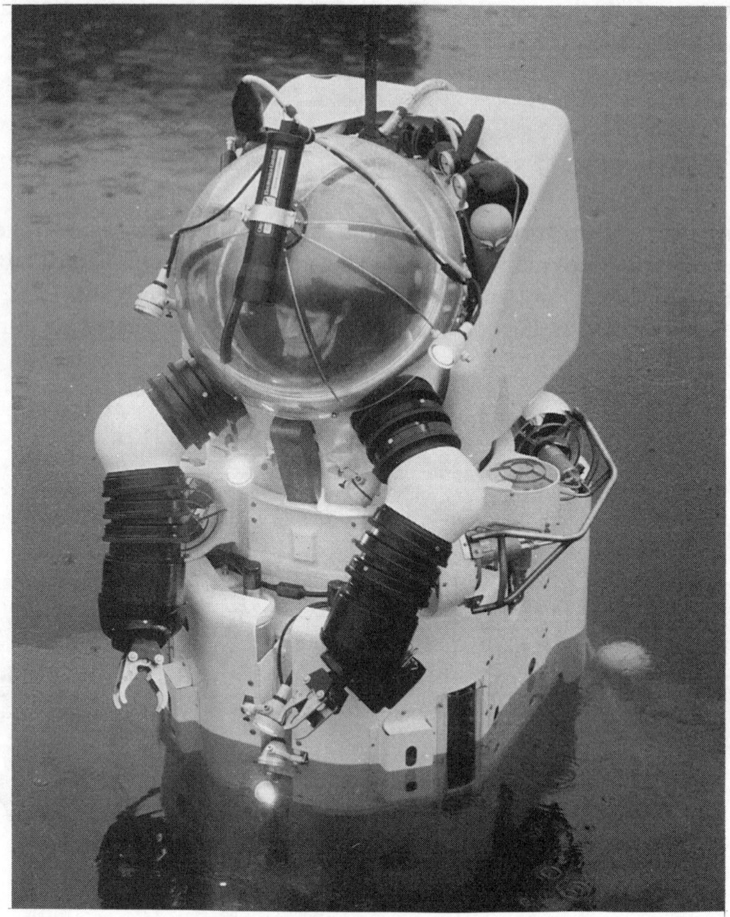

Figure 1.3 'Spider' diving suit (exoskeleton). (Courtesy Slingsby Engineering Ltd)

is completely surrounded by his device and it responds to signals transmitted by the limbs of the operator. The remote manipulator arm shown in Figure 1.4 is used by the Space Shuttle to deposit and retrieve items from orbit. This is not a true robot device as it is operated from within the Shuttle cabin by the astronauts. Prosthetic devices, such as artificial limbs, that respond to stimuli from the muscles or voice of the wearer are also nonrobotic devices but can be very similar in appearance and construction to robotic arms.

Tasks in, for example, underwater exploration, that at the moment require teleoperators or remotely controlled vehicles, will in the future increasingly utilize robots. This will be for various reasons ranging from hazardous working conditions for humans, to the problems experienced with signal propagation delay, signal corruption due to noise, and the restrictions caused by being attached to an umbilical cable.

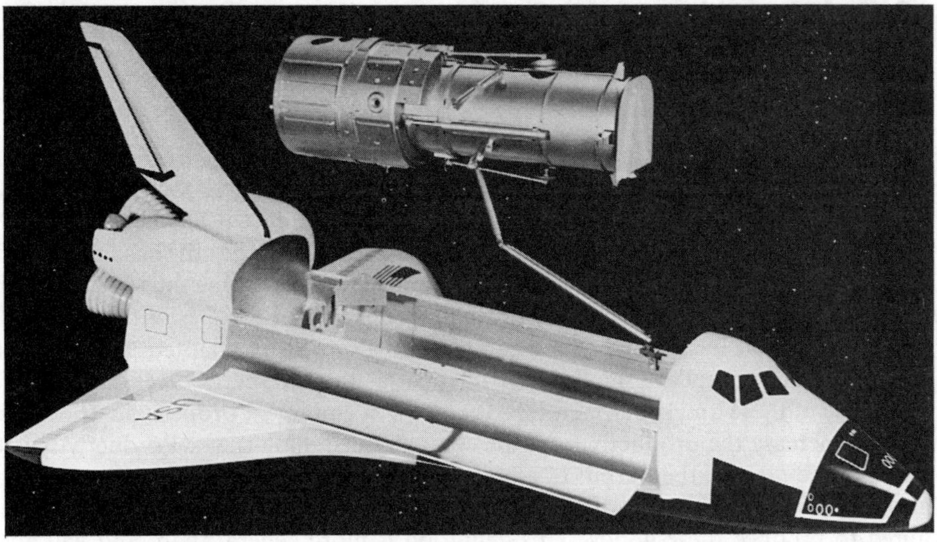

Figure 1.4 Space shuttle remote manipulator (telechiric device). (Courtesy NASA)

WHY INDUSTRIAL ROBOTS ARE USED

On consideration of robots as machines, in comparison to human labor, some features are immediately apparent.

- They do not become fatigued or distracted, this means that they are capable of producing a job of consistent quality, at a steady rate, with a greatly reduced probability of incurring scrap or the need for rework.
- Continuous working, almost twenty-four hours per day, seven days per week, is possible. The only stops necessary being for maintenance or repair.
- They can work in areas either unsafe or unpleasant for humans, thus reducing health and safety risks and boredom problems. This in turn reduces absenteeism and labour turnover. It also reduces the cost of meeting health and safety regulations.
- Industrial robots do not demand wage increases, fringe benefits, pensions, holidays, sick pay, or other burden costs.
- Material utilization is improved, e.g. less waste in spray painting.
- Consistency of quality and regularity of cycle times gives improved control of the manufacturing process. This can result in a number of improvements in efficiency such as reduced work in progress and faster throughput times.
- The capital cost of a robot is essentially in the form of a once-only payment, whereas labor costs are always present and increase every year.
- Robot costs, due to advances in electronics and computing, are increasing at a slower rate than labor costs. Therefore as each year passes robots become a more attractive investment in relation to human labor.

- In comparison to humans they can be much stronger and are therefore able to lift heavier weights and exert larger forces.
- Some robots can be very precise in their movements.

As well as advantages over humans, robots also offer advantages over the specialized, dedicated equipment designed for the mass production systems described later. This type of equipment normally becomes obsolescent when the product it has been designed to produce reaches the end of its original application. This means that the cost of the design and manufacture of the equipment has to be recouped in a short time thus reducing profit margins. Industrial robots, however, can be reprogrammed for a new task after their original application has ceased. They can also be more easily adapted than dedicated equipment to accommodate small design changes as the product evolves.

Because of these and other factors, robots, when employed properly in appropriate tasks, will increase the productivity of the company installing them. Productivity is the measure of the ratio of the output of a business, in terms of value of goods provided, in relation to the input, in terms of the overall cost of manpower, machines and materials required to produce these goods. This definition is more general than, and will give a different ratio to, that of labor productivity. Labor productivity only relates output or units manufactured, to the amount of labor hours required to achieve that output. It is obvious that employing robots instead of people must improve labor productivity, however it is improvements in general productivity that must be aimed at.

A number of surveys on the reasons for using industrial robots have been carried out in recent years, notably in the USA, Germany and Japan. The results have been roughly summarized as follows in descending order of popularity of reply:

- To reduce labor costs.
- To improve working conditions.
- To increase flexibility for accommodating design changes, etc. This is becoming more popular as product life cycles reduce in length, especially in the electronics industry.
- To improve product quality.
- To increase output rate.
- To reduce material wastage.
- To ease compliance with health and safety regulations.
- To remedy high labor turnover problems.
- To improve production control and work flow.

There may also be other reasons for the installation of a robot. The firm may urgently require to increase its competitiveness and the introduction of robotization could be a way to leapfrog the competition. It is also good for the company image to be seen to be embracing the most up-to-date, and presumably efficient, technology. The company may also simply want to gain experience in this field, and explore the capabilities of a robot in relation to their own particular requirements.

ROBOTS AND REPROGRAMMABLE AUTOMATION IN MANUFACTURING

To understand fully where robotics fits into the overall manufacturing picture it is helpful to examine the various groupings that manufacturing is normally divided into. The classification adopted here is the most common, i.e. Process, Mass, Batch, and Jobbing production.

In process production the product is commonly a powder, liquid or gas. The process will be continuous, or as continuous as demand allows, although there may be occasional pauses as a change is made to the type of material being fed through the system. The investment in capital equipment will be very high as the process will be almost totally automated using special purpose equipment dedicated to the one product. Conversely, labor costs will be low and will comprise mostly technical, administrative and managerial personnel. Typical process plant would be an oil refinery, or a continuous casting mill for producing steel slabs.

Mass production involves the same principles as Process production except that in this case discrete items are the finished product. These may be either industrial components such as conrods or transistors, or complex assemblies of components such as automobiles or television sets. Characteristics of this type of production are high capital investment in dedicated special purpose machinery, specialization of the workforce, and a deskilling of operations to allow easy interchangeability of workers between jobs. The variety of product is usually reduced to an absolute minimum, and there is a high level of organization of plant and personnel to ensure a smooth flow of work through the factory. All of these characteristics are an attempt to ensure the lowest possible cost per unit produced.

Batch production occurs when specific quantities of a product are required on a noncontinuous basis. Although batches may not always be repetitive, the requirement often recurs on a regular basis, e.g. an order for five hundred items appearing every three months. Because of this type of demand the equipment used will be much more flexible than in the previous case. Also the shopfloor workers will tend to be more skilled and hence more highly paid than the workers in the Process plant. In a large factory engaged in this type of production, many batches of different components will be moving through the plant following various routes, often requiring time on common machines, and creating a complex control problem. For these reasons, and because batch production contributes so much to the national economy, much attention has been devoted to the application of computer technology to machine and material control systems. This has given rise to numerical control machining, robotics, computer aided design and manufacture, flexible manufacturing systems, etc. (these terms will be explained more fully in Chapter 17). This has been done in an attempt to attach the benefits of mass production to batch production, thus reducing the unit costs of goods produced. Typically components may be produced for the aerospace and other light or medium engineering industries.

Finally jobbing production occurs most often when a single item is made specifically for a customer. This requires much flexibility of equipment, skilled operators and craftsmen, and usually results in a very high cost per unit produced. Typical products would be ships, bridges and oil rigs.

The relative positions of these four classes of production are shown in Figure 1.5. The three axes shown represent quantity of items produced, variety of components or products, and relative cost per unit produced. Figure 1.6 shows the possible area of application and effect of reprogrammable automation systems, including robots, within the manufacturing realm. As can be seen they can be applied throughout mass, batch, and jobbing production with a consequent reduction in unit cost in each case. This reduction in cost is achieved through higher efficiency by, for example, improving processes, reducing the amount of work in progress, and by reducing the labor cost per unit produced.

CONCLUSION

Industrial robots are now to be found in a wide variety of applications such as spot welding, arc welding, material handling, and assembly. An analysis of robot population distribution, application statistics, and growth rates will be found in Appendix 1, the applications themselves are covered more fully in Chapter 13. Potential users of robots are many and already applications have been found outside the manufacturing industry for nonindustrial robots, for example in outer and inner space with planetary exploration craft and undersea submersibles. The main problem with many potential nonindustrial applications, such as domestic chores, is the unstructured environment in which they take place. This necessitates increased sensing and computational abilities, and these are still in their development stages.

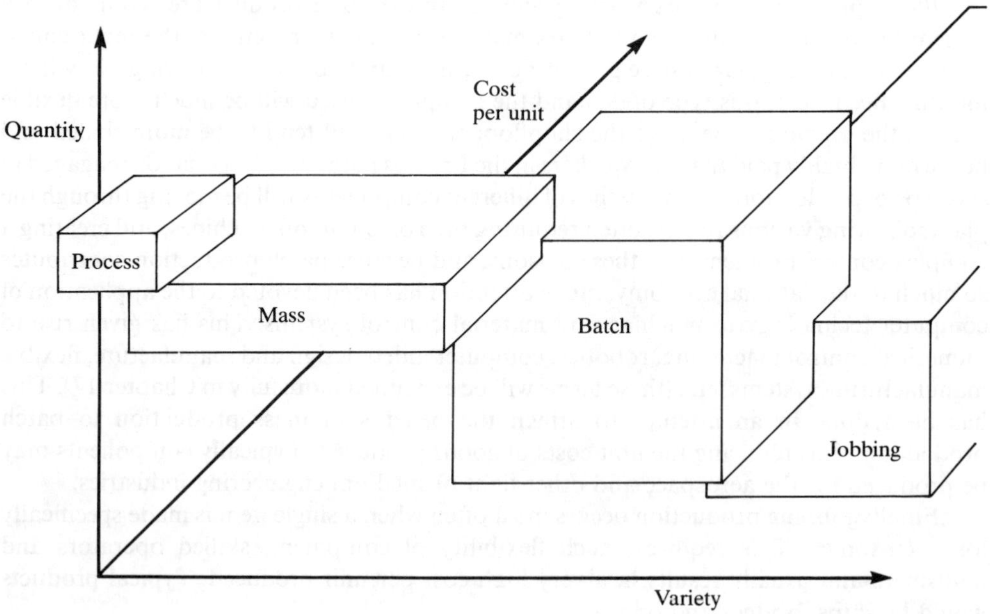

Figure 1.5 Four classes of production

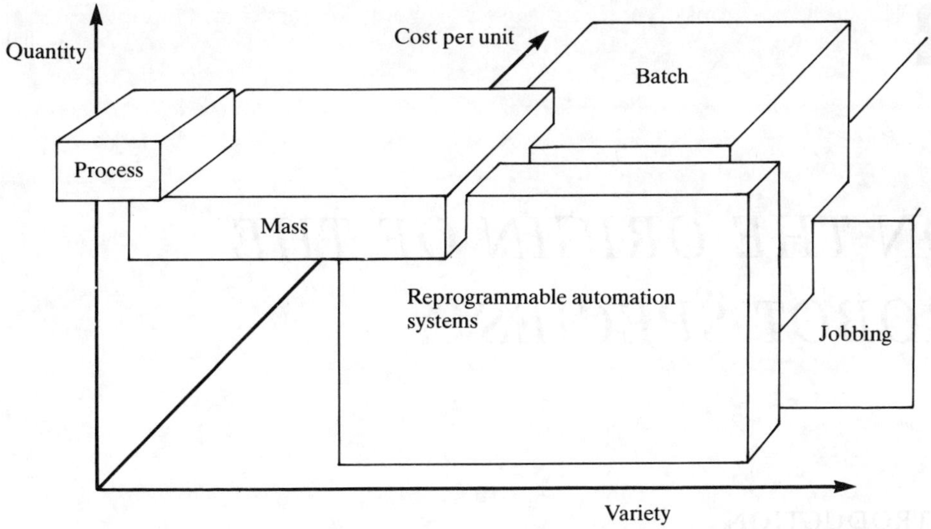

Figure 1.6 The area of application of robotics and other reprogrammable automation systems

It is apparent that industrial robots are having a steadily increasing effect on society. Human workers have been, and will continue to be, displaced by robots and other automation systems. This may mean retraining to work with a robot, movement to another job within the firm, or at worst redundancy. This latter possibility is naturally the one that generates the greatest apprehension. Flexible automation will reduce the number of people employed in manufacturing in the long term and the redistribution of labor required will have to be carefully handled. However, although the social implications must always be remembered, and if possible controlled, the following chapters will show that industrial robotics is an enthralling subject with many challenges for the broadest spectrum of people involved in industry.

2

ON THE ORIGIN OF THE ROBOT SPECIES

INTRODUCTION

The practicality of producing self-replicating robots for extraterrestrial applications is already under consideration. Imagine one of these machines engaged in mining and self-reproduction on some far away asteroid a century hence. Imagine also that contact with humanity has long been lost. The robot may turn its artificially intelligent mind to the heavy task of contemplating where did it all start, how did robots come to be, did they appear spontaneously, or did they evolve in some way? Who created them and with what ultimate purpose? Fortunately these questions applied to robots are rather easier to answer than if applied to man, they also of course contain less awesome implications. Should the robot concerned have this chapter loaded in memory it will hopefully have its curiosity satisfied to some extent.

In the previous chapter the essential elements that comprise an industrial robot were shown. In this chapter the developments that have contributed to the emergence of the robot will be examined thus enabling the current state of evolution to be viewed in true perspective technically and socially. As it exists today the industrial robot is a coalescence of a great deal of man's scientific progress and technical invention, and epitomizes his technical and social aspirations. Some day factories will provide the material wants of western society's consumerism with little input of human labor, indeed manufacture and assembly will be carried out completely automatically. The concept of a world or country where everyone is provided for without the need of drudgery by the citizens is an ancient dream indeed.

We begin at a point just over two centuries ago in the last half of the eighteenth century. A new freedom was being experienced in thought, the Reformation had taken place and the Bible was freely available in English to all who could read. However, people were experiencing a new bondage to work. In the early textile mills women and children worked fourteen hours per day in extremely poor environmental conditions, although most workers were still employed in agriculture. In considering technical achievements relevant to robotics up to this point we may note the following.

In the realm of computation the abacus had been around for some thousands of years;

in 1614 John Napier had invented logarithms. William Oughtred utilized this principle to devise the slide rule in 1621, and in 1642 Pascal constructed his Arithmetic Machine – the first digital calculator. In the field of power Otto Von Guericke demonstrated an electric 'machine' in 1647 and an air pump in 1684, and Newcomen invented an engine using steam in 1712. In the area of control the names of Bouchon and Falcon are associated with the application of punched cards to control patterns produced by looms around 1720.

Using the above inventions as a guide, and considering the construction of a robot, we can group technical discoveries into two basic areas: (1) those relevant to the production of power for the robot, and (2) those relevant to the control of that power through the communication and analysis of all information related to the task being performed.

THE AGE OF REVOLUTIONS

The last quarter of the eighteenth century saw the American Revolution and Declaration of Independence in 1776, the French Revolution and the Declaration of the Rights of Man in 1789, and what has been called the English Revolution, i.e. the Industrial Revolution which really began in Britain around 1760. It started at this time with improvements in steelmaking but was really boosted with the invention of the first rotating steam engine by James Watt and Matthew Boulton in 1783. Edmund Cartwright built a loom capable of being driven by steam power in 1784 thus heralding the introduction of true mechanization of manufacturing. Efficient control of this new steam power came in 1788 with James Watt's governor used on the Boulton/Watt steam engine. This represents the first commercially successful and practical feedback control device used on a machine and is explained in Figure 2.1.

We have therefore in the space of fifteen years, from 1775 to 1790 the beginning of man's emancipation from soul destroying labour, and also his freedom to choose his own government and how he would live his own life. The Industrial Revolution, however, caused profound changes in society. As industry grew in specific localities so the labor to supply it came from the countryside around, thus fewer people were employed in agriculture while the industrial population grew and gathered in larger and larger cities. It is with this revolution and its effects on society that analogies are drawn when the implications of the introduction of today's microprocessors are considered.

A popular theme in fiction is the enslavement of man by his own mechanical inventions. This has, in an indirect way, always been true in that man becomes dependent upon his tools and machines and in this way becomes bonded to them. However, in 1794 a mechanical invention directly caused the prolongation of real human bondage. This was Eli Whitney's Cotton Gin which separated the cotton fibers from the seed, or in other words 'cleaned' the cotton. The Negro slaves in the southern states of America were despondent and had little spirit for work, they were hardly worth their keep in the cotton fields as they only cleaned a few pounds of cotton per day. Public opinion was against slavery and it may have been that the time was ripe for freedom. Unfortunately for the Negroes the cotton gin suddenly meant that one slave could clean one thousand pounds of cotton per day, thus greatly increasing the slave's productivity and hence his value. The system of slavery was therefore set for the then foreseeable future.

By 1796 we have the first practical application of a power system widely used in

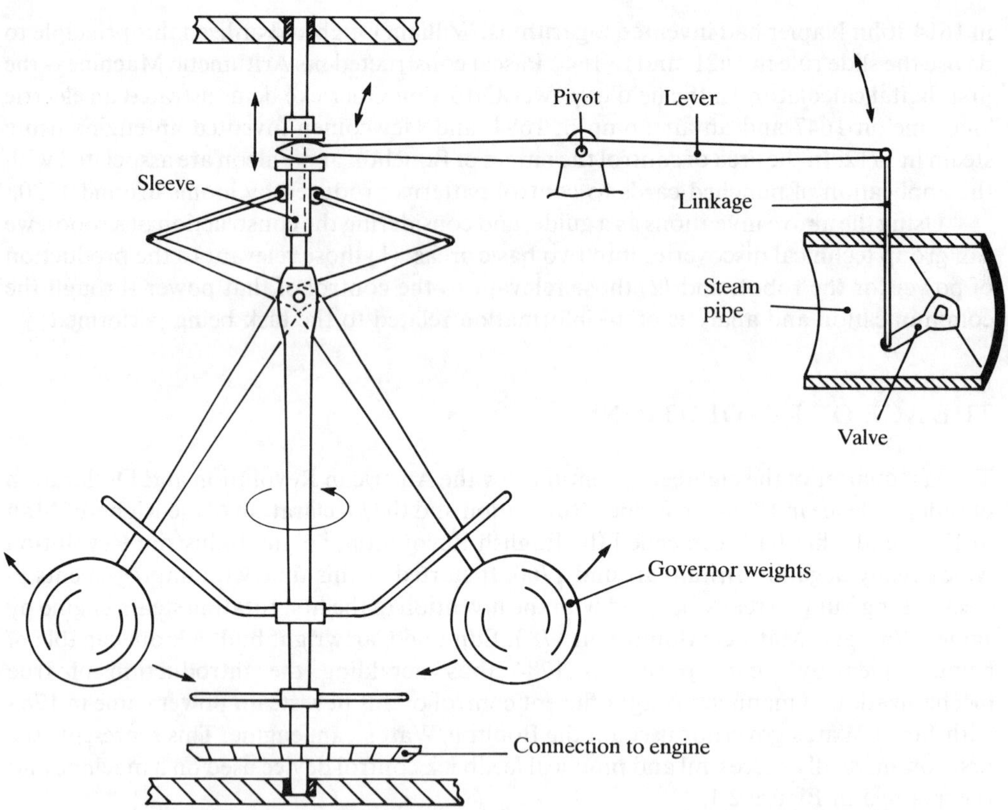

To maintain the speed of a steam engine constant, a means to adjust the flow of steam to the engine in response to a variation in load on the engine or in steam pressure was necessary. This was achieved by connecting the engine spindle to the flywheel of the governor via gears and pulleys. This caused the governor assembly to rotate at a speed directly proportional to the speed of the engine. If the load on the engine increased, the governor weights would drop downwards and inwards due to the decrease in speed and centrifugal force. This would cause the sleeve to move upwards and push the linkage at the right downwards, hence increasing the valve aperture and allowing more steam to flow, thus increasing the engine speed. As the engine speed increased the weights would fly upwards and outwards thus pulling the sleeve downwards again and decreasing the aperture of the valve. Eventually the system would reach a balanced position corresponding to the desired speed, preset by the operator's adjustment of the linkages.

Figure 2.1 James Watt's centrifugal governor (1788)

robotics today, hydraulics. This was the hydraulic press invented by Joseph Bramah. As with Watt's governor this should be considered in some detail as the concept is relevant to later discussions. In 1647 Pascal had deduced from his experiments that, 'in a fluid at rest pressure is exerted equally in all directions'. He constructed a machine consisting of a large piston and cylinder connected to a small piston and cylinder, the intervening space being filled with fluid. From this it was observed that when pressure was applied to the fluid by the piston in the small cylinder this pressure would act uniformly on the underside of the large piston. Therefore assuming the area of the smaller piston was one unit of area and the area of the larger piston one hundred units of area, then if one unit of force is applied to the small piston a resulting one hundred units of force will act on the large piston. But it was not until Bramah's invention that an effective practical machine was constructed. This is shown diagrammatically in Figure 2.2.

The ideas of Bouchon and Falcon to control a loom by punched cards were perfected in 1804 by Joseph Marie Jacquard. These cards were punched with a pattern of holes and arranged to pass over a drum in sequence in a continuous fashion. By synchronizing the feeding of the cards, the movement of the shuttle and the loom rods which were pushed against the cards and their holes, a pattern could be produced in accordance with the instructions contained in the punched cards. By using large numbers of cards it was possible to produce a highly complex pattern. For example, Jacquard had a portrait of himself produced that required 24,000 cards which was said to produce a weave so intricate that most observers took the finished pattern to be an engraving. This principle of storing information by means of punched holes on cards remained in use for many applications until very recently. It also provided another stepping stone along the path of control techniques that led to the robotic control systems of today.

One necessity for the efficient manufacture and application of any modern machine including robots is the standardization and interchangeability of components. The first person to initiate this system was Eli Whitney in 1805 in the manufacture of rifles for the American government. The influence of Whitney's inventions on the history of the United States was remarkable, the industrialization of the northern states was accelerated, and the dependence on the southern states on the agriculture of cotton was increased due to his

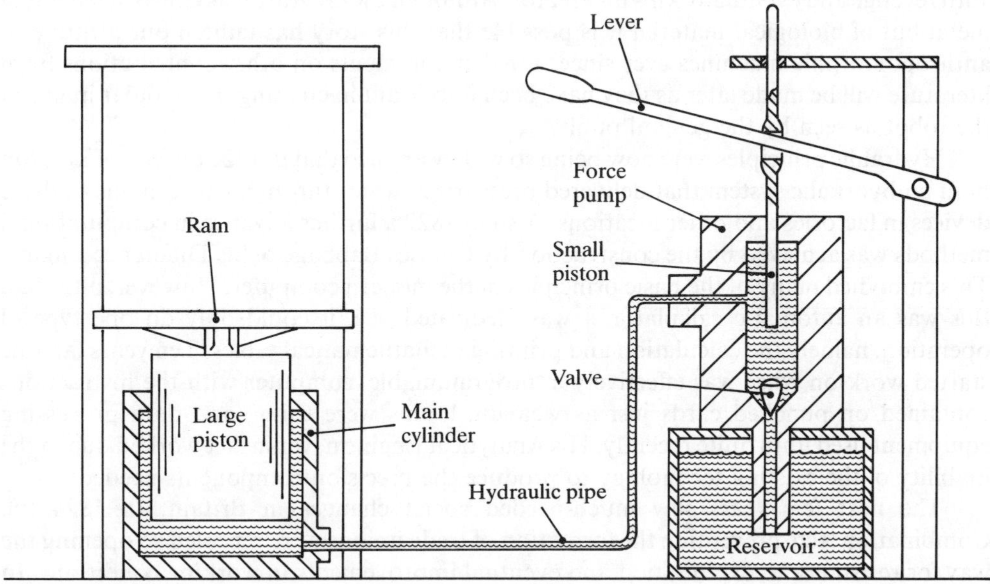

The press consisted of a force pump which, on the upward stroke of a lever, drew water from a reservoir through a one-way valve. On the downward stroke of the lever the water was forced through the pipe and into the main cylinder. The force applied to the lever, the lever length, and the proportion of the two pistons determined the pressure exerted on the ram.

Figure 2.2 Bramah's hydraulic press (1796)

cotton gin. This difference in interests was eventually to be a contributing factor in the occurrence of the Civil War. The increasing volume of cotton was also fuelling the textile mills of Britain adding a further incentive to the industrial revolution already underway.

Just as industrial robots displace workers in modern industry, so mechanization in the textile mills at the beginning of the nineteenth century displaced many textile workers. Unfortunately this does not seem to have been carried out in quite as diplomatic a manner as would be undertaken today. Also the implications of being out of work would be much more unpleasant than they would be in our present society. Toward the end of 1811 in the area of Nottingham in England the Luddite movement was formed, their purpose being to riot against the new machinery. They destroyed equipment but abstained from violence against persons. A band of Luddites was shot down in 1812 and continuing repressive measures coupled with an increased general prosperity caused the end of the movement shortly afterwards.

Mary Shelley started writing *Frankenstein* at the age of eighteen and completed it before she was twenty. The book was published in 1818 and provided all subsequent science fiction writers with the original 'man destroyed by his own creation' theme that was to be elaborated upon so often in the future. The monster has the 'vital spark' of life imparted to him then his creator, Frankenstein, flees from him in horror. Later, when almost putting the finishing touches to a female companion for the monster, he considers the 'race of devils' that might result from the two creatures and destroys this second creation. The original monster, in the beginning innocent and benevolent, becomes filled with revenge and eventually kills his creator. Although the creature was constructed not of metal but of biological material it is possible that this story has tainted our attitude to anthropomorphic machines ever since. Further comments on other contributions from literature will be made later as they have been important in creating the popular image of the robot as seen by the general public.

Hydraulic principles were now being so widely applied that in 1822 the city of London built an hydraulic system that delivered pressurized water through street mains to drive devices in factories and other locations. Also in 1822 a further advance in computational methods was achieved by the construction by Charles Babbage of his Difference Engine. This embodied many of the basic principles of the modern computer. However, although this was an automatic calculator it was 'dedicated', i.e. it could only do one type of operation, namely the calculation and printing of mathematical tables. Ten years later he started work on what was effectively a 'programmable' computer with the instructions contained on punched cards just as weaving looms were using, and data processing equipment used until quite recently. His Analytical Engine never in fact worked due to the inability of the existing technology to produce the precision components needed.

The next few years saw much-needed social changes in Britain. In 1824 the Combination Acts preventing the formation of trade unions were repealed so opening the way for resistance to exploitation and eventual improvements in working conditions. In 1833 slavery was abolished throughout the British Empire. Between these two dates the stage, or compound, air compressor was patented in 1829. Compressed air had been used ever since bellows had been utilized for stimulating fires, but this invention saw the beginning of the use of compressed air for many industrial applications and eventually as a drive system for industrial robots.

Electricity, today the most popular drive for industrial robots, has not yet been mentioned. This is because as yet no real practical application had been found for it. The work of Galvani, Volta, Faraday, and Henry was essential, but the first use of electricity to drive a motor was in 1834 by Moritz Jacobi. Eventually in 1839 he built a battery powered, passenger carrying, electric boat.

Modern automation equipment such as the controllers for machine tools and robots, require their activities to be coordinated and sequenced in a logical manner. As digital 'ON' 'OFF' techniques are employed a system is required to organize the circuits involved. In 1854 George Boole published *An Investigation into the Laws of Thought* which was a mathematical approach to logical deduction. This was based on the assumption that a statement can be in one of two states, 'true' or 'false', (1 or 0, ON or OFF). His work has since been developed to provide a basis for much of the logic systems in use in automation and robotics today.

Advances in machine tools had been making steady improvements in the quality and accuracy of components produced, but it was not until after the American Civil War that the first evidence of true automation in manufacturing is seen. This was witnessed by the development in 1873 by Christopher Spencer of a fully automatic lathe. The instructions required by the lathe were contained in the form of adjustable cams mounted on two large drums mounted beneath the lathe bed. These drums, termed 'brain wheels', were in fact the memory of the lathe just as the memory of a modern robot is contained on, for example, a floppy disk. One of these drums controlled the operation of the part of the lathe that held the material to be cut, and the other drum governed the sequence of operations carried out by the machine's cutting tools. A cam is a device usually mounted on a spindle or drum that rotates. It is designed so that it imparts movement to a rod, or rod and roller arrangement, in a manner suitable to the application. Some examples of cams are shown in a later chapter. Cam mechanisms were used in early automata such as novelty toys and are still in use today in many types of automated machinery.

The nine hour day arrived for engineering workers in Britain in 1872, working conditions having improved considerably due to the widespread formation of the trade unions. By 1880 electric motors were beginning to be used to drive machines and machine tools. Other forms of power were still popular as is evidenced by the installation of a city-wide pneumatic system during the 1880s in Paris. This was the period when electronics was being born as Edison observed electronic conduction in a vacuum in 1883. The next refinement in the automatic processing and computation of information occurred in 1890 and was due to Herman Hollerith. He was a statistician on the staff of the US Bureau of Census and it was evident to him that some sort of mechanical means of handling the vast amounts of data being gathered was necessary as manual methods were far too slow. His Pantograph Punch and Electrical Tabulator enabled the operator to record the data for each person in the census by punching holes in a card. The information was then processed electrically and tabulated automatically.

THE CENTURY OF THE ROBOT

As the century of the robot began the basic drive systems were in existence albeit in primitive form, the control techniques were beginning to be established, and the social aspirations and demands for greater productivity were almost as they are now. Most of the developments relevant to robotics that remained to be made were in the realm of electronics and control of machines. Working conditions were steadily improving and the management and organization of labor were becoming more scientific. These improvements in work efficiency were notably due to the work of F.W. Taylor and Frank and Lillian Gilbreth in the field of Work Study.

The pace was set in 1904 with the invention of the thermionic valve by J.A. Fleming and in 1906 with the invention of the triode by De Forest which made it possible to amplify signals electronically. Also in 1906 it is notable that an oil hydraulic system was used to raise and lower the guns of the USS Virginia, an indication of the progress being made in fluid power. Mass production techniques were introduced in 1912 by Ford in the manufacture of his 'Model T'. This principle was to prove a stimulant towards the creation of more sophisticated automation devices for manufacture and assembly of goods. The first of the century's two world wars lasted from 1914 to 1918 inevitably stimulating further technical progress in many areas.

Karel Capek, a Czechoslovakian playwright introduced the word 'robot' to the world in his play *Rossum's Universal Robots* in 1922. The word itself roughly means 'worker'. The story is set 'somewhere between 1950 and 1960', and considering future developments Capek's dates have shown themselves to be significantly appropriate. In the play placards that hang from the general manager's factory office could well be hung in a present-day robot manufacturer's office. Some of them read, 'Cheap Labour, Rossum's Robots', 'Robots For The Tropics, 150 Dollars Each', 'Do You Want To Cheapen Your Output? Order Rossum's Robots'. Many other statements made in the play are unexpectedly accurate, for example, 'One robot can replace two and a half workmen'. The play had strong parallels with Frankenstein, however, in that the robots were not mechanical constructions. They were first created by Old Rossum when he 'attempted by chemical synthesis to imitate the living matter known as protoplasm, until he suddenly discovered a substance which behaved exactly like living matter although its chemical composition was different . . .'. The end result was a robot whose running expense was 'three and fourpence per hour' for 'food and all costs'. In Capek's story, as in Shelley's, the creature eventually eliminates its creator, only in this case the whole of mankind is eradicated.

In 1937 Howard Aiken thought of utilizing the principles used in the previously mentioned card controlled machines to produce a fully automatic calculating machine. He then contacted the IBM corporation and they started work on the 'automatic sequence controlled calculator'. However, World War Two started in 1939 and the developed world was thrown into terrible conflict. Both sides in the struggle made leaps of technological progress as each fought for supremacy. The pressures for higher production in industry and for better control of weaponry and radar systems saw progress in areas that were later to have many commercial industrial applications. One example particularly relevant to robotics was the combining of hydraulic systems with increasingly sophisticated control techniques for servo mechanisms used in various applications. The manpower that trained in these areas and was involved in the search and development projects provided a pool of talent for commercial exploitation later when secrecy was lifted.

Meanwhile the young Isaac Asimov was writing stories in the USA concerning advanced intelligent robots. These stories proved to be highly popular and considered the robot from a scientific viewpoint rather than simply a vehicle for a melodrama. He introduced the now familiar term 'robotics' to convey the study of robots and he also anticipated the age of solid state microelectronics by giving his robots a 'positronic' brain. In 1942 in the story *Runaround* he proposed the 'Three Laws of Robotics' which are quite appropriate to present-day robots, although they have not yet achieved the levels of intelligence of Asimov's machines. These laws were: (1) a robot may not injure a human being, or, through inaction, allow a human being to come to harm; (2) a robot must obey the orders given it by human beings except where such orders would conflict with the First Law; (3) a robot must protect its own existence as long as such protection does not conflict with the First or Second Law. These laws would be programmed into every robot to ensure absolute safety, and although we know that if something can be programmed it can usually also be reprogrammed, they do serve as a practical reminder of the desirable characteristics modern robots should exhibit.

During the war Bletchley Park in Britain was the scene of intense efforts at code breaking. In 1943 Mr T.H. Flowers, who was responsible for research into switching problems for the Post Office, was asked to assist in this task. Using as a basis work already done on electronic and relay based machines he constructed, at the Post Office establishment at Dollis Hill, a new, all-electronic code-breaking device. This used 1500 vacuum tubes and was operational by December 1943, probably the world's first all electronic computer. Because of its relative power it was nicknamed Colossus. This was installed at Bletchley Park and so also was a second version in time for the Allied invasion of Europe in June 1944. Further machines were built which could effectively handle 25 000 characters per second.

By 1944 the previously mentioned IBM sequence controlled calculator was completed. By present standards it was extremely ponderous but it was another rung upwards on the latter towards achieving a computer suitable for controlling a robot. The machine had over three-quarters of a million parts and over five hundred miles of wire, to add two numbers it took 0.3 seconds and to divide numbers took around ten seconds per operation.

ENIAC, probably more similar to today's concept of a computer than Colossus, was completed in the USA in 1946. ENIAC was an acronym for Electronic Numerical Integrator And Calculator. As might be expected at this time it had been designed for military purposes and to calculate such things as shell trajectories. It could add numbers in one five-thousandth of a second, but required 15 000 square feet of floor space, weighed thirty tons, and contained 18 000 vacuum tubes. In Britain by 1950 possibly the world's first commercial computer was produced by Ferranti, the Mark 1 Star. This only required about 130 square feet of floor space and contained 4000 valves; however it required 27 kW of power and air conditioning, still a long way from the small fast minicomputer required to control today's robots.

The vacuum tube or valve type computers, as well as being large and slow were also unreliable. It was really the invention of the transistor by Shockley, Bardeen and Brittain in 1947 that acted as the catalyst for further progress. This electronic component was many times smaller than a valve, had no moving parts, was reliable, and could be used as a switch or a solid state amplifier. Thus computers and electronic controls for machinery could now be reduced massively in size and their reliability increased. In the same year the General Electric Company in the USA demonstrated how it was possible to control a machine tool using magnetic tape. This was termed the 'record playback control' system as the machine movements to make the initial workpiece were recorded on the magnetic tape and then played back when each subsequent duplicate workpiece was automatically machined. This system was a precursor of modern robotic control systems.

Also in the USA about this time J.T. Parsons devised a method to control a milling machine using the now familiar punched cards. These cards would supply the required information to a controller which would then drive the cutter along the stipulated path. In collaboration with the US Air Force and the Massachussetts Institute of Technology the method was used to produce a prototype three axis control on a milling machine in 1952. This same year saw the GEC technique developed further at MIT and introduced commercially by the Giddings & Lewis Machine Tool Co., it was called the Numericord system and was the first commercial NC (numerical control) system. This incorporated tape preparation equipment for transferring the relevant data to punched paper tape which was then used to produce multichannel magnetic tape. The tape was subsequently used on the control system of the machine tool. Savings of fifty percent in machining time were claimed as compared to manually controlled cutting operations. About this time in Britain, Ferranti had also developed a three-dimensional magnetic tape control system. Thus with the progress in computation methods, electronic techniques, and machine tool systems the time was almost ripe for the emergence of the robot. Until now the idea of the robot had been mused over but the technology to turn the dream into a reality had not been available.

George Devol devised his concept of a system for 'programmed article transfer' in 1954, and later in collaboration with Joseph Engelberger produced the first industrial robot as we recognize the term today. Devol's control patents were bought up by the CONDEC Company which set up a subsidiary called Unimation. This marked in a small way the coming together of the numerical control technology already described and the

Figure 2.3 The Cincinnati Milacron T3 with minicomputer control. (Courtesy Cincinnati Milacron)

concept of a robot existing in popular literature. Much talk was apparent at this time of the 'automatic factory' as people generally realized the potential of applying mechanical and electronic control systems to manufacturing. This was evidenced by the construction in 1954 of the first large scale automated production line, i.e. Project Tinkertoy, devised to produce and assemble electronic articles. It was not, however, the 'programmable' automation familiar to robotics but comprised 'dedicated' machinery that was mainly suited to the one task for which it had been designed.

Unimation installed their first robot in 1961 in a die casting application. These robots were still not controlled by a computer as many of the modern robots are. Their actions were governed via a control unit on the machine which was set to the appropriate sequence by the operator. Unimation originally produced a polar coordinate robot of which an updated version is shown in Figure 13.28. To enable computers to be used economically for control of machines such as robots, further advances had to be made in electronics. This happened in the early 1960s when it became possible to produce many transistors and diodes on the one 'chip' of silicon, i.e. produce an integrated circuit (IC). Progress continued and in 1967 the first large scale integrated circuit (LSI) was produced containing hundreds of components. In 1970 the Intel Company announced the 'microprocessor', a device that could interpret and carry out instructions and have the ability of carrying out arithmetic operations on a single chip. Due to this rapid progress in reducing cost and size and in increasing the ability and reliability of computers, we see in 1974 the introduction of the first minicomputer-controlled robot, the Cincinnati Milacron T3 (Figure 2.3). The appearance of the true computer-controlled robot marks the end point of our history. Although this event happened well over a decade ago the improvements and refinements since then are still ongoing and indeed these are the subjects of study in the chapters that follow.

3

CLASSIFYING AND DESCRIBING INDUSTRIAL ROBOTS

When describing an industrial robot various terms are utilized by suppliers and users. The precise meaning and implication of these terms should become apparent as this text progresses. At this point an overview of the terms is given as a framework for the following chapters. Often the terms are used to classify robots and although this is quite valid there are many gray areas where subgroupings overlap. Here seven commonly used classification groupings and subgroupings are described: geometric configuration and work envelope; drive method; servo control system; path control system; programming method; application; level of intelligence.

GEOMETRIC CONFIGURATION AND WORK ENVELOPE

Robot arms are constructed of members, or links, that move in relation to each other on rotational and/or linear joints. These joints are considered more fully in Chapter 4. The arrangement and combination of these joints define the *geometric configuration* of the robot. A large variety of configurations is therefore theoretically possible although in practice robot designers have confined themselves to a relatively small number of permutations.

The volume of space created within the virtual surfaces swept by the robot arm at maximum and minimum reach is termed the robot *work volume*. The shape of this volume is the *work envelope* and this, in conjunction with the configuration, classifies the physical appearance of the robot. There are now seven commercially available configurations, each of which is obtainable in a slightly different form in a variety of models from a large number of manufacturers. These configurations and their associated work envelopes are shown in Table 3.1 and discussed briefly below. As is evident in the sketches the maximum theoretical work envelope is restricted in each case due to the physical construction and constraints of the robot. Each robot will have a wrist at the end of the arm with an additional one, two or three axes to allow orientation as well as positioning of a tool about a point.

Table 3.1 Robot classification by geometric configuration and work envelope	
Configuration	**Work envelope**
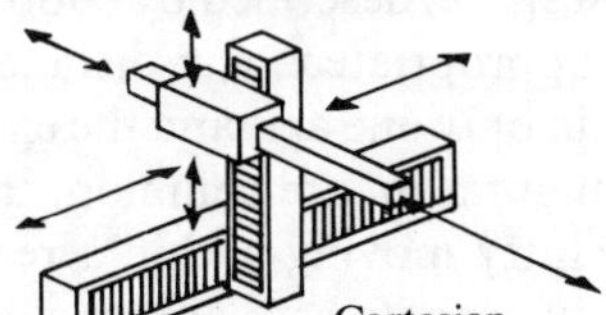 Cartesian	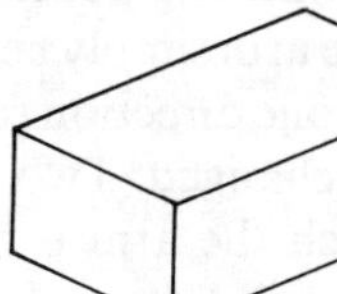 Rectangular
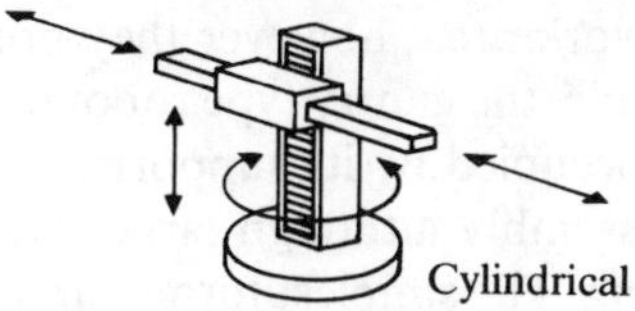 Cylindrical	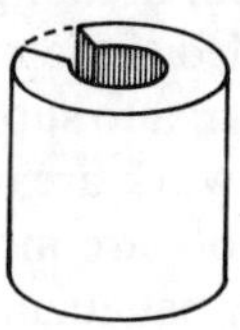 Cylindrical
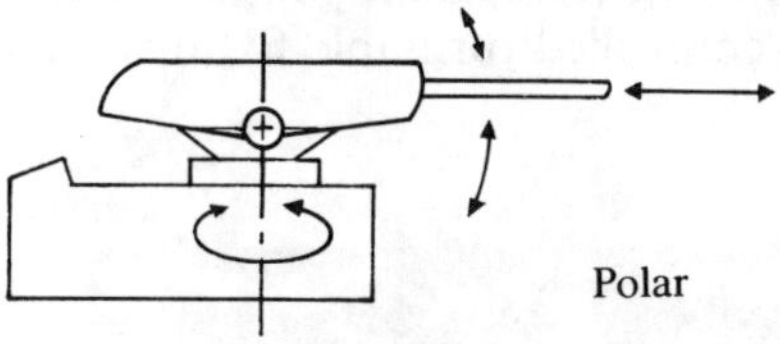 Polar	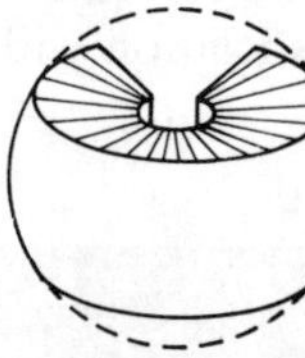 Spherical
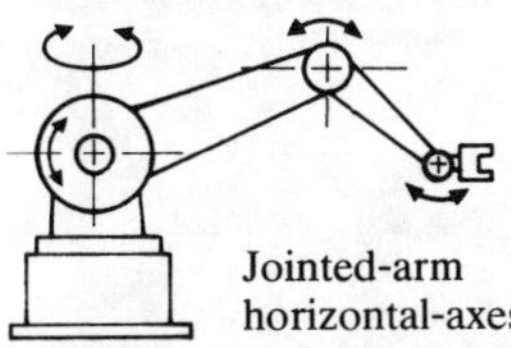 Jointed-arm horizontal-axes	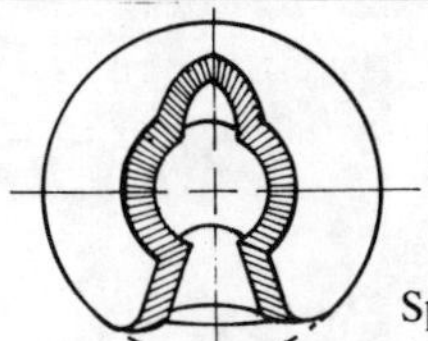 Spherical
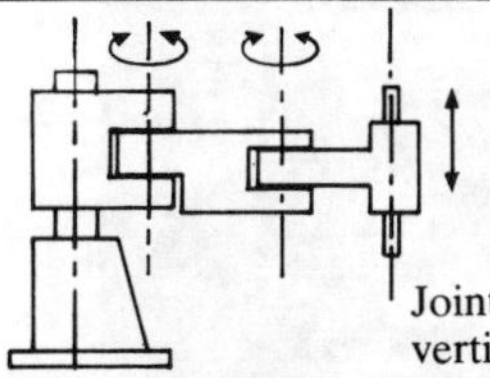 Jointed-arm vertical-axes	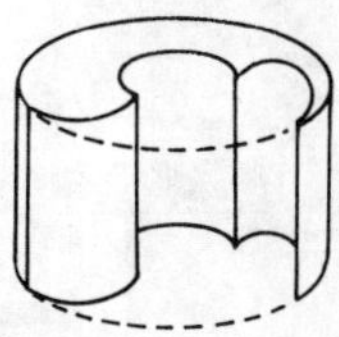 Cylindrical
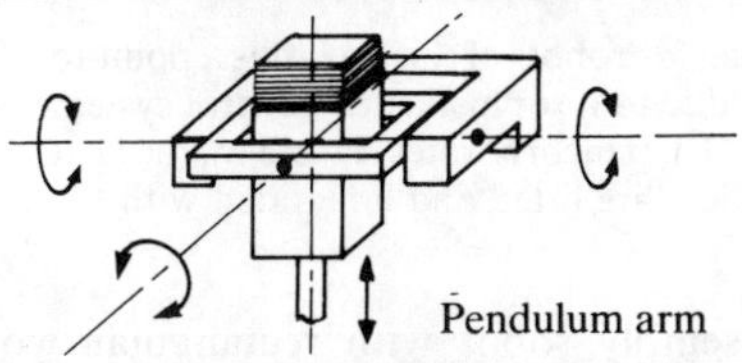 Pendulum arm	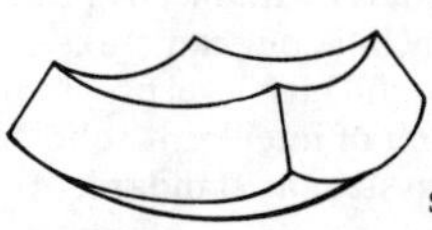 Partial spherical
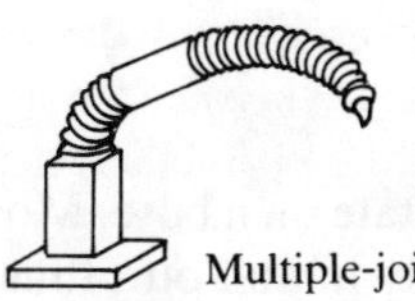 Multiple-joint arm	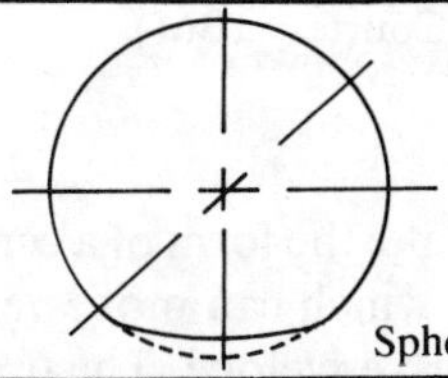 Spherical

Cartesian

This configuration provides a rectangular work envelope. The arm's three major axes are rectilinear and provide movements which are relatively simple to control. To move the end effector of the arm to a point in three-dimensional space, described by coordinates x, y, z, each axis of the arm simply needs to be moved the appropriate x, y, z, distance. Thus a movement in only one direction requires a movement in only one axis and the end effector orientation is not changed. Two varieties of this configuration are common: the gantry type robot in which the arm is suspended from a gantry moving within a rectangular framework, and the slideway mounted arm which may have its base and one linear axis mounted on the floor or along the side of a workbench. Due to the ease with which a linear axis can be extended these robots can cover a large work area, however they normally have a poor work area to support area ratio. For example, the gantry type robot must, by definition, have a work area less than the floor area occupied by its supports.

Cartesian robots are most commonly used for assembly although larger types are sometimes used for palletizing and machine tool loading. The simplest forms can be used for injection moulding machine servicing. Some very large complex examples are used in a gantry form in association with integrated servo controlled turntable fixtures for seam welding.

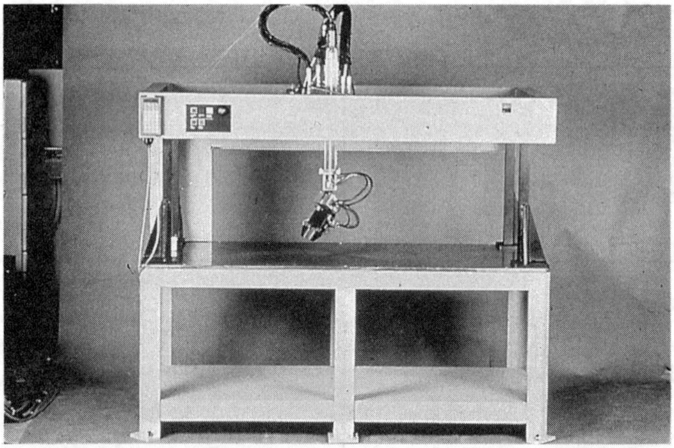

This robot is hydraulically powered, which is unusual for robots of this size. It is a point to point machine but can make complex movements due to a sophisticated control system and powerful high level programming language (AML). It incorporates features of the next generation of intelligent robots in that sensing devices are fitted and integrated with the control system as standard.

Figure 3.1 IBM 7565 gantry type cartesian assembly robot with rectangular work envelope. (Courtesy IBM)

Cylindrical

These robots take the form of a central pillar which can rotate on a base. Mounted on the arm is a pillar which can move vertically, and horizontally in and out, thus providing a cylindrical work envelope. The up–down and in–out movements require control of only

This robot has programmable sequence control with manually set stops for position control. The robot shown can be constructed from individual modular units. These units can be arranged to give a variety of configurations. Robots such as this are simple, fast and relatively inexpensive; their control is simple point to point between mechanical stops.

Figure 3.2 Martonair cylindrical pneumatic robot

one axis in each case. Similarly a rotational movement about the central vertical axis also requires control of only one axis. However, should a straight line movement in a left to right direction be required, then coordination of the rotational and in–out axes is necessary. Should it also be desired to maintain a tool on the end effector at a constant orientation then control of a third wrist axis is necessary. Thus the control system must be capable of calculating the individual axis movements necessary to allow coordinated straight line movement. Should desired points be fed to the controller from a terminal keyboard in rectangular coordinates, then the robot processor will have to be capable of carrying out mathematical transformations to achieve the appropriate joint movements necessary.

Cylindrical robots are robust and have a good work area to floor area ratio. Their long reach suits them to loading and unloading machine tools and palletizing. A particularly efficient work layout for them is to have them positioned in the center of radially arranged machine tools and a conveyor system. One robot can then remove work from an input conveyor, transfer it from one machine to another, then deposit it on an output conveyor. Small robots specially designed for attachment to NC machine tools are produced by GM Fanuc and are described as having a 'dual cylindrical coordinate' configuration.

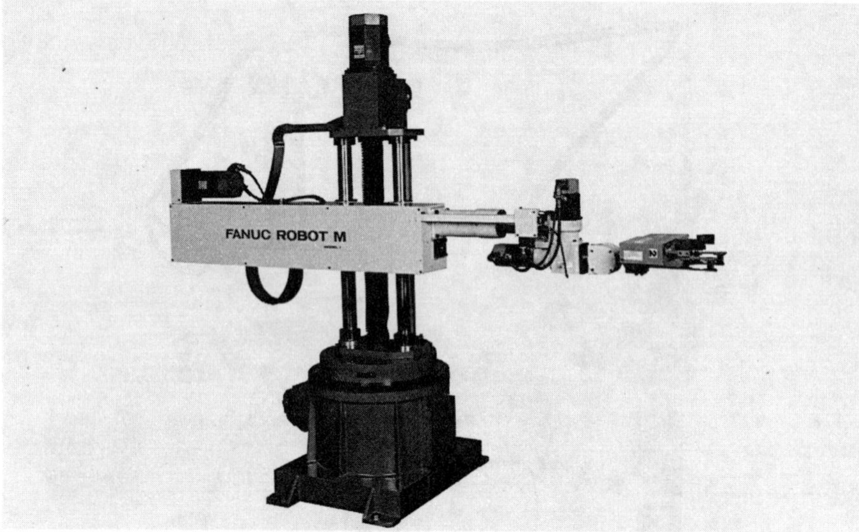

A cylindrical coordinate robot driven by electric AC servo motors. Suitable for material handling and servicing a number of machine tools.

Figure 3.3 FANUC robot M – model 1. (Courtesy FANUC)

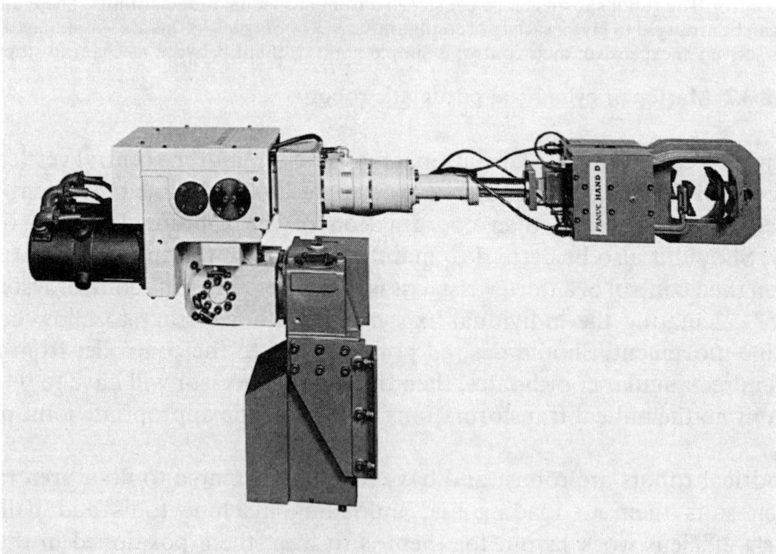

A dual cylindrical coordinate robot driven by electric DC servo motors. It is specially designed for mounting on a machine tool for loading and unloading operations.

Figure 3.4 FANUC robot MF – model 020. (Courtesy FANUC)

Polar

Alternatively called 'spherical' this is the earliest type of configuration and is therefore still widely used, although few new robots are now designed in this manner. The first Unimate robots by Unimation took this 'gun turret' form. Although these robots are robust, they do not have the flexibility of the jointed-arm robots and yet still require complex control if straight line movements or off-line control is required. Their work envelope is a partial sphere and they are used for spot welding, palletizing, and manipulation of heavy loads.

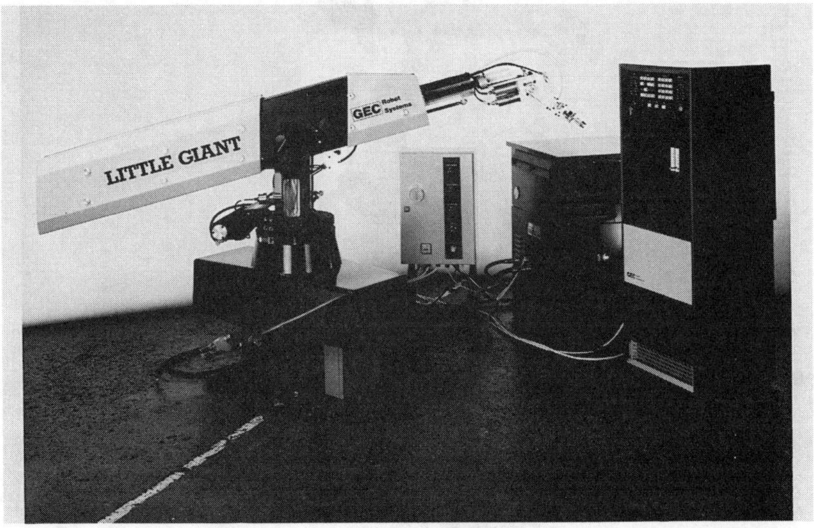

A hydraulically powered polar robot. Control is point to point with linear interpolation.

Figure 3.5 GEC Little Giant. (Courtesy GEC)

Jointed-arm horizontal-axis

Sometimes called 'revolute', 'jointed-arm spherical', or 'anthropomorphic'. They normally have a rotational vertical 'waist' with two arms connected by rotational horizontal 'shoulder' and 'elbow' joints. This type most closely resembles the human arm configuration and with a three-degree-of-freedom end effector they are adaptable to a very large variety of tasks. The rotation of the base coupled with the movements of the other two major axes produces a hollow spherical work envelope. The base of the robot takes up very little area in comparison to the work volume that can be encompassed by the arm. Straight line motion in any direction requires coordination of all joint axis movements, thus this type also requires complex control programs.

They are as equally suited to reaching down into bins to pick components as they are to lifting loads to high levels for stacking purposes. Their versatility allows them to be used for such diverse tasks as paint spraying, seam welding and spot welding, adhesive application, assembly, and heavy materials handling.

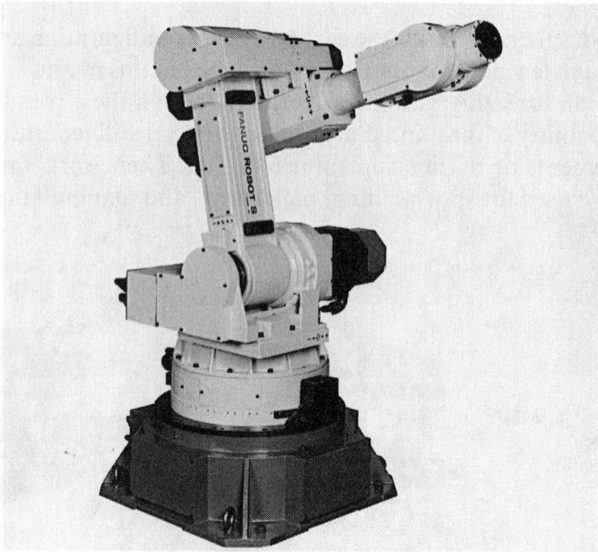

A jointed-arm horizontal-axes robot driven by AC servo motor. Control is point to point.

Figure 3.6 FANUC robot S – model 500. (Courtesy FANUC)

A jointed-arm horizontal-axes robot driven by DC servo motor. Point to point with controlled path.

Figure 3.7 Cincinnati Milacron T3-776 robot. (Courtesy Cincinnati Milacron)

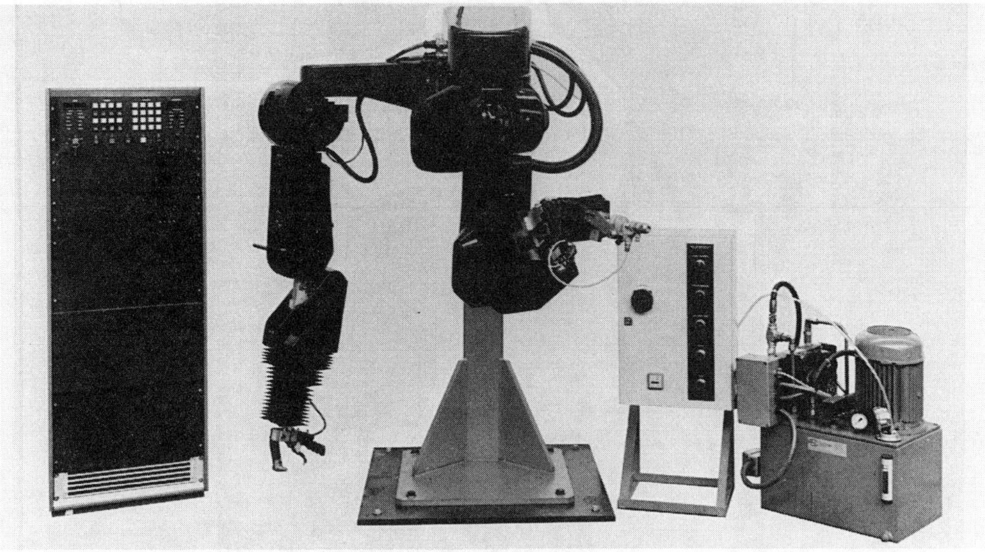

A jointed-arm horizontal-axes robot; hydraulically powered and with continuous path control. The control panel is on the far left; the teaching arm is left of the pedestal, and the hydraulic power pack is on the far right.

Figure 3.8 GEC Comparm robot. (Courtesy GEC)

Jointed-arm vertical-axis

The most common example of this type is the SCARA design (Selective Compliance Assembly Robot Arm) for assembly. The construction allows a certain amount of compliance (or movement) in the horizontal plane when attempting to assemble objects in a vertical direction. This is particularly important when inserting one component into another as the necessary precision of the arm is reduced, providing there is an adequate 'lead-in' or chamfer on one of the components. The assembly robots, being normally only used for lightweight tasks, are usually of simple construction and do not normally require full servo control on every axis. There are always exceptions, however, and one such is the Lamberton heavy duty robot. This is a massive machine with a vertical axis construction. It can be used in areas such as forges, foundries, and other places where heavy workloads are met. On all of these robots a partial cylindrical work envelope is produced.

Pendulum arm

Specially designed by ASEA for assembly work this configuration is also now used for gantry mounted welding robots. It provides a low inertia structure suitable for high acceleration and high speed motion. The work envelope is a partial sphere.

Multiple-joint arm

This configuration provides excellent flexibility and can be used for a number of applications. The type shown in Table 3.1 is that produced by 'Komatsu Ltd' and is specially designed for spray painting, spot welding and seam welding. It can reach into car

A jointed-arm vertical-axes robot. Electrically or hydraulically powered with point to point control. It is specially designed for heavy lifting tasks.

Figure 3.9 Lamberton AA series heavy duty robot. (Courtesy Lamberton Robotics Ltd)

body shells to carry out operations in areas that would be difficult to access with robots of a more conventional construction. Other forms of this type of arm have been produced one of which has been designed to reach into a nuclear reactor for inspection and maintenance purposes. The work envelope produced by these robots is almost a complete sphere with a flat base and hollow core.

At the time of writing these last two configurations are unique to their respective manufacturers. Illustrations of them can be seen in Chapter 13.

DRIVE METHOD

Drive methods and their relative advantages and disadvantages are covered more fully later, however a few introductory comments are given here. Industrial robots are driven by either *electric power* or *fluid* power, the latter category being further subdivided into *pneumatic* and *hydraulic* power. Today the most common drive method is electric with various types of motor being used, e.g. DC servo, stepper, and brushless servo. Electricity is clean, quiet and ideal for electronic control. Hydraulic robots are still common in existing installations but fewer new models are being produced except for heavy duty

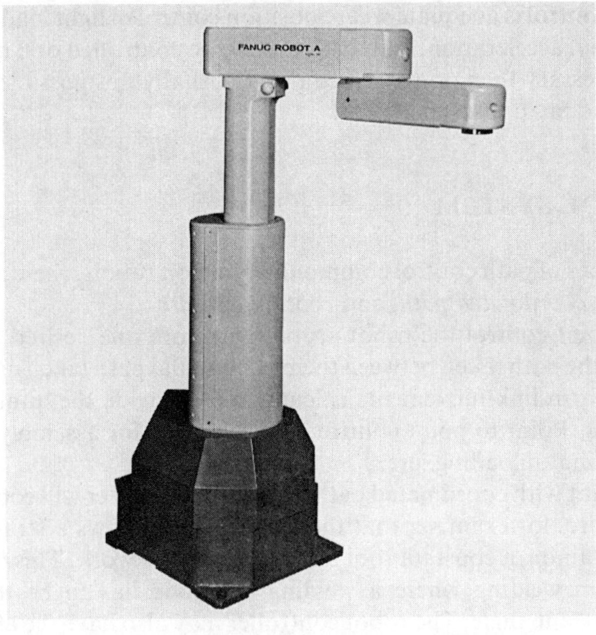

A jointed-arm vertical-axes robot with direct drive motors. It is specially designed for high speed assembly.

Figure 3.10 FANUC robot A – model 500. (Courtesy FANUC)

applications. They are capable of providing high power to size ratios, however they are relatively difficult to control, expensive, and their power pack is noisy. Improvements in electric motor design are gradually reducing the power to size ratio advantage. Pneumatic robots are fast, relatively inexpensive, but usually not servo controlled. They are ideal for light assembly or packing work but are not usually suitable for heavy duty tasks or where speed control is necessary.

SERVO CONTROL SYSTEM

Robots are either *servo controlled* (usually closed loop) or *non-servo*-controlled (usually open loop or with simple feedback using limit switches). To gain full advantage of microprocessor control, achieve good precision under heavy load conditions, and carry out complex tasks with confidence, full servo control is necessary. In this method of control commands are sent to the arm drives to move each axis the requisite amount. The actual movement is monitored for both displacement and velocity and compared with the command signal. The difference between the command and the action is the error, this value is used as feedback to the controller to enable further commands to be modified accordingly. Most electric and hydraulic robots are servo controlled. Pneumatic robots are usually non-servo-controlled. In this case a command signal is sent and it is assumed that the arm reaches its intended position, usually a fixed or programmable mechanical stop. Non-servo-control is adequate where position control of light loads only is required. However, if velocity, acceleration, and torque are to be controlled or if movement against heavy loads is necessary then non servo control is usually not possible. The majority of industrial robots today use servo control.

PATH CONTROL SYSTEM

There are three types of path control commonly found on robots, these are: *point to point*, *point to point with coordinated path*, and *continuous path*.

In point to point control the robot arm moves from one desired point to the next without regard to the path taken between them. The actual path taken may be the result of a combination of arm link movements calculated to provide the minimum travel time between the points. Point to point control is widely used for assembly, palletizing, and machine tool loading/unloading, etc.

In point to point with coordinated path the robot controller will coordinate each axis of the arm to ensure, for example, that the end effector follows a straight line between points while maintaining a constant tool orientation to the work. This has advantages in tasks such as seam welding where a welding electrode has to be traversed along a workpiece at a constant angle. The robot controller may also have the ability to carry out circular interpolation, enabling it to move the end effector in a circle once taught three points on the circumference or a center point and radius. A further example of this type of control is in line tracking. This is commonly used in spot welding on car assembly lines. In this application a pattern of spot welds is applied to a car body while the body moves along an assembly line. The robot controller receives the information used for coordination from feedback instruments monitoring the velocity of the line.

Finally in continuous path control the robot moves along a continuous taught path. In this method the arm is usually led through the path required, for example when paint spraying, and signals from the feedback devices in the joints are constantly monitored by the robot controller. In the earliest examples of this type of control continuously varying signals, i.e. analog signals were recorded on magnetic tape and these signals used to replicate the movement. Today the feedback sensors are usually sampled at a very high frequency and the signals stored as a very large number of discrete points in the controller.

PROGRAMMING METHOD

Industrial robots can be programmed by various means and they are described in more detail later in the text. Essentially they can be programmed either *on-line* or *off-line*. On-line methods require direct use of the robot and will utilize teach pendants for point to point programming and slave arms or pistol grip attachments for continuous path programming, these will be explained later in Chapter 12. The more-primitive robots may employ cam drums and mechanical stops to allow limited sequence programming, but these are now very unusual. More recent robots have the ability to be programmed off-line, i.e. the robot can continue working on a particular task while a program for a new task is prepared on a computer terminal.

APPLICATION OF FUNCTION

As more robots are designed for specific tasks this method of classification becomes more relevant. For example many robots are now designed for assembly work in particular, they are termed 'assembly robots' as they are not readily adaptable to other applications. For seam welding some suppliers now provide complete welding systems incorporating the robot, the welding equipment, and the other material handling facilities such as turntables, as an integrated unit. Thus in this case the robot is sometimes termed a 'welding robot', although as a discrete unit many of these robots could be adapted to a variety of tasks. Some robots are specifically designed for heavy load manipulation, and are labeled 'heavy duty robots'.

INTELLIGENCE LEVEL

As briefly mentioned in the introductory chapter, the Japan Industrial Robot Association uses a particular set of definitions and classifications for industrial robots. This method incorporates definitions that reflect the intelligence level of the robot and it is therefore useful to quote them here from the Japanese Industrial Standard JIS B 0134-1986.

After defining the terms 'manipulator' and 'robot' as discussed in Chapter 1, the standard goes on to give a General Classification of robots.

Operating robot. A robot having controllers, and actuators for mobility and/or manipulation, remotely controlled by human operator. (Into this category would fall the telechiric, nonrobotic devices, mentioned in Chapter 1.)

Sequence control robot. A robot that operates sequentially in compliance with preset information on sequence, conditions, and positions.

- Fixed sequence robot. A sequence robot for which the preset information cannot be easily changed.
  (It is only from the category below onwards that these statements could be covered by the definition adopted in this text for an 'industrial robot'.)
- Variable sequence robot. A sequence robot for which the preset information can be easily changed.

Playback robot. A robot that can repeat an operation on the basis of instructions concerning sequence, condition, position and other information imparted by moving the robot under operator control.

Numerically controlled (NC) robot. A robot that can execute the commanded operation in compliance with information on, for example, sequence, conditions and position, loaded numerically or by program, without being moved.
(The above category takes us to the level of the conventional commercial industrial robot of today, sometimes referred to as first generation robots.)

Intelligent robot. A robot that can determine its own actions through artificial intelligence.

- Sensory controlled robot. A robot whose operation is controlled by sensory information.
- Adaptive controlled robot. A robot having an adaptive control function. Note: adaptive control signifies control in which the characteristics of the control system change in response to change in the environment in order to satisfy the prescribed conditions.
- Learning controlled robot. A robot having a learning control function. Note: learning control signifies control in which appropriate operation is obtained reflecting operating experience.

Eventually, intelligent robots will be able to work autonomously to achieve pre-programmed task objectives using sensing and artificial intelligence, these would be called third generation robots. They are only part-way to being seen today in the form of welding robots with adaptive control for seam tracking, and assembly robots aided by vision systems. These robots that are presently utilizing various types of sensing for adapting their programmed instructions, albeit in a limited way, to suit their external environment, are sometimes termed second generation robots.

II

ROBOT DESIGN AND CONTROL

4

PRINCIPLES AND PROBLEMS IN ROBOT DESIGN AND CONTROL

INTRODUCTION

This first chapter of Part II is intended to introduce the basic concepts upon which robot design and control are founded. It has been noted that an industrial robot is a manipulator powered electrically, hydraulically, or pneumatically and usually controlled by an electronic or electromechanical system. The following chapters will discuss these features in more detail.

The block diagram in Figure 4.1 illustrates the approach adopted here. The input to the system comes from a human operator using a teach pendant, computer terminal or other teaching method. The input may also come from a supervisory computer where the robot is operating as part of a hierarchical system. Thus a programming technique is required often involving the use of a programming language, this is covered in Chapter 12. The information from the input is processed by the control system which sends signals to the robot actuators and receives feedback signals from the internal and external sensors. The control system, control hardware components and sensors are examined in Chapters 5, 9, and 10. The various actuators are discussed in Chapters 6 and 7, the transmission components in Chapter 8, and the end effectors in Chapter 11.

Essentially an industrial robot, by virtue of its definition, must move its members through space, this means that a number of factors concerning the arm must be considered if the system is to be efficiently controlled. These fall into two main categories, namely *kinematic* and *dynamic* factors, both of which can be analysed mathematically. If, for example, it is required to move a robot end effector through a path within the robot work envelope, then the position and direction of the end effector's movements necessary to achieve this can be represented mathematically. In turn the torques or forces required to be generated by the actuators can be calculated. Knowledge of these desired forces will then allow the appropriate control signals to be evaluated, e.g. an increase or decrease in current or voltage. Thus control is being effected by modifying an input signal to an

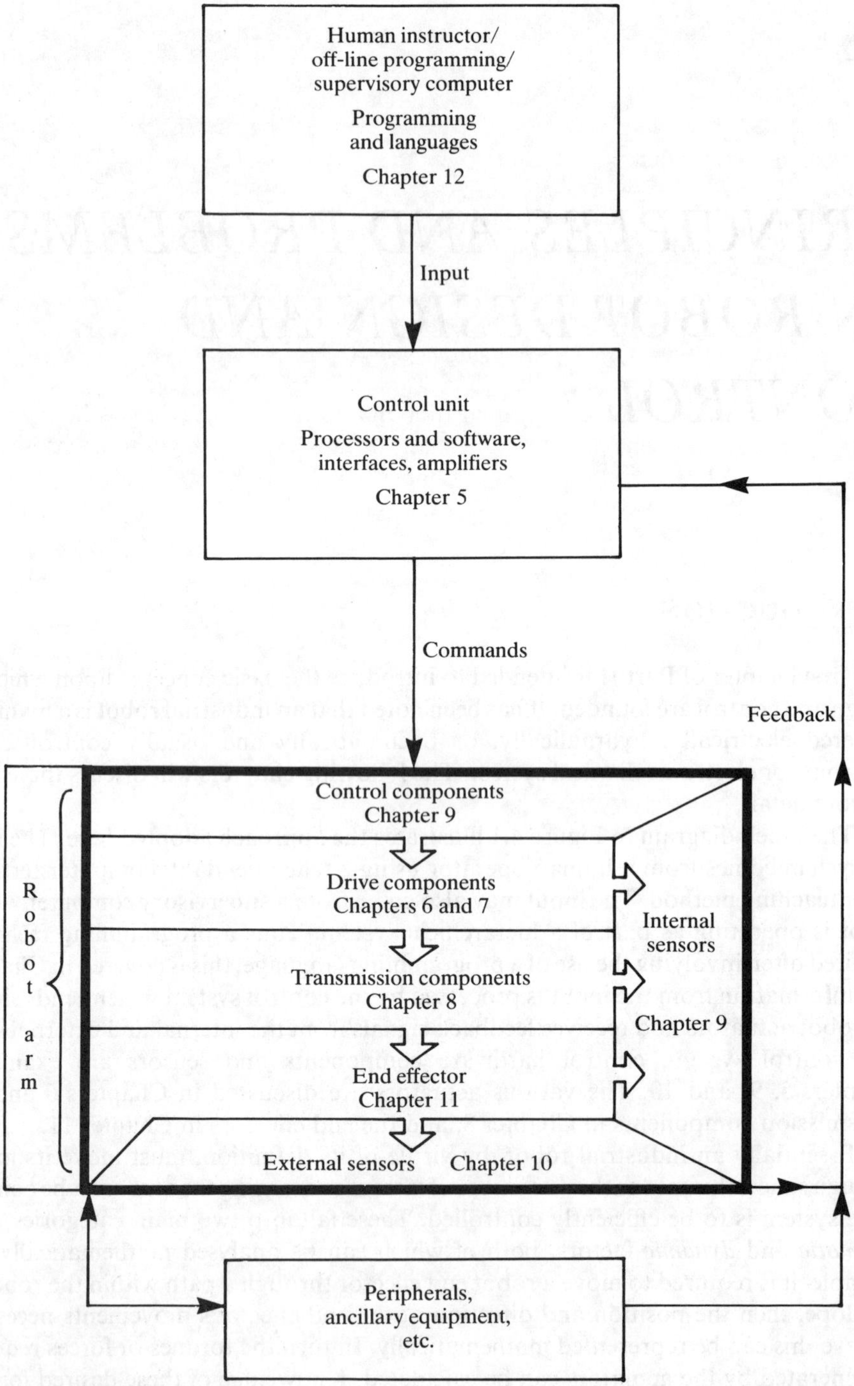

Figure 4.1 Block diagram illustrating the interrelationship between subject areas covered in this section

actuator, thus controlling the force which alters the joint positions, which then determine the final path of the end effector. Each of these stages can be defined mathematically, enabling full computer control to be maintained on a real time basis. These quantities mentioned above will all be manipulated within the control system in the form of vectors (a vector is any physical quantity that requires a direction to be stated in order to define it completely) as they will have both a value and a direction.

KINEMATIC CONSIDERATIONS

Kinematics is a branch of mechanics that is concerned with motion without reference to mass or force. In the next section dynamics are considered, and there it is the effect of the laws of motion on the arm that are of concern. Here it is the measurement and description of the motion, i.e. in displacement and time that is examined. Kinematics is concerned with how the component parts of a mechanism move with respect to each other, e.g. a mechanism for lifting the hood of a car bonnet, a mechanical excavator, or an industrial robot. Thus, in an industrial robot system kinematics can be used to model and derive programs that will allow the robot end effector to be moved, positioned, and oriented in three-dimensional space in a controlled manner. This is done by analyzing the movements of the combination of the robot joints and links and so determining the resulting movement of the end effector.

There are essentially two problems to be considered here: the forward kinematic problem, and the inverse kinematic problem. In forward kinematics the joint angles of each axis of the robot are known and the position and orientation of the end effector is calculated in relation to some coordinate frame. This is the simpler of the two problems. It is experienced when leading a robot by hand to a point or through a path, then commanding the robot to remember that point or path. In inverse kinematics the desired position and orientation of the end effector is known and all the possible combinations of joint angles necessary to achieve this are calculated. This is a more difficult problem. By utilizing geometric information on the robot, held in a database, programs, or algorithms, can be constructed. These algorithms can then be used to program the robot off-line or enable graphical simulations to be carried out.

The number of degrees of freedom of the robot influences its ability to position and orientate an object in space. A free rigid body in space has six degrees of freedom, that is it can move horizontally back and forward say along the X–X' axis, horizontally from side to side along the Y–Y' axis, and vertically up and down along the Z–Z' axis, it can also rotate about each of these three axes. This is shown diagrammatically in Chapter 12 on Programming. Each joint of a robot must have at least one degree of freedom, although the construction is usually such that one link is constrained so as to be able to move in only one manner relative to the other link. Figure 4.2 shows examples of some articulations or joints and the number of degrees of freedom obtainable from each one. If more than one degree of freedom is given to a joint or articulation this can complicate control and increase the processing time. It is sometimes said that each axis of a robot constitutes a degree of freedom. This is not necessarily true. Obviously a seven axis robot cannot have seven degrees of freedom as there can only possibly be a maximum of six. It follows that it is possible to have a robot with six individual joints or axes but much fewer degrees of

Comments	Number of degrees of freedom	Type
Revolute Rotation about one axis Requires only one parameter to specify position in space	One	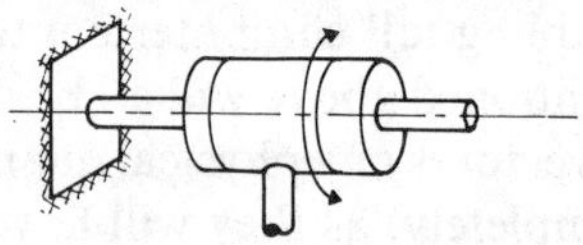
Prismatic Linear movement along one axis Requires only one parameter to specify position in space	One	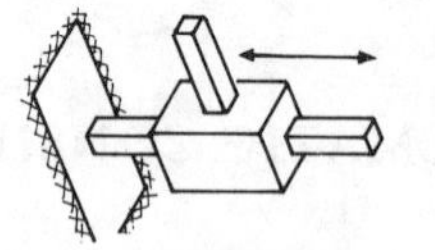
Screw Composite linear and rotational movement along one axis Translation defined by pitch and rotation Only one parameter need be specified	One	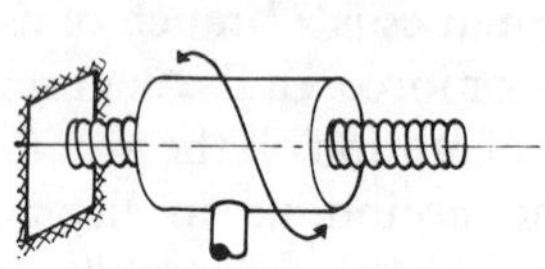
Cylindric Linear and rotational Independent movement along and around one axis Two parameters to be specified	Two	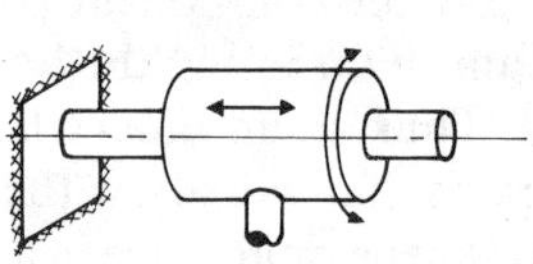
Spherical Can rotate about three axes (precession, nutation and spin) Three parameters to be specified	Three	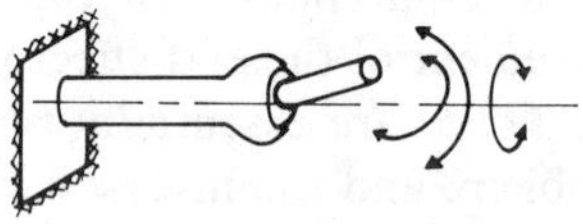
Planar Linear movement along two axes in one plane and rotation about one axis Three parameters to be specified	Three	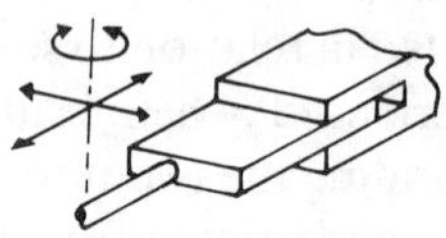

Figure 4.2 Some articulations, joints or kinematic pairs

freedom. For instance a revolute robot with no wrist or waist movement can have any number of revolute joints, but if the joint axes are parallel then it can have no more than three degrees of freedom (see Figure 4.3). Additional axes can, however, give greater flexibility which may be necessary for reaching around corners into car bodies, machine tools, pipework, etc. Even with true six degrees of freedom constructions it is possible for the degrees of freedom available for control to degenerate to singularity in some configurations or postures. For example, when the robot is at full reach it may be impossible to achieve the required position and orientation of the robot wrist, this can also apply for other postures depending on the individual arm construction. These special configurations may also give dynamic problems where excessively high torques and accelerations may be required to achieve desired velocities and paths. Basically, therefore, the flexibility of a robot system is a function of its geometry. If the geometry is specialized with, say, parallel axes the flexibility of the system will be reduced.

The above discussion shows that in order to achieve full computer control, or off-line programming, of the robot it is necessary to be able to translate the world coordinates of

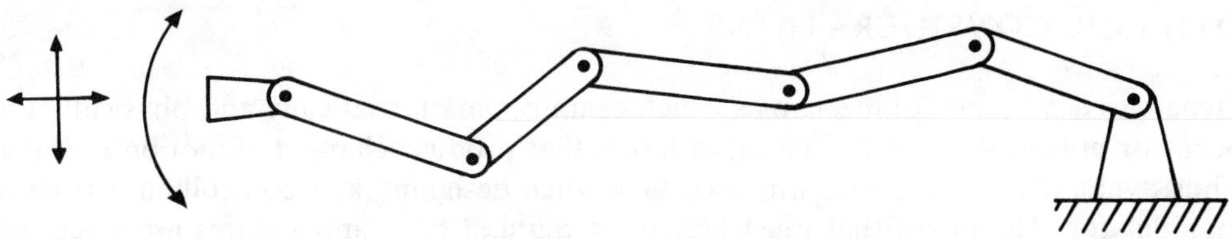

Figure 4.3 A six-axis manipulator but with only three degrees of freedom as all six revolute pair axes are parallel

the end effector into a number of corresponding joint coordinates for the robot wrist and arm. In the case of a jointed arm robot a mathematical model is constructed that is essentially that of a moveable open polygon with one of its ends fixed to the ground (i.e. the robot base) and a free end (i.e. the end effector). See Figure 4.4. The polygon is thus composed of the links and joints or kinematic pairs corresponding to the appropriate limb members of the robot. A program is then constructed that can carry out coordinate transformation from the world coordinates of, say, point $P=(x, y, z, \alpha, \beta, \gamma)$ to joint coordinates say $\theta_1+\theta_2+\theta_3+ \ldots$, etc. All this must be carried out in real time, involving possibly thousands of calculations per second, especially when the input from sensors is being monitored. The algorithms used for this must therefore be very efficient. The equations used to conduct the mathematical transformations are usually constructed with matrices. Further information on coordinate systems is included in Chapter 12 on Programming.

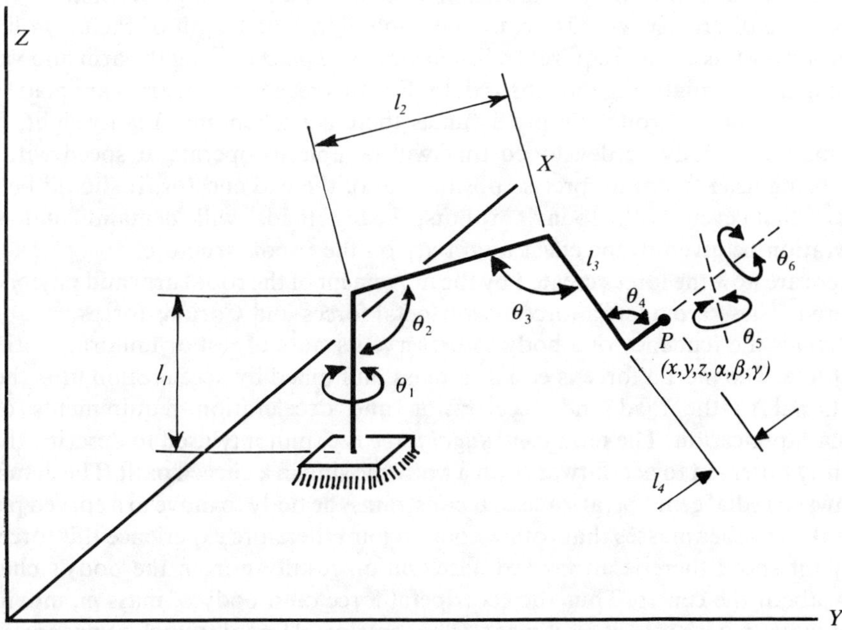

Figure 4.4 Open polygon representation of robot for kinematic analysis

DYNAMIC CONSIDERATIONS

Dynamics is a branch of mechanics which examines mathematically and physically the behavior of bodies under the action of forces that produce changes of motion in them. Thus dynamic considerations are necessary when designing and controlling a robotic manipulator. The forces that must be considered and their implications are discussed below.

Firstly, assuming the robot is being used on Earth, the arm, end effector, and payload are subjected at all times to the force of gravity. This means, for example, that if the arm is being held horizontally then its own weight will tend to pull it downwards. Depending on the robot structure there must be some means provided to compensate for this, the main methods being as follows. Firstly the robot joint construction may incorporate some sort of ratchet mechanism which mechanically prevents return movement. Or, if the robot is hydraulic then the flow of hydraulic fluid can be blocked by closing a valve. This means that the incompressible fluid locks the robot arm into whatever position it was in when the valve was closed, i.e. the weight of the arm is supported by the fluid held in the actuators. However, if the robot is pneumatic, some sort of clamping mechanism is necessary since air is compressible. If this is not done then any load on the actuators will tend to compress the air held within and so allow loss of position of the arm. To hold an arm outstretched against gravity in electric robots, the arm position must be constantly monitored and the electric servo motors continuously compensate for position deviation. To minimize load on the motors the arm is usually fitted with balance weights, or is of a parallelogram construction, so providing balance for various configurations. It is also normal to have fitted electromechanical brakes to allow the arm to be locked in position when necessary. This is, of course, important for safety purposes in the case of a power failure.

The force of gravity will also cause bending along the length of each arm link. To compensate for this would require strain sensors to be placed along the arm and valuable processing time in analyzing the sensor data. For this reason robot arms are made strong and rigid and the controller simply assumes there is no bending. Lightweight, flexible robots may eventually be developed that will be able to operate at speed with much sensing being used to ensure precise positioning of the end effector. It should be noted, however, that even with today's robots, heavy loads will demand that special consideration be given to the effect of gravity on the robot structure.

There are now the forces created by the movement of the robot arm and payload to be considered. These are: inertia forces, centripetal forces and Coriolis forces.

Inertia is the tendency of a body to preserve its state of rest or uniform motion in a straight line. The inertia force is equal to mass multiplied by acceleration, it is therefore proportional to the load and acceleration and deceleration requirements of each particular application. The term *centrifugal force* is commonly used to describe the force apparently observed to act outwards on a body moving in a curved path. The actual force is the inward radial centripetal force that constrains the body to move in a curved path. All robot links or other masses that rotate about a joint therefore experience this force. Even at constant speed there is an inward acceleration resulting from the body's change of velocity about the center. Thus the centripetal force on a body of mass m, moving in a circle of radius r, with velocity v, is mv^2/r. The centripetal force is therefore proportional to the velocity squared. *Coriolis forces* are apparent forces used to simplify calculations involving rotating segments such as the coupling between the segments of a robot arm. In a novel by Arthur C. Clark, observers in a hollow cylindrical vessel in outer space are standing on the inner wall of the vessel. The cylinder is spinning to produce artificial gravity and the observers are looking towards one end of the cylinder which of course takes the form of a disk. From the center of the disk a waterfall tumbles 'down' to the cylinder wall. This waterfall rather than coming vertically down traces out a distinctly curved path. This curved path is caused by the Coriolis forces which are the apparent forces used to account for the tangential acceleration. In the robot arm the Coriolis forces will be proportional to the products of the velocities of the various articulations.

A further consideration is that related to the natural frequency of the robot. The natural frequency of a system is that frequency at which free oscillation occurs. If the robot system therefore has a natural frequency equal or close to the operating frequency of some device associated with the robot, or in the robot vicinity, then it is possible for unwanted vibration to be set up in the system through resonance. When a small periodic force is applied to a system capable of oscillation, such as an industrial robot, the system is forced into oscillations of small amplitude. However, as the frequency of the applied force approaches the natural frequency of the system, the amplitude of the oscillations builds up. This amplitude reaches a maximum when the frequency of the applied force equals the natural frequency of the system. The system is then said to be in resonance. The most famous example of this is that of the Galloping Bridge in the USA. This was a bridge that was constructed such that its natural frequency allowed it to be excited by naturally applied forces until it was in resonance. The bridge literally shook itself apart. For this reason marching troops fall out of step when crossing a bridge to ensure there is no one, large, exciting frequency applied to the bridge by their footsteps. In an industrial robot the same situation applies. The robot structure must be designed to ensure that its natural frequency is not likely to allow resonance to occur. For example, an electric motor with an out-of-balance shaft, or a hydraulic pump causing pulsing in the hydraulic fluid, rotating at 3000 rpm, will set up excitation forces at the operating frequency, 50 Hz. If the robot system has a natural frequency of 50 Hz then serious problems can arise through unwanted vibration occurring throughout the entire robot. If the natural frequency of the robot is too low, however, excitation forces may be picked up from the surrounding ground and general environment, again causing problems.

There are, therefore, forces produced by gravity, the robot actuators, the movement of the robot articulations, and forces due to resonance. All should be anticipated and compensated for in the original robot design. Usually assumptions are made to facilitate modeling the robot system and constructing the operating programs. For example, there will of necessity be clearances between moving parts at joints, it will be assumed that no movement occurs here. In practice this will be minimized by careful mechanical design and anti-backlash devices. The friction forces to be overcome at the joints may be compensated for by estimating values based on experience, or if small may be ignored. Also, as mentioned previously, if the robot is of rigid construction then the elasticity of the arm members will be assumed to be negligible.

THE CONTROL SYSTEM CONCEPTS

Any control system is a means by which a variable quantity or a group of variable quantities is made to conform to a desired value. In the case of an industrial robot the absolute aim is to manipulate the end effector carrying a tool or other load in a prescribed and controlled manner in space. This necessitates calculation of the joint coordinates and subsequent control of these joints by controlling the applied forces, this being achieved by controlling the voltages and currents applied to motors or valves. The following paragraphs look briefly at the general control concepts involved.

Industrial robots may have open loop or closed loop control. Only the simplest are of the open loop type and indeed under some definitions may not be termed robots at all. In an open loop system if an input signal or command is given to the robot to make a particular movement, the arm should move to the specified position and stop. However, there will be no feedback signal to indicate to the controller that this has in fact been done. The control loop has not been closed, a signal has been sent to initiate a movement, but no feedback signal has been received to verify that movement has occurred. An early example of open loop control was the Jacquard Loom mentioned in Chapter 2. The information contained in the punched cards controlled the pattern woven by the loom but no information from the process was used to control the machine's operation. The cam operated machine tools used in today's mass production factories are open loop systems. The pneumatically operated industrial robot that has its positions controlled by setting mechanical stops is an open loop system. Simple teaching robots using stepping motors may also be open loop, the position being determined by a specified number of pulses being applied to the motor.

Most industrial robots utilize closed loop servo controlled systems. The terms 'closed loop control system' and 'feedback control system' are used synonymously here. The definition adopted by the American Institution of Electrical Engineers states, 'a feedback control system is a control system which tends to maintain a prescribed relationship of one system variable to another by comparing functions of these variables and using the difference as a means of control.' The British Standards Institution defines a closed loop control system as, 'a control system possessing monitoring feedback, the deviation signal formed as a result of this feedback being used to control the action of a final control element in such a way as to tend to reduce the deviation to zero.' Feedback is defined as, 'the transmission of a signal from a later to an earlier stage', and monitoring feedback as, 'the feedback of a signal representing the controlled condition along a separate path provided for that purpose, for comparison with a signal representing the command signal to form a signal representing the deviation.' James Watt's governor described in Chapter 2 is a good example of a closed loop device whereby the speed of the engine was set by adjusting the length of the vertical link connecting the valve and governor levers. Similarly an industrial robot with a closed loop system requires feedback components, or sensors, to measure the response of the system to the command signal, and provide a feedback signal. Industrial robots are also usually servo controlled as the high power to the actuators is controlled by low power signals. In a servo system a low powered signal is received, either as the command signal or as the feedback signal, and this is used to give control over the high powered output. For instance, in the hydraulic servo valve shown and explained in Chapter 9 on Control Components, the flow of the low pressure fluid in the pilot valve moves the control spindle in the high pressure chamber thus allowing high pressure fluid to flow to the actuator.

To describe a control system the graphic symbols of a block diagram are used. There are four symbols employed in the notation; these are shown in Figure 4.5. The arrow

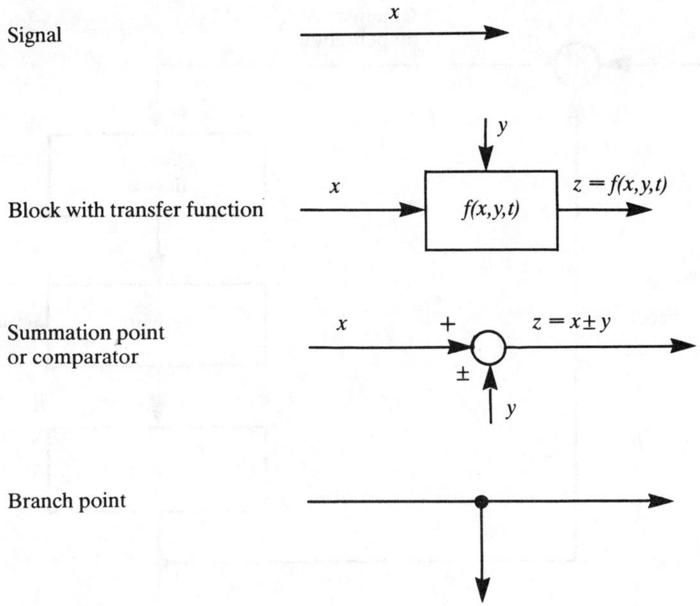

Figure 4.5 Notation for control, block diagrams

designates the signal which will be a quantified physical variable acting only in the direction indicated. The block designates the functional relationship between the signals. The signals that enter the block are independent variables, i.e. inputs or causes, and the signal leaving the block is the dependent variable, i.e. the output or effect. Inside the block is often written the transfer function, it is this that explains or determines the relationship of the output to the inputs. In other words the output from the block is the input multiplied by the transfer function. The circle describes a summation point or comparator. Here two signals are added or subtracted, as determined by the plus or minus signs next to the arrows. The branch point signifies one signal simultaneously causing two separate effects.

Figure 4.6 shows a simplified closed loop control for one actuator of a robot. The command signal is sent to a control element, say a servo valve in an hydraulic system. This control element will have its own transfer function, as will all the other blocks in the system depending on their physical construction and operation. This control element will cause operation of the actuator, one of the disturbances to the response of the actuator will be due to the load being manipulated. The actual operation of the actuator will be monitored by the feedback element or sensor and the resulting signal sent round to the comparator. This feedback signal will be subtracted from the command signal giving an error value. Thus a compensating signal will then be sent to the control element, and so the loop is closed. Subsequent chapters include more detailed discussion on these control system components.

There are many factors to be considered when designing a control system. For example, the stability of a control system is determined mainly by its response to a suddenly applied signal or transient. If this signal causes the system to overcorrect itself

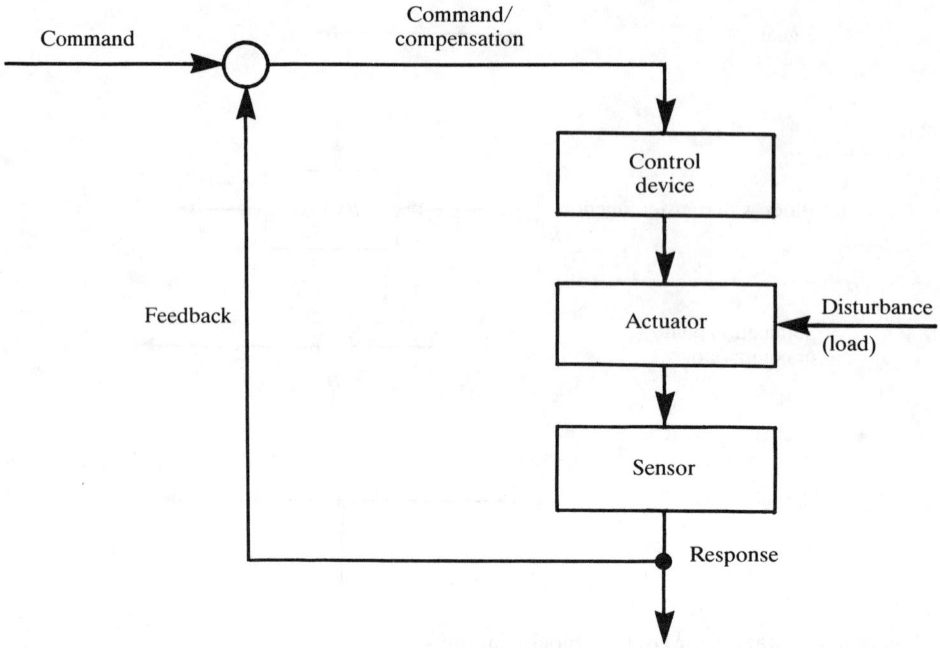

Figure 4.6 Simplified control, block diagram for one axis (closed loop system)

then 'hunting' may occur. Hunting is the term given when the system first overcorrects itself in one direction and then overcorrects itself in the opposite direction. This is undesirable, and corrective measures are usually taken in the form of 'damping' which slows down the system response to avoid overshooting. When damping is designed into the system such that correction produces exactly the desired result then the system is said to be critically damped. If the damping is designed so that the system is too sluggish then it is said to be overdamped, and if the damping allows the system to oscillate a few times before it settles then it is said to be underdamped. The damping may be achieved by, for example, forcing oil through a small orifice as in a shock absorber, or in the form of electrical resistance as in an electronic circuit. In the lower cost teaching robots it may be found that hunting is avoided by a creating a 'deadband' around the desired point. The axis is reckoned to have achieved a satisfactory position if it comes to rest at any point within the deadband, this removes the possibility of hunting but of course reduces the final precision of the robot.

Adaptive control is a concept of direct relevance to robotics as this signifies the capability of the control system to modify its own operation to obtain optimum operating conditions to suit the task in hand. Adaptive control essentially involves three principles: (1) continuous assessment, monitoring, or sensing of present conditions, e.g. arc condition and position monitoring in seam welding, or sensing of forces experienced in component assembly; (2) deciding on the basis of the information from (1) what changes to make in the system to achieve optimum conditions; (3) modifying the system on the basis of the decision made in (2). Adaptive control can therefore be used in many instances in robotic work and will become much more widespread as the use of sophisticated sensing increases and artificial intelligence techniques are employed for information processing.

TYPES OF MOTION CONTROL

The control system can be designed to allow operation of the robot in one of three ways, i.e. point to point, point to point with coordinated path, and continuous path. These are illustrated in Figure 4.7 and a brief description now follows.

Point to point motion control

With this type it is only the start and stop positions of the end effector that are of importance and are stored in system memory. The path traced by the end effector as it moves from point to point is deemed to be unimportant. Usually this type of control is designed to allow the fastest speed to be achieved between the points. The disadvantage is that the arm movements are unpredictable and therefore may cause problems in cluttered work areas.

Coordinated path motion control

This is in effect a combination of point to point and continuous path movement. Here the robot is taught the point locations and they are held in memory. The robot then when making the movements between the points, works out the joint movements and speeds to allow coordinated movement. This enables the end effector to trace a predictable path between the taught points. This coordination may allow a straight line or circular movement of the end effector depending on the interpolation method selected by the operator. A more sophisticated controller is necessary for this, to allow all the necessary calculations to be carried out in real time. The time between points will probably be longer than with simple point to point as each joint may not be able to reach maximum velocity when working in concert with the others.

Continuous path motion control

The path traced by the end effector in this case can be a very complex three-dimensional one. This method differs from the previous as each incremental movement of the end effector is as important as the start and end points. How this is achieved is best described by considering how the robot can be programmed. In the teach by lead through method, the arm or slave arm is led through the necessary movements and the positions of each of the joints are monitored continually, i.e. they are sampled at high frequency and stored in memory. When these coordinates are played back they are fed to the actuator controls and so a duplicated movement is obtained. There are three main methods by which the real time control of the robot in continuous path control is effected. The first is simple servo control, and the remaining two really adaptive control methods. In the first, which is the most common, the necessary signals to achieve the required positions are fed to the actuators. The actuator movements are sensed by feedback components and signals for corrective action are sent back round the control loop. Although the full path is held in memory only past and present data are used for servo control. The second method is to use a preview sensor such as a vision system mounted on the robot arm to anticipate any

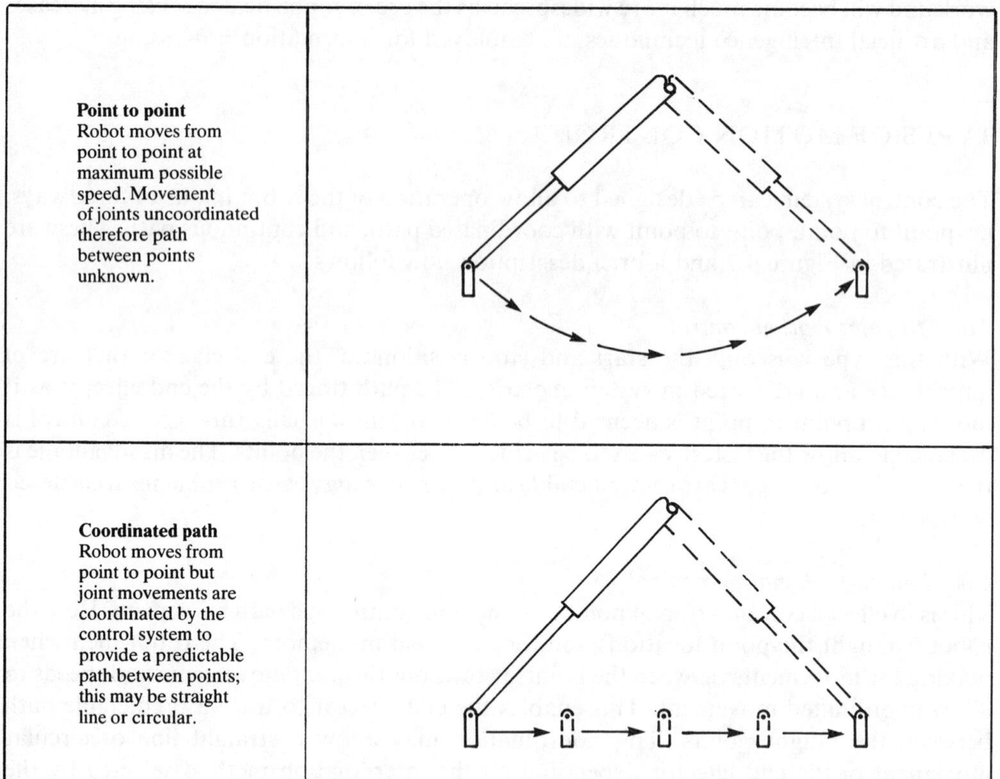

Figure 4.7 Robot motion control

changes likely to affect the preplanned progress of the end effector. This is typical of the control method best-suited to seam tracking and welding. The third method is to use a full trajectory calculation approach in which all of the joint coordinates and accelerations have been precalculated to achieve the required profile. The movements required both present and future to achieve this profile are being continually modified in the light of the feedback data from various sensors.

5

THE CONTROL UNIT

INTRODUCTION

The concepts of open loop control and closed loop servo control should now be familiar to the reader. In this chapter the operation of the control unit and its components will be examined briefly, as a full discussion and description on the operation of microprocessors and computer hardware and control is beyond the scope of this text. However, this chapter will provide an appreciation of the hardware and software involved in the control unit of an industrial robot.

Figure 5.1 illustrates the control unit functions and their relationship to the robot arm components and other external information sources.

Inputs to the control unit will be of the following categories. Programming instructions; these may come from various sources, e.g. the robot programmer via a teach pendant. Inputs will also come from the system tape drive, disk drive, or bubble memory. These mass storage devices will contain the robot operating system software, initialization procedures, and operating language. Instructions can also be input to the control unit from a computer terminal, downloaded from an off-line programming and simulation terminal via a postprocessor, or from a supervisory computer as part of a factory-wide hierarchical control system. The control unit will also receive input from the internal displacement and velocity sensors in the robot arm, external systems such as vision systems, and external process sensors such as conveyor belt start and stop signals, and finally there will often be input signals from programmable controllers used to control various actions within the work area.

Outputs from the control unit will be in the form of communications to visual display units, printers and, if present, a supervisory computer. These communications will take the form of status statements, fault diagnosis, or data for further analysis such as number of jobs completed or information from external sensors. The unit must also send out the appropriate command signals to the control components for the robot actuators.

Within the control unit the input signals have to be made compatible with the control computer hardware and data handling procedures, and the outgoing signals from the computer system will have to be amplified and conditioned to suit the control components and peripheral equipment. This task is handled by the interface hardware which incorporates analog to digital and digital to analog conversion, signal conditioning, and signal amplification including, for example, power supply control to electric servo motors.

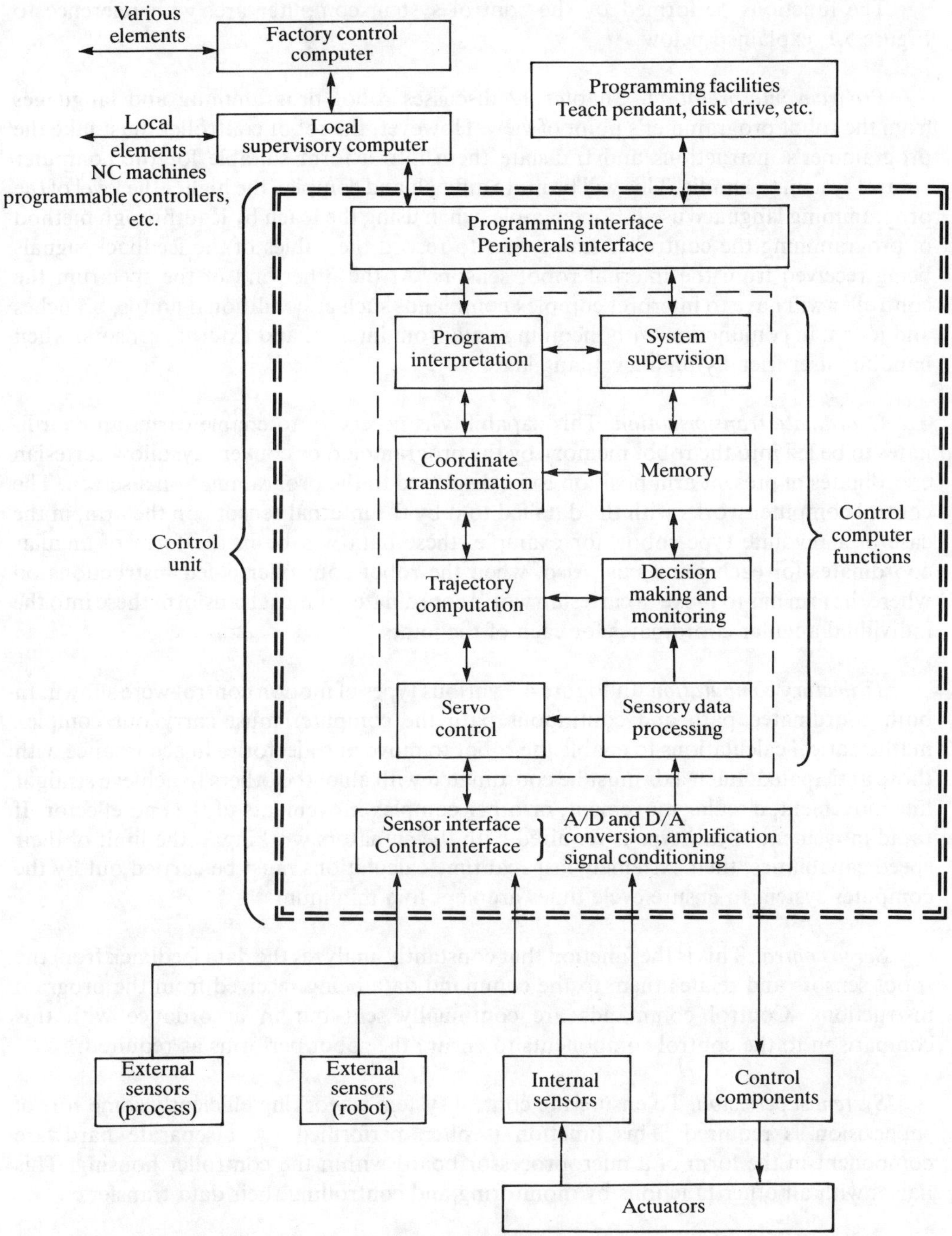

Figure 5.1 The control unit (functional diagram)

The functions performed by the control system computer are, with reference to Figure 5.1, explained below.

Program interpretation. Chapter 12 discusses robot programming and languages from the robot programmer's point of view. However, the robot controller must take the programmer's instructions and translate them into a form suitable for the computer control system to handle. This will be more difficult and complex the higher the level of the programming language used. For example, when using the teach by leadthrough method of programming the controller simply has to record the values of the feedback signals being received from the internal robot sensors. At the other end of the spectrum the controller will have to interpret complex commands such as conditional jumps, branches and loops, in conjunction with incoming data from internal and external sensors, when handling user friendly high level languages.

Coordinate transformation. This capability is necessary to enable cartesian coordinates to be fed into the robot memory by the programmer, or conversely, allow cartesian coordinates of present arm position to be displayed to the programmer on a screen. The control computer works with the data fed to it by the internal sensors on the arm, in the case of a revolute type robot, for example, these data will be in the form of angular coordinates for each joint axis. Also, when the robot controller is fed instructions on where the arm has to move in cartesian world coordinates, it must transform these into the individual angular coordinates for each of the joints.

Trajectory computation. In Figure 4.7 various types of motion control were shown. In both coordinated path and continuous path the computer must carry out complex mathematical calculations to enable the robot to move in trajectories in accordance with those anticipated. Each axis must be coordinated with all of the others to achieve straight line movement, circular movement, or other complex movements of the end effector. If rapid movement of this type is required with the actuators working at the limit of their speed capabilities, then extremely fast real time calculations must be carried out by the computer system to ensure cycle times are kept to a minimum.

Servo control. This is the function that constantly analyzes the data feedback from the robot sensors and relates them to the command data being received from the program instructions. Control commands are continually sent out in accordance with this comparison to the control components to ensure the robot performs as required.

System supervision. To ensure the control system is working efficiently some sort of supervision is required. This function is often performed by a separate hardware component in the form of a microprocessor board within the controller housing. This liaises with all other functions by monitoring and controlling their data transfer.

Memory. Information on how the system is to perform is held in this area. Often the memory will be loaded at the beginning of the workshift of the robot. Data will be loaded from disk on the operating system and control language being used. If running a previously-created work program this will also be loaded in from disk. The memory may also contain the data necessary for initializing the system and the unique 'signatures' of the various characteristics, such as the exact length of links between joints, relevant to the particular robot being used. The memory will be constantly updated by information from the sensors and may be used to store data before sending it downline to data logging facilities or a supervisory computer.

Decision making and monitoring. This is the logical area within the system that monitors the sensor data, compares them with the instructions from memory and relates them to the computed trajectory. This function, in conjunction with the system supervisor, may also decide on what data should be passed for communication to external devices and systems.

Sensory data processing. After incoming sensor signals have been modified to suit the computer control system by the interfacing function they will be analyzed before passing the information to the servo control or decision making function. For example, the information from the internal sensors such as resolvers will be processed and sent to the servo control function, whereas the more complex information from a vision system, having already been preprocessed, will be passed to the decision making function.

The above description, including Figure 5.1, is not intended to be representative of any computer architecture that may be employed in control. Its purpose is to illustrate the various functions that are fulfilled by the control unit in an industrial robot system. An appreciation of the electronics involved in achieving this is covered in the remainder of this chapter. Some fundamentals of computers and computer control are covered and the basic components of computers and electronic systems are described. Concepts of logic and digital control are considered and finally interfacing is examined. For some readers already familiar with these subjects the following pages can be omitted; however, they are included for completeness as an understanding of them is necessary for the later chapters.

ELECTRICAL BASICS

An introduction to the terms used is given below.

Electric current. When electrons are in organized motion along a conductor an electric current is said to exist. This motion is caused by the existence of an electric charge. A source of charge is therefore required and this may be provided by a mechanical generator or a chemical battery. In the industrial robot situation a robot will normally utilize electricity from the mechanically generated mains supply. In the event of a mains failure there is usually a chemical battery backup to ensure the control system does not lose essential information. The standard unit of current is the ampere and its symbol is the letter I or i. The strength of current in a conductor is directly proportional to the emf applied and inversely proportional to the conductor resistance.

Emf, potential difference, and voltage. The charge source mentioned above will have a degree of capability of producing electron flow. This capability is indicated by the electromotive force (emf) which is measured in volts, symbol V or v. When an emf is applied to a circuit a potential difference (pd) will exist across it; this also is measured in volts. When the circuit is composed of a number of circuit elements separate pds will exist across each element. In a circuit, current flows in a direction from the higher to the lower potential, from the positive to the negative, or to 'ground' (or 'earth') if so connected. This is purely convention as the electron flow is in fact in the opposite direction.

Resistance. This may be considered as the opposition of a circuit or circuit element to current flow. If the emf remains constant then the higher the resistance the lower the current. The resistance present is related to the material, cross-sectional area, length, and temperature of the conductor. The unit of resistance is the ohm, symbol Ω. The relationship between pd, current, and resistance in a circuit is given by Ohm's law, $V = IR$. A basic circuit is shown in Figure 5.2.

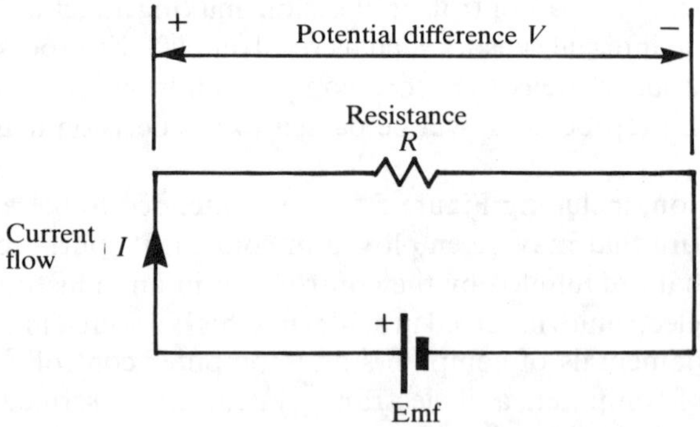

Figure 5.2 Simple circuit

DC (direct current) and AC (alternating current). Direct current is current which flows in one direction only and is usually of constant value. One source of DC is a battery although the value decreases over time as the emf decreases. Another source is a power supply which takes its own power from the AC mains supply and converts it to DC. Alternating current is current which flows first in one direction and then the other. For example, the UK AC mains supply changes its direction fifty times per second (50 Hz or 50 Hertz). The voltage/time graph has a sinusoidal shape as its value is proportional to the sine of an angle. In the USA the mains frequency is 60 Hz. Figures 5.3 and 5.4 illustrate DC and AC voltage/time graphs, the graph obtained for the AC is termed a 'waveform'.

Waveforms. The concept of plotting voltage against time can be applied to any cyclically changing voltage or current to obtain a time related graph or waveform. See Figure 5.5.

Frequency. This is the number of complete cycles of a wave that occur in one second; it is measured in Hertz.

Amplitude. This is a measure of the 'height' of a waveform from a given datum. In a

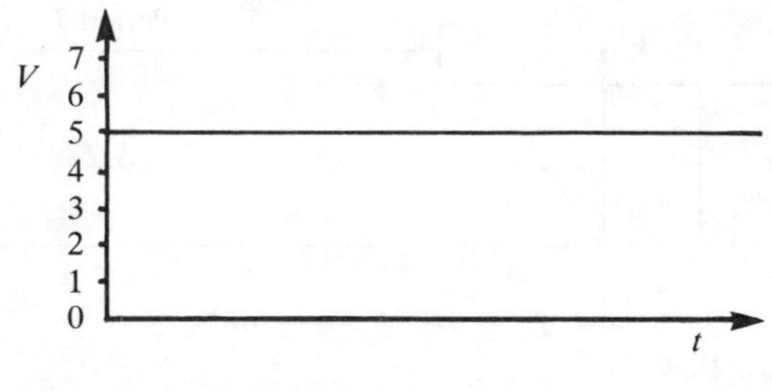

Figure 5.3 Ideal voltage/time graph for DC supply

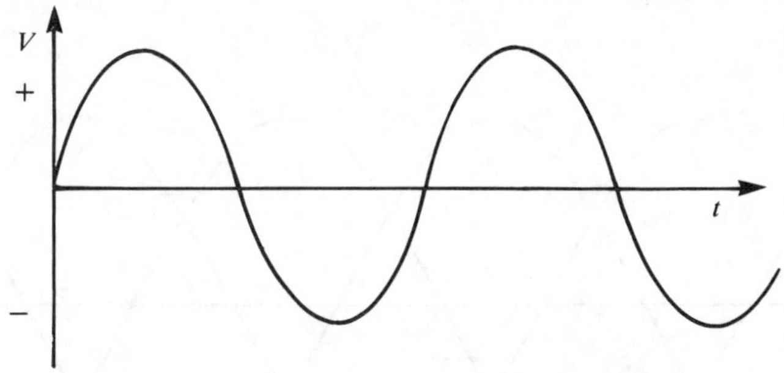

Figure 5.4 Voltage/time graph for AC mains supply (sine wave)

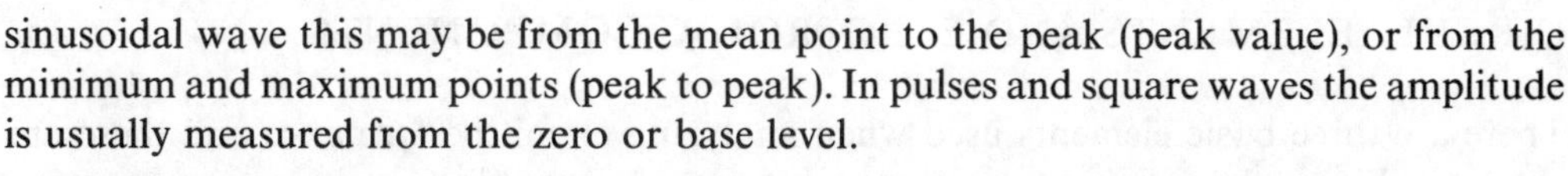

Figure 5.5 Some other waveforms

sinusoidal wave this may be from the mean point to the peak (peak value), or from the minimum and maximum points (peak to peak). In pulses and square waves the amplitude is usually measured from the zero or base level.

Period. When a quantity, such as voltage or current in a cyclical waveform, is a function of time the quantity is said to be periodic. The period of the function is T and is equal to l/f, where f is the frequency. Thus the periodic time is the time taken for one complete cycle. Figure 5.6 illustrates amplitude, period, and frequency.

Phase. If two points are considered as lying on the path of a waveform, then they will be points of equal phase if their displacements at a given moment in time are exactly similar. In the two-phase and three-phase supplies that will be mentioned later in the book the alternating voltages provided by these supplies are displaced relative to one another by one quarter and one-third of a period respectively. Figure 5.7 illustrates the waveform of a three-phase power supply commonly used in industry.

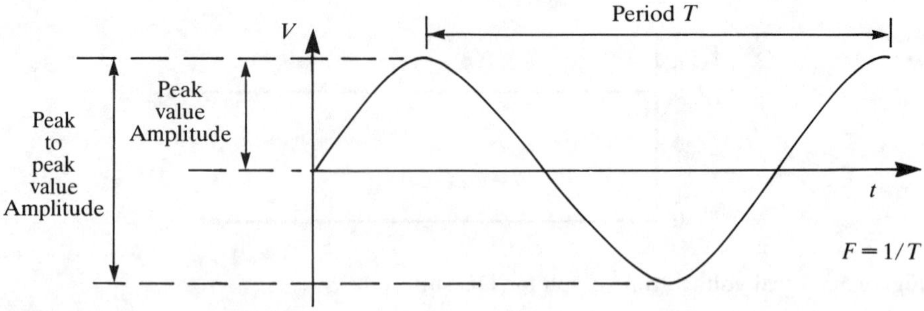

Figure 5.6 Amplitude, period and frequency

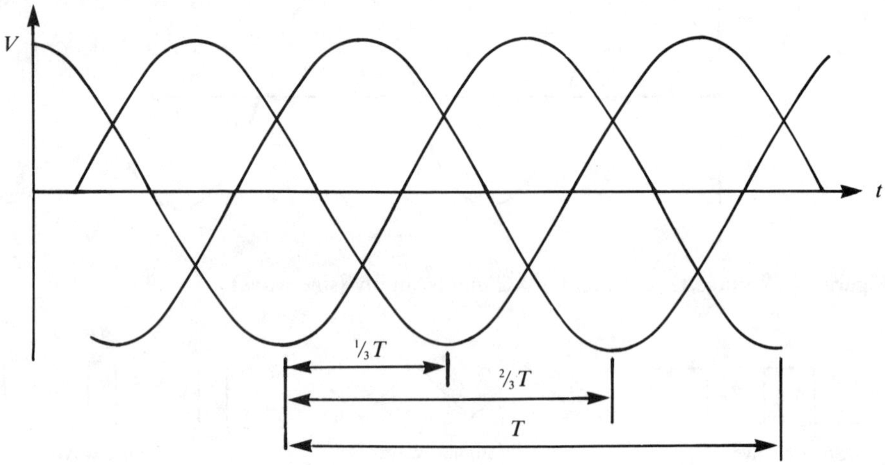

Figure 5.7 Waveform of three-phase power supply

CIRCUIT ELEMENTS AND ELECTRONIC COMPONENTS

There are three basic elements used when analyzing circuit performance and these are shown in Figure 5.8. The first, resistance, has already been discussed.

Capacitance. This is a measure of the ability of a device to store energy in the form of an electric field, it is measured in farads (F). A capacitor is a component which can be used to store electrical energy. The voltage across a capacitor cannot change instantaneously and it will not pass a DC current. Capacitors can therefore be used to perform various functions, such as smoothing out ripples in a power supply or blocking out DC signals.

Inductance. This is a measure of the ability of a device to store energy in the form of a magnetic field, it is measured in henrys (H). Where there is a conducting coil of material present in a circuit it will have the property of inductance. The current across an inductance cannot change instantaneously.

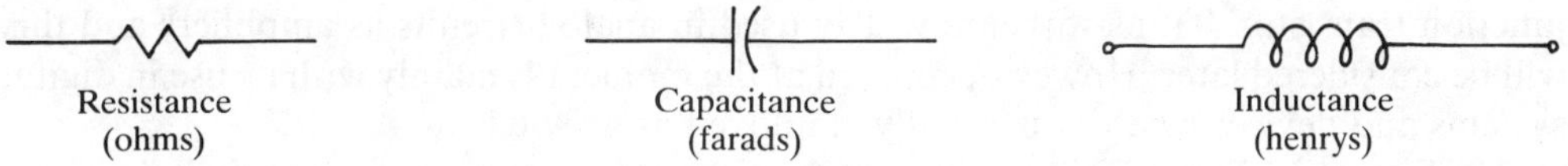

Figure 5.8 The three basic circuit elements

Discrete components used in electronic control circuitry are resistors, capacitors, diodes, transistors, and integrated circuits (ICs). The first two have been explained above, the remaining components are now discussed. All five components and their circuit symbols are shown in Figure 5.9.

Diodes. Diodes are components that have the characteristic of conducting electric current in just one direction. Thus they can be used as 'one way valves' to direct the movement of signals within the control system. They can also be used in circuits to change alternating current into direct current. Diodes are 'semiconductor' devices in that they are solid state and their electrical conductivity can be determined by the manner in which a voltage is applied across their contacts. The diode was the first type of semiconductor discovered and was thus the forerunner of today's solid state electronics industry. In the sketch a 'forward biased' diode is shown. When a voltage is applied across the diode with the positive reference terminal on the left and the 'bar' on the diode to the right, the current flows freely through the diode as shown. If the voltage is now reversed no current will flow. Another characteristic of diodes is that they maintain an almost constant voltage drop across themselves when switched on. This means that the voltage across a diode remains constant and independent of the current flowing through it. For the diode to be 'switched on' the voltage across it must be greater than this constant or 'forward' voltage.

Transistors. Transistors are essential parts of any electronic control system. Discrete transistors have three terminals although there are a number of different types available. In the common bipolar junction transistor (Figure 5.9) these terminals are the base (b), the collector (c) and the emitter (e). In the less widely used field effect transistor (FET) the terminals are the gate (G), the drain (D) and the source (S). FETs require a very small input signal current, but their high frequency response is poorer than the more common

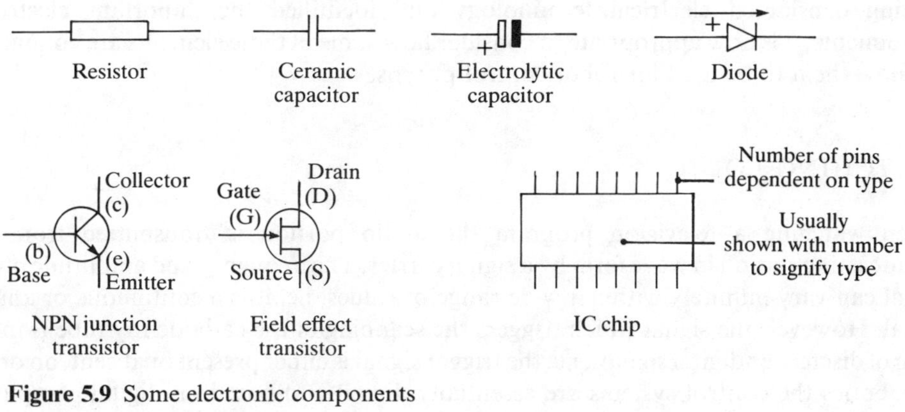

Figure 5.9 Some electronic components

junction transistor. Transistors are widely used in analog circuits as amplifiers and this will be considered later. However, concern at the moment is mainly with its use in digital systems and devices where it is usually employed as a switch.

In Figure 5.12 an NPN transistor is shown within a commonly used digital circuit. The NPN designation refers to the semiconductor structure of the device and in this case signifies that the collector has to be positive with respect to the emitter for the device to work. If a signal is applied to the base it will switch the transistor on or off. The emitter is connected to ground. If input V is 0 then the transistor is off and the output is at the value of the supply voltage +V. However, if a small signal is applied to the base raising the input voltage to a high enough value, the transistor will be switched on, allowing the output to be connected to ground. This gives a 0 or off signal at the output.

Transistors and diodes are the basic building blocks of the circuits used in digital control. The majority of these components will be found within a robot control unit in miniaturized form as the elements of an integrated circuit chip.

Integrated circuits. ICs were first developed in 1958 and are now applied in virtually all cases where electronic control is required. By using technology which allows carefully controlled layers of semiconductor material to be diffused or deposited in regions governed by patterns produced by a photolithographic process, solid state integrated circuits can be fabricated. This technology is continually advancing, and at the moment hundreds of thousands of discrete components such as transistors can be produced on 'chips' of about one centimeter square. All the ordinary electronic circuit elements and components, with the exception of inductance, can be fabricated in semiconductor form. These can then, in miniaturized form, be integrated into purpose-built IC chips in large numbers to create devices for control, sensing, decision making and computation. In fact, as all the functions of digital computers are obtained by simple on/off signals it is possible to integrate so many elements into a monolithic structure that the 'computer on a chip' has now been available for some time. Moving a step further, these single chip computers can be interconnected on a single printed circuit board to form a new breed of very fast and powerful computers capable of performing tasks of considerable complexity. Thus as advances are made in 'microelectronics' so the structures or architectures of control systems also change.

Having considered electrical terminology and identified the important electronic components, it is now appropriate to consider how some of these elements are configured to allow them to be used for robot control purposes.

SWITCHING LOGIC

When watching a television program the audio portion is transmitted from the broadcasting station in wave form by a signal carrier. The frequency and amplitude of this signal can vary infinitely within a wide range of values, i.e. it is a continuous or analog signal. However, the signal which triggers the scanning of the cathode ray tube is in the form of discrete or digital signals, i.e. the trigger signal is either present or absent, on or off. In robotics the control systems are essentially digital and based on digital computing facilities. Therefore some attention is given here to switching logic which is the basis of digital computer operation. It should be noted, however, that the robot will also have need to handle analog signals, e.g. it may have to transmit a constantly varying voltage signal to a servo valve, or it may have to receive a constantly varying voltage signal from a potentiometer. The digital computer can handle these signals by using digital to analog and analog to digital conversion devices. These are considered later.

Switching logic is used not only by digital computers but also by other automation and instrumentation systems. The elementary logic operations used to perform the logic functions are AND, OR, and NOT. These three elements are all that is needed to perform all the logic and computational functions of the computer and robot control system. The circuits that carry out these functions are called 'gates' as they open and close the logic paths within the system. There is a special type of algebra used to handle these operations. It is termed Boolean algebra after its inventor George Boole who set down its laws in 1854. There are sometimes limitations in using the AND and OR elements, and therefore it is possible to invert them into NAND and NOR functions. Indeed when these two elements are properly combined they can perform all necessary control functions. The circuits used to create the gates are shown here in Figures 5.10 to 5.19,* as though composed of discrete semiconductor components. This of course may be the case in some instances, but normally many millions of these gates are incorporated in miniaturized form throughout the large scale integrated circuit chips used in the robot controller. The operation of these gates is now discussed.

Figure 5.10 shows an AND gate in the form of its symbol, a diode circuit, and the associated truth table. The Boolean algebra function for this gate can be written as $C=A \cdot B$ or C=A AND B. In Boolean algebra the $\cdot$ signifies AND. The symbol shown for the gate is used on logic circuit diagrams and will normally be interconnected to many others to show how a system operates. The AND gate shows that there will be an output at C only when there is an input at A AND B. There may be any number of inputs to the gate but C will only be ON when *all* of the inputs are ON. The circuit diagram shows how this type of gate can be constructed using diodes. When the inputs at A and B are positive (1 or ON) voltage pulses then the diodes are reverse biased, no current can flow through them, and the output will be positive. If any or both of the inputs go to 0 then the diode(s) will become forward biased, current will flow freely through them, and the output will now be 0. The 'truth table' is a method of expressing exactly the function of a logic circuit. As

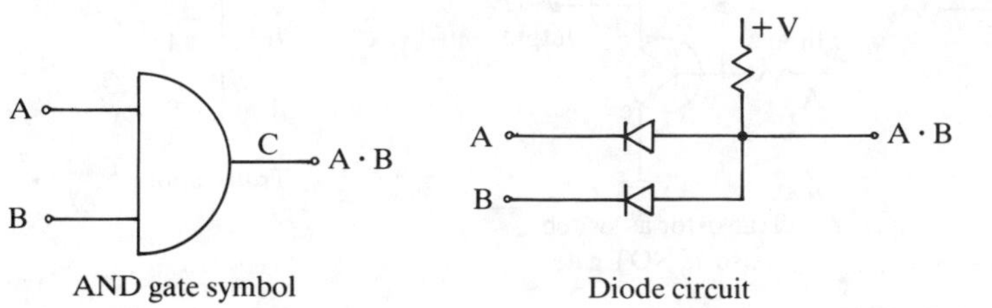

A	B	A · B
0	0	0
0	1	0
1	0	0
1	1	1

Truth table

Figure 5.10 A two-input AND gate using two diodes

*Symbols for representing gates vary considerably. The British Standard for example uses rectangular boxes to represent the gates. The symbols employed here, however, are probably more widely recognized and also give an indication as to the direction of the logic 'flow'.

can be seen for this circuit the output A · B is 1 only when input A *and* input B are 1. (These truth tables show the output conditions for all possible combinations of input signals. The number of possible combinations is given by 2^n where n is the number of variables, i.e. inputs. The convention for writing truth tables is that 0 represents no signal present and 1 that a signal is present.)

Figure 5.11 illustrates the symbol, circuit and truth table for an OR gate. In Boolean algebra the + signifies OR. As the truth table shows, the output will be 1 if input A *or* input B is 1. The circuit diagram shows that if there is an input at A or B or both, then the diode(s) will be forward biased, current will flow, and the output voltage will rise to give a signal. For no input no current flows and the voltage is 0.

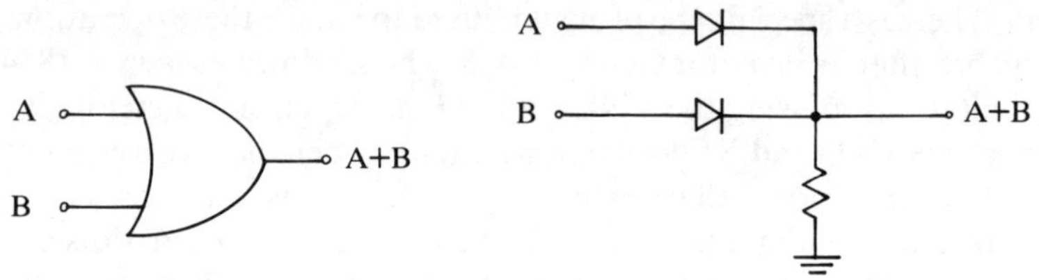

OR gate symbol

Diode circuit

A	B	A+B
0	0	0
0	1	1
1	0	1
1	1	1

Truth table

Figure 5.11 A two-input OR gate using two diodes

Figure 5.12 shows a NOT gate. This is based on the transistor switch previously described. This circuit 'inverts' the input signal. An input of 1 at A will give an output of NOT A, signified by the convention $\bar{A}$ in the diagram. Thus the NOT gate is an 'inverter' giving the 'complement' of the input signal. With no input signal, 0, the transistor switch is held open by the negative base voltage and the output is 1. A positive input voltage 1, forward biases the emitter junction, the collector current will flow, and the output voltage will drop to virtually 0. The small circle on the NOT element symbol, in a similar manner to the bar above the output, indicates the inversion process.

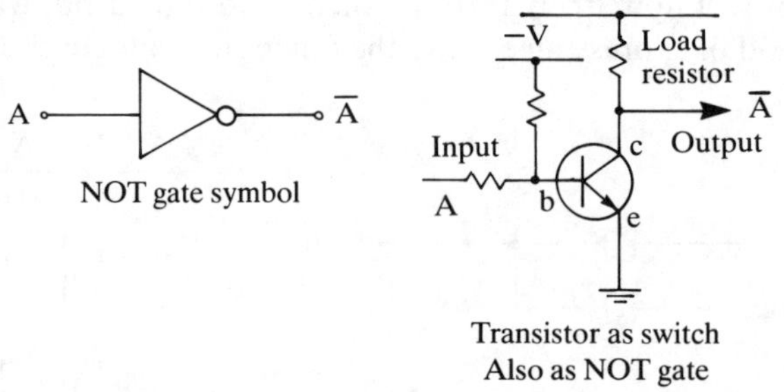

NOT gate symbol

Transistor as switch
Also as NOT gate

A	$\bar{A}$
0	1
1	0

Truth table

Figure 5.12 A NOT gate using a transistor

The NOR gate shown in Figure 5.13 is composed of a transistor and two diodes. This is in effect a combined OR and NOT function or NOT OR gate. Thus a NOR gate is a device which has two or more inputs such that the output is only active when no inputs are

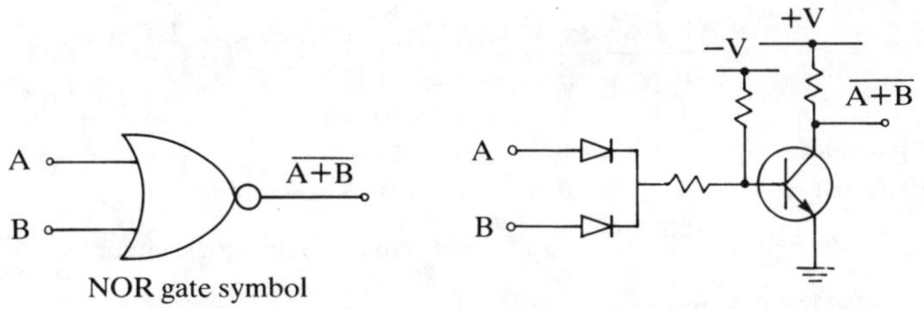

NOR gate symbol

Diode and transistor circuit

A	B	$\overline{A+B}$
0	0	1
0	1	0
1	0	0
1	1	0

Truth table

Figure 5.13 A NOR gate using two diodes and a transistor

present. In the OR gate shown earlier the output signal is less than the input because of the diode voltage drop. This means that if the gates were connected to each other in a series, or 'cascaded', then the signals would rapidly decrease to an unreliable level. By using the NOR gate the transistor restores the signal level but of course inverts the signal. In the circuit shown the transistor is held OFF by the negative base voltage and the output is 1. A positive voltage signal at A or B raises the base potential, forward biases the emitter junction, turns the transistor switch ON, and drops the output virtually to 0. Other advantages of this circuit are that the NOR output can be used to supply inputs to many other gates, and also it can be used as the building block for all the basic logic operations.

The NAND gate in Figure 5.14 has all the advantages of the NOR gate and is easy to fabricate in integrated circuit form, it is therefore the most common. This gate represents an AND gate whose output has been complemented to give a NOT AND function. A NAND gate is therefore a device with two or more inputs and one output, so that the output signal is positive except when all the inputs are positive. In the circuit diagram with positive inputs at both A and B, both the parallel diodes are reverse biased and no diode current flows. The positive base current causes a high collector current and the output is virtually 0. When one or both of the input signals drops to 0 the diode(s) conducts to ground, thus no base or collector current flows and the output will be 1.

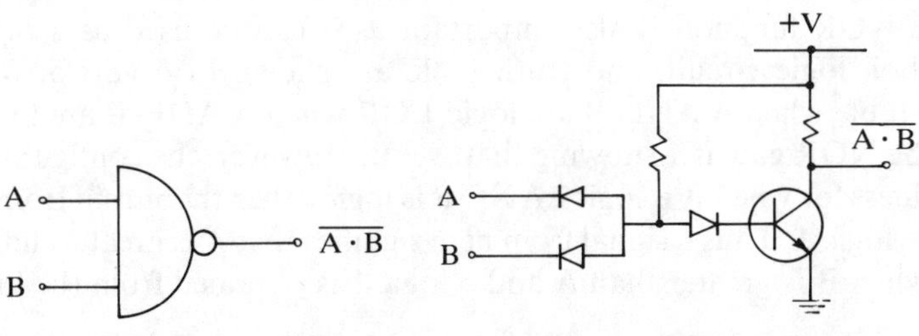

NAND gate symbol

Diode and transistor circuit

A ·	B	$\overline{A \cdot B}$
0	0	1
0	1	1
1	0	1
1	1	0

Truth table

Figure 5.14 A NAND gate using two diodes and a transistor

In the EXCLUSIVE OR or XOR function the output is 1 only if the inputs are different. The symbol, truth table, and logic circuit comprising two NOT gates, two AND gates, and one OR gate are shown in Figure 5.15. This element is extremely useful and forms the basis of the logical function in the computer system arithmetic adders. How it does this will be explained on page 66.

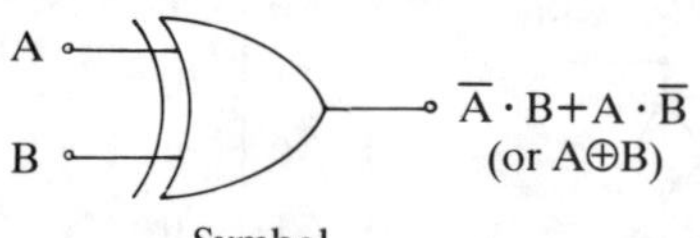

Symbol

A	B	$\overline{A}\cdot B+A\cdot\overline{B}$
0	0	0
0	1	1
1	0	1
1	1	0

Truth table

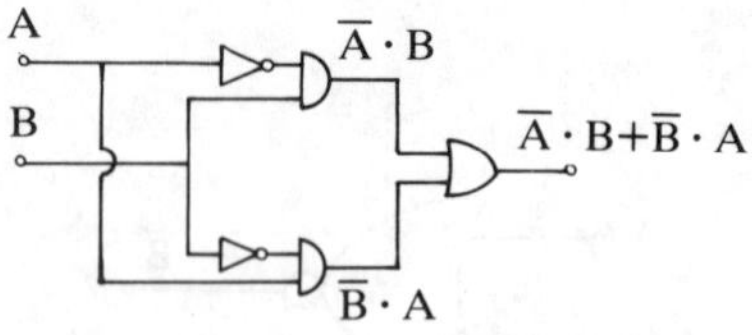

An equivalent logic circuit

Figure 5.15 The EXCLUSIVE OR function

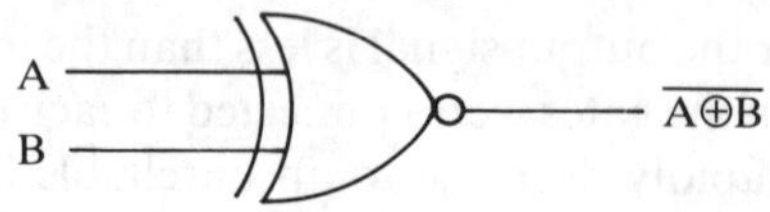

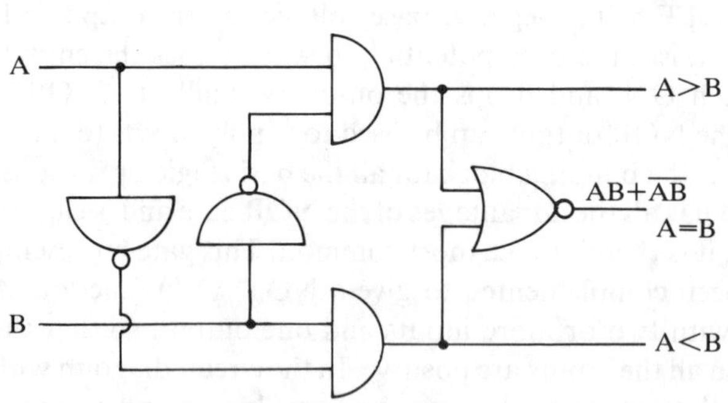

A	B	A>B	A=B	A<B
0	0	0	1	0
0	1	0	0	1
1	0	1	0	0
1	1	0	1	0

Figure 5.16 EXCLUSIVE NOR function and truth table

The EXCLUSIVE NOR function is also important as it can be used as a digital comparator. The symbol, logic circuit, and truth table are shown in Figure 5.16. As evidenced by the truth table when A AND B are logic 1 OR when A AND B are logic 0 then the output from the NOR gate is 1 showing that A = B. However, the configuration shown has added usefulness for when A is logic 1 AND B is logic 0 then the output from the upper AND gate will be logic 1. Thus a signal from here signifies that A is greater than B. The converse applies when B is greater than A and a logic 1 is obtained from the lower AND gate.

LATCHES

Within the digital control system it is necessary to have the ability to allow binary information to be held temporarily for use in memory or computation. The operations that allow this are implemented by 'bistable' logic circuits, i.e. circuits that have only two stable states. These are called 'flip flops' or 'latches'. How one type of flip flop operates is shown in Figure 5.17. In this case it is shown in a practical application where it is sending a

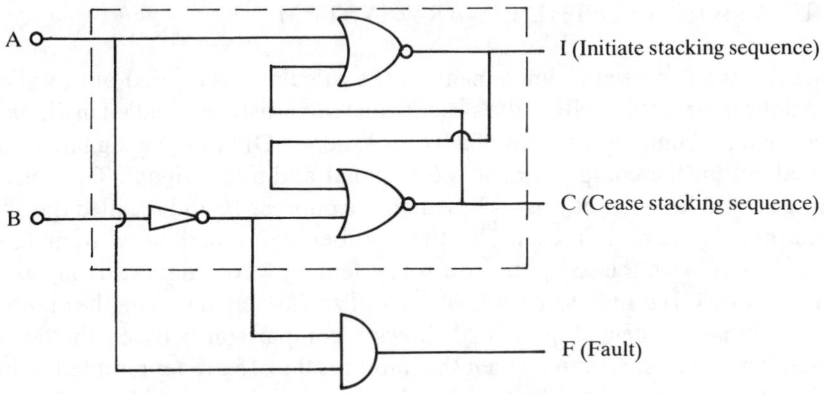

Figure 5.17 NOR flip-flop or latch (incorporating additional AND gate)

signal to a robot. The robot has to maintain a stack of components being fed to another operation, within a given height range. The maximum and minimum height of the stack is sensed by two photoelectric sensors that send a value of 1 to the flip flop when occluded by a component.

Assume that the sequence begins with no components in place. There is therefore a 0 signal from the sensors on both the A and B lines. The 0 signal on B passes through the NOT gate and becomes a 1 into the lower NOR gate. Thus the signal from the lower gate is 0. This signal is fed into the upper gate together with the 0 signal from the upper sensor A. These two signals provide a 1 output from the upper gate telling the robot to initiate the stacking sequence. As the stacking progresses the lower sensor becomes occluded causing the signal on B to go to 1. This is inverted to 0 before being input to the lower gate. This means that there is now a 0 signal, and a 1 signal from the upper gate output, going into the lower gate. The lower gate continues therefore to produce logic 0 and the stacking process is maintained. When the robot stacks the components high enough to occlude the top sensor the input from A changes to 1. This means that there is now a 1 and a 0 into the top gate giving an output of 0. This output is fed into the bottom gate together with the existing 0 signal from B. The output from the lower gate will now be 1 telling the robot to stop the stacking sequence and possibly go to a second task.

The circuit will now operate in reverse. The robot will continue to work on its second task even after input A goes to 0, and will continue to do so until the sensor providing input to B goes to 0 so resetting the system. Should at any time sensor A be occluded and sensor B not occluded then this will signify a fault, e.g. if the components were being stacked in a magazine jamming may have occurred. This fault will be recognized by utilizing the AND gate as shown.

Although demonstrated here in a simple control application thousands of these flip flops or latches are used in integrated form to construct memory chips. Registers are one type of memory device formed in this way. They are used to store multiple bits of information which can be processed in a variety of fashions, as for example in the successive approximations register used in the analog to digital converter mentioned on page 77.

COMPUTATION AND THE BINARY SYSTEM

Logic circuits are also used to implement all the calculations carried out by the control system. All the information within the system must obviously be handled in digital signals, each piece of data being in the form of a 0 or a 1, i.e. an ON or OFF signal. This may be represented within the computer as a +5 V signal and a 0 V signal. The conventional numbering system for everyday use is based on the number 10 and is called the decimal or denary number system. For example, the number 252 could be written $(2 * 10^2)+(5 * 10^1)+(2 * 10^0)$. In the computer control system, however, data can only exist in one of two stages, 0 or 1. It is therefore necessary to utilize a system based on the number 2, this is called the binary system. Figure 5.20 shows a comparison between the decimal and binary number systems. As can be seen the numbers 0 to 15 are represented in binary by four digits which can be either 0 or 1. Using this system the number is written right to left, the first number written being termed the least significant bit (LSB), the last number written being termed the most significant bit (MSB). Each column therefore represents a power of 2 and each time this is reached a new column is started, moving right to left. The numbers 0 and 1 are called binary digits or 'bits' for short. Typical microprocessor-based systems such as computers usually manipulate data in blocks of 8, 16, or 32 bits. The number 252 is therefore written as 11111100, i.e. $(1 * 2^7)+(1 * 2^6)+(1 * 2^5)+(1 * 2^4)+(1 * 2^3)+(1 * 2^2)+(0 * 2^1)+(0 * 2^0)$. The MSB, the bit on the far left is a 1 in this case, and the LSB is a 0.

To carry out addition in binary the same principles as used in decimal arithmetic apply, i.e. the numbers are added column by column and the numbers carried where necessary into the next higher column. For example, adding decimal 15 and decimal 11: decimal 15 in binary is 1111 and decimal 11 is 1011. Adding these in binary gives:

```
   1 1 1 1
   1 0 1 1
 ---------
 1 1 0 1 0   i.e. decimal 26.
 ---------
```

As addition is the basis of the arithmetic functions in the computer it is worth noting the logic circuits used for this purpose.

Figure 5.18 shows the half adder circuit. This circuit accepts inputs on A and B. As can be seen from the truth table if both inputs are 0 then the outputs are 0. If A is logic 1 and B is logic 0 or vice versa, their sum (S) will be 1. If both are logic 1 then their sum (S) will be greater than the capacity of that bit and the value will have to be carried (C) to the next higher bit column. This circuit is called a 'half adder' as it is only capable of accepting two inputs. If a complete circuit for addition is required a full adder circuit is necessary as shown in Figure 5.19. The full adder can accept the two numbers to be added plus the carry input from the previous operation. It comprises two half adders and an OR gate. Thus when adding two binary numbers of more than one bit we require a half adder for the first addition column and full adders for subsequent columns.

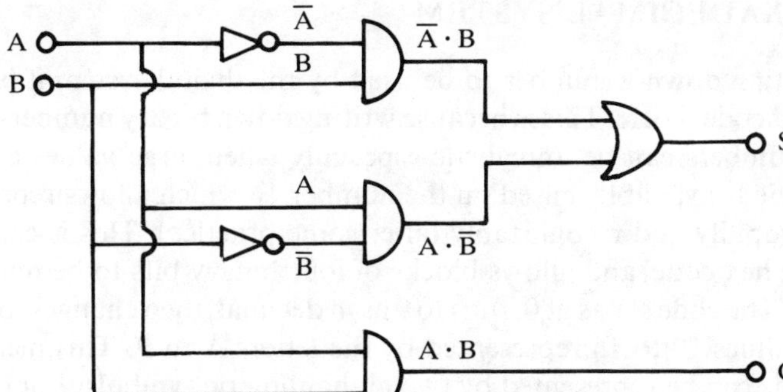

A	B	S	C	A plus B = sum
0	0	0	0	0 plus 0 = 0
0	1	1	0	0 plus 1 = 1
1	0	1	0	1 plus 0 = 1
1	1	0	1	1 plus 1 = 0 with carry 1

Truth table — Binary addition

Figure 5.18 Half adder

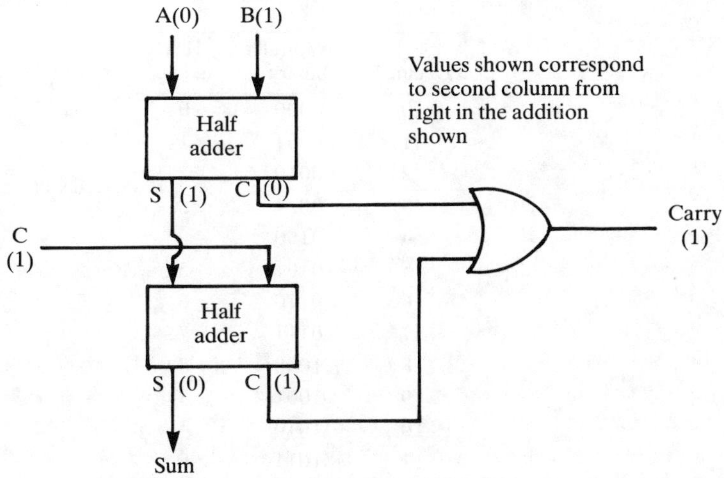

	A	0	0	1	0	1
plus	B	0	1	0	1	1
		1	0	0	0	0

Figure 5.19 Full adder

THE HEXADECIMAL SYSTEM

When writing down a number to be used by the digital system it is necessary to find a convenient code to use. This is because writing down binary numbers is tedious and using decimal numbers can be unrealistic especially when large values are being considered. Thus a code is available, based on the number 16 which allows representation of binary numbers rapidly and recognizably (after some practice). This is called the hexadecimal system or 'hex code' and allows blocks of four binary bits to be represented by one hex character. The code starts at 0, runs to 9 as in decimal, then changes to alphabetic notation with the values 10 to 15 represented by the letters A to F. This means that an eight bit binary code can be represented by two alphanumeric symbols. For example, the decimal value 7 in binary is 00000111. In hex this would be translated into blocks of four bits 0000 and 0111 and be represented by 07. Similarly decimal 20 is binary 00010100, which is split into 0001 and 0100 to give 14 in hex. Decimal 205 is binary 11001101 which would be CD in hex. For reference the decimal, binary and hex codes are shown in Figure 5.20.

The hex code is useful for humans to use when representing binary number values. It is also possible to input hex code directly to a control system if proper translation facilities are included in the computer. However, with industrial robots high level languages or codes are almost always used and these will be considered in Chapter 12.

THE COMPUTER SYSTEM

Having considered the basic ingredients of digital systems it is now appropriate to look at the computer more generally. The discrete components, gates, and their assemblies are now treated as 'black boxes' within the larger 'box' of the robot controller.

Decimal	Natural binary	Hex code
0	0000	0
1	0001	1
2	0010	2
3	0011	3
4	0100	4
5	0101	5
6	0110	6
7	0111	7
8	1000	8
9	1001	9
10	1010	A
11	1011	B
12	1100	C
13	1101	D
14	1110	E
15	1111	F

Figure 5.20 Comparison of number systems and codes

In most servo controlled industrial robots the control system is based on one or more microprocessors. To describe a microprocessor it is necessary to first consider the operation of a digital computer. The computer has essentially five basic functions as shown in Figure 5.21. These are:

1. The input function which allows communication from the outside world to the computer.
2. The output function which allows communication from the computer to the outside world.
3. The memory function which stores information and also the instructions as to what to do with the information.
4. The arithmetic and logic function, or the ALU (arithmetic/logic unit). This carries out the mathematical and logical functions of the computer.
5. The control function which sequences and controls the computer operations.

Just as with the overall robot control system described earlier these are functional categories only, and do not necessarily correspond to discrete hardware components. However, the categories contained within the large box in Figure 5.21 will be found in the digital computer as a physical area termed the CPU or central processing unit.

THE MICROPROCESSOR SYSTEM

A microprocessor is a large scale integrated circuit (LSI) device, that embodies the functional abilities of a digital computer contained in the CPU as discussed above. It will

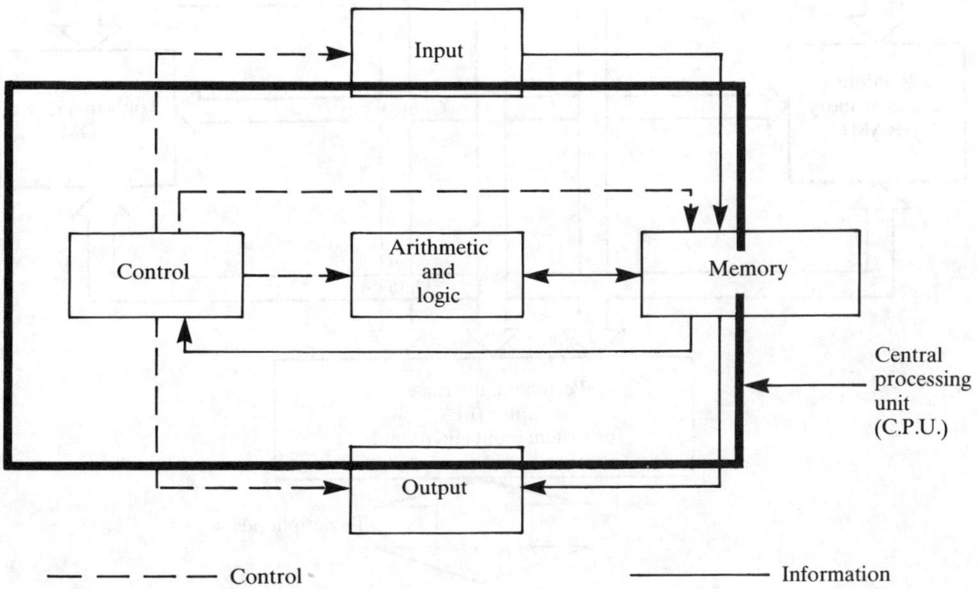

Figure 5.21 Digital computer (functional diagram)

usually contain therefore arithmetic and logic capabilities, an ability to perform the control function, some memory capacity, and facilities for input and output, i.e. communication with the outside world. Physically it will be comprised of the equivalent of thousands of discrete transistors and diodes creating the gates necessary to perform the appropriate computer functions. The microprocessor does not operate in isolation but as a part of a larger system. A typical minimum system configuration is shown in Figure 5.22. This comprises the microprocessor chip, a RAM chip, a ROM chip, and a PIA chip.

The ROM chip contains the 'read only memory' function. This contains the program of instructions necessary for the system to be able to perform, i.e. to sequence the microprocessor through the necessary operations. The information on this chip will have been created by the manufacturer to suit the system configuration being used, this information is fixed and cannot be altered.

The RAM chip contains the 'random access memory', this area can be loaded with instructions by the programmer to suit the particular application to which the system is being applied. This area can also store data temporarily, transmit (WRITE), and receive (READ) data as and when required by the program.

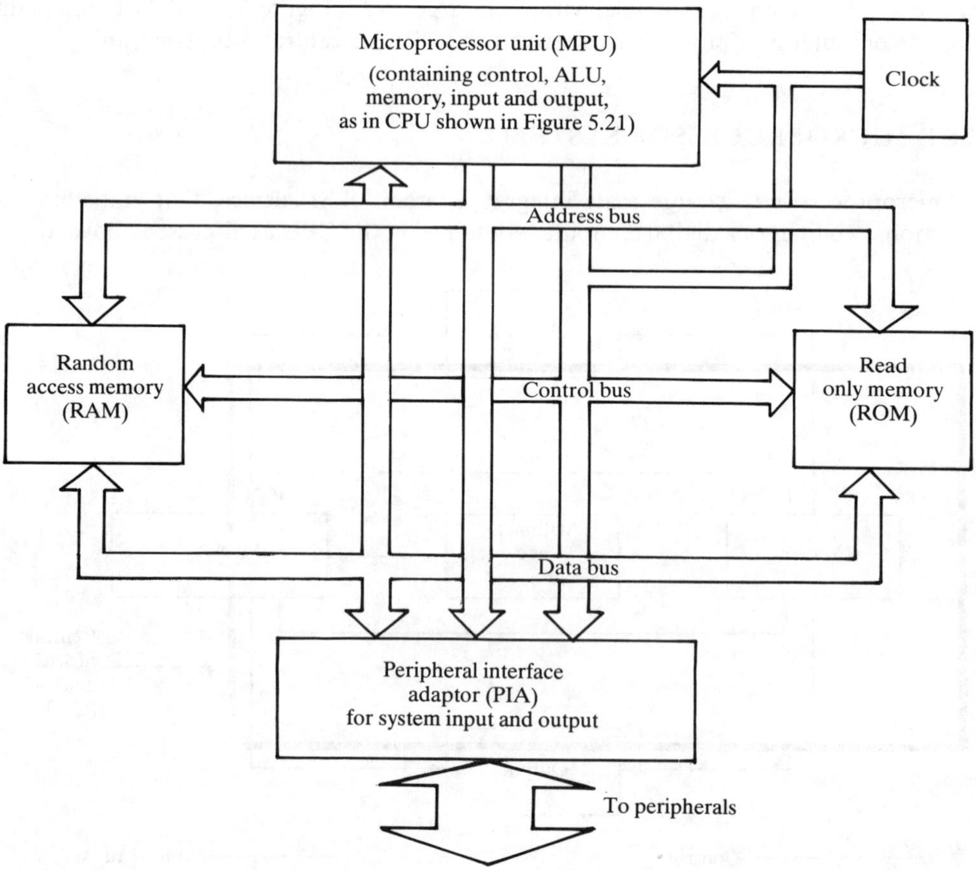

Figure 5.22 Basic microprocessor system

In the system shown in the figure a 'peripheral interface adaptor' or PIA chip is being used to interface between the microprocessor system and the outside world. A PIA is used where parallel data are being handled. In parallel communications there is a wire for each bit of information being transmitted. Another device that could have been used here is the USART, i.e. the 'universal synchronous/asynchronous receiver transmitter' which is used where serial data input and output are being handled. Serial communication requires one wire along which the bits can be transmitted in series. Both serial and parallel systems require an additional ground wire, and the parallel system also requires a further wire for control purposes, these systems are considered again later. The input/output chips are complex pieces of hardware under software control, handling communications both into and out of the system.

Information between the system chips is sent in packets or blocks of binary data (bytes). Depending on the system used a byte may contain eight, sixteen, or thirty-two bits. These data are transferred around the system on groups of electrical connections called 'bus' lines. A bus provides a communication path between all sections of the system and there is generally one for each of the control, address and data communications functions as shown in Figure 5.22. There will also be a power bus to drive the system.

To ensure the information is transferred properly and that packets of data do not interfere or corrupt each other, the timing of the system flow is controlled precisely by the system clock. The clock is a chip based on a piezo-electric crystal which sends out a continuous stream of timing pulses. When a voltage is applied to a piezo-electric crystal it vibrates at a precise frequency. It is this vibrational frequency that is used to provide the timing pulses. These pulses are utilized by the control function to 'enable' the appropriate system components at precisely the correct time to transmit or receive data to or from the bus structure.

It is now common for an industrial robot controller to have multiple microprocessor systems. For example, there can be one microprocessor system to control each axis, one to control peripheral equipment inputs and outputs, and one to act as a supervisor over the total system. This type of distributed control offers advantages of speed as each subsystem can make decisions on its own without consulting the supervisory system. This principle can be applied to 'add on' equipment, for example it is possible to purchase 'intelligent' grippers which contain their own microprocessor system. This type of gripper incorporates tactile sensors and can analyze the sensory data before transmitting only the essential information to the robot controller. Thus the overall robot control system may contain seven or eight microprocessor-based 'sub' systems, this is shown schematically in Figure 5.23.

This concludes an appreciation of the microprocessor system and from now on this will also be treated as a black box capable of handling and manipulating input and output data from the rest of the overall robot system.

PERIPHERALS

Deviating slightly from the course steered so far, it will be helpful at this stage to step outside the robot controller and note very briefly the various pieces of peripheral equipment associated with the control system.

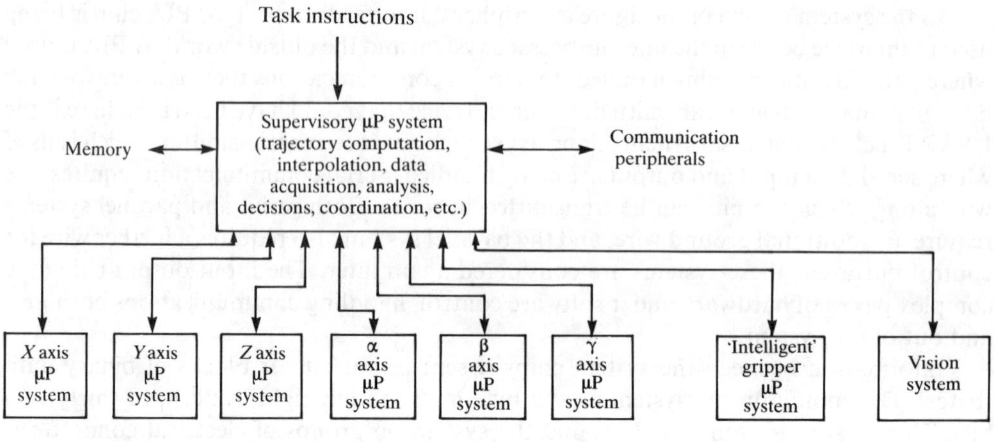

Figure 5.23 Two-level hierarchical control

To store information on the robot operating system, high level language information, and the work task programs, floppy disks are often used. These are magnetic disks containing all the necessary data to run the robot. They are loaded into disk drive units which read the coded information on the disks and transfer it to the robot control system memory. Other storage media may be used such as magnetic tape (now uncommon) and bubble memories which are still expensive.

A visual display unit (VDU or CRT) is used to display information on how the robot is operating, i.e. what stage it is at in a program, what malfunctions have occurred, messages from the robot sensors, diagnostic messages, etc. To provide a permanent record, or hard copy, of this information a printer will also be required. This may generate the print in a variety of ways, e.g. a daisy wheel giving good quality but slow speed, a dot matrix giving slightly poorer quality but higher speed, or a laser printer which gives good quality and high speed but is more expensive. Dot matrix printers are usually appropriate for industrial robot applications.

To enable the robot programmer to input information to the system, either a teach pendant or computer keyboard is used. The teach pendant is a small custom keyboard, sometimes with a joystick, that can be held in the hand for convenient use when programming. This will be considered further in Chapter 12.

INTERFACING

Serial and parallel data transmission

The microprocessor system must be able to transmit control commands to the outside world, i.e. motor controls, servo valves, solenoids, VDUs, printers, etc. It must also be able to receive signals from the arm feedback sensors and from other peripheral equipment such as the teach pendant. To do this a method of interfacing the microprocessor system to the other devices is required. The microprocessor system has been considered up to the input/output stage where data were being either received or transmitted between the microprocessor chip and the outside world via a special purpose chip. Data within the system will be handled in parallel. It may be that the other devices will be in close proximity, i.e. less than fifty meters, to the controller and therefore parallel transmission may be used. However, if the communication was to be over a longer distance then serial transmission would be necessary.

Within a conventional robot system the robot controller, the control components, the feedback sensors and most peripherals will all be in close proximity and therefore parallel data transfer used. This is obviously much faster than serial transfer as all the bits required for one byte are transferred contemporaneously, each along its own wire. Thus to transmit an eight bit word, eight wires would be required for the data, plus one wire for a ground signal to provide a reference value, plus one or two wires carrying control signals. To ensure compatibility between equipment communicating in parallel various standards have been adopted. One of the most common is the IEEE 488 parallel standard interface. This has twenty-five wires and a twenty-five pin connector.

When it is necessary to transmit data over longer distances serial data transfer is necessary to reduce the cost of wiring. It is also necessary when using, for example, a telephone line to carry data, as may be done when carrying out remote diagnostics. When data are transferred serially only two wires are needed, one to carry the data, and one to act as a ground so providing a datum voltage level. The equipment compares the value on the ground wire with that on the signal wire to determine whether the bit is a 1 or a 0. Serial data transmission may be either synchronous or asynchronous. The latter is the more common. With this method the receiver recognizes the the first 0 in the flow of information as a 'start' bit then proceeds to accept and decode the following information according to a predetermined system, which will be discussed in the next section. This method takes longer than the synchronous as start and stop bits are required to identify each character. With synchronous transmission clock pulses are sent by the transmitter to synchronize the transmitting and receiving of the data, this allows a higher density of information to be transmitted in a given time. The RS232-C is a standard hardware interface for serial data transmission in either a synchronous or an asynchronous manner.

To convert parallel data to serial data, a 'shift register' is used. This register accepts the parallel input from the data bus as shown in Figure 5.24. Each pulse from a clock will cause the bits in the register to move down one place and the bottom bit will be output. For an eight bit word applied to the register, eight clock pulses will be required to output it serially. When receiving serial data once again a register is used. In this case the serial data are entered at the top and each clock pulse enters one bit of data into the register again moving a bit one step downwards with each pulse. Shift registers are configured with other registers and control logic into LSI chips, e.g. the USART mentioned earlier.

The ASCII code

For transfer of data between the robot system controller and other computers and peripheral equipment some sort of standardized code must be adopted to represent the characters to be communicated along the data lines. The data will be either numeric, alphanumeric, special characters such as brackets, or control characters which will be used by the equipment at both ends of the communication lines to control the data flow. The most widely used code is the one originally put forward by the American National

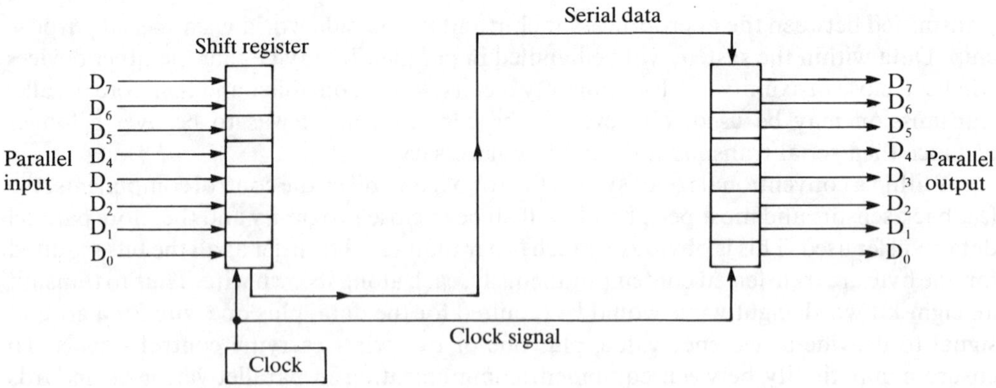

Figure 5.24 Simplified sketch showing the use of shift registers for parallel to serial and serial to parallel conversion

Standards Institute called the American Standard Code for Information Interchange or ASCII. Another code, the EBDIC, Extended Binary Coded Decimal Information Code, is used by IBM. However, it is the ASCII code that is examined here. This is a seven bit code and can therefore represent 128 different characters, often an eighth bit is added as a method of error detection. Figure 5.25 shows this code in tabular form together with examples of implementation. The control characters in the first two columns are easily identifiable by the computer as bits seven and six will both be zero, any other combination will mean the data refers to a character suitable for processing. Thus the rows specify the bits one to four, while the columns along the top specify the bits five to seven. To translate ASCII code to hex code the row and column values shown on the right and along the bottom of the table can be used. For example, the control character SUB is in ASCII code equal to binary 0011010. This can be split into 001 and 1010 which is hex 1A.

Baud rate

The rate at which data are transferred between equipment is also important. Speed of transmission must be equal to speed of reception otherwise communication would be impossible. This speed is measured in serial bits transferred per second and is termed the 'baud' rate. Thus a baud rate of 1200 signifies that 1200 bits of information are being transferred per second. Both the transmitter and the receiver must be compatible. Usually the baud rate can be set on the equipment either by hardware using small switches or by software using a keyboard. Once the transmitter and receiver are set to the same baud rate transmission can begin. The number of characters being sent per second will equal the baud rate divided by the total character length including start, stop, and parity bits used for error detection.

As a general comment here, it may be noted that in very large systems using equipment supplied by a variety of manufacturers, interfacing problems become more complex. One approach to this is to use communications protocols, this will be discussed in a later chapter.

Bit number												
			Seven	0	0	0	0	1	1	1	1	
			Six	0	0	1	1	0	0	1	1	
			Five	0	1	0	1	0	1	0	1	
Four	Three	Two	One									
0	0	0	0	NUL	DLE	SP	0	@	P	‘	p	0
0	0	0	1	SOH	DC1	!	1	A	Q	a	q	1
0	0	1	0	STX	DC2	″	2	B	R	b	r	2
0	0	1	1	ETX	DC3	#	3	C	S	c	s	3
0	1	0	0	EOT	DC4	$	4	D	T	d	t	4
0	1	0	1	ENQ	NAK	%	5	E	U	e	u	5
0	1	1	0	ACK	SYN	&	6	F	V	f	v	6
0	1	1	1	BEL	ETB	’	7	G	W	g	w	7
1	0	0	0	BS	CAN	(	8	H	X	h	x	8
1	0	0	1	HT	EM	)	9	I	Y	i	y	9
1	0	1	0	LF	SUB	*	:	J	Z	j	z	A
1	0	1	1	VT	ESC	+	;	K	[	k	{	B
1	1	0	0	FF	FS	,	<	L	\	l	:	C
1	1	0	1	CR	GS	—	=	M	]	m	}	D
1	1	1	0	SO	RS	.	>	N	^	n	~	E
1	1	1	1	SI	US	/	?	O	_	o	DEL	F
				0	1	2	3	4	5	6	7	Row / Column

E.g. 9 is 0111001 or 39 (hex)
SUB is 0011010 or 1A (hex)

Figure 5.25 American Standard Code for Information Interchange, or ASCII code

Digital to analog and analog to digital conversion (DAC and ADC)

The microprocessor-based controller of the robot is a digital system using binary numbers to represent values. Conversely the servo motors in an electric robot and the servo valves in an hydraulic robot are analog devices and they respond to analog control signals. The transducers or sensors used for feedback to the control system are also often analog. For instance, the tachogenerator is an analog device providing an analog signal, as too are potentiometers, resolvers, and strain gauges. When communicating with these analog components the robot controller must therefore convert digital signals to analog values when issuing commands, and convert analog to digital when receiving information. This necessitates the use of digital to analog and analog to digital converters. Once again these are usually in the form of integrated circuit chips.

The principle of one type of digital to analog converter is shown in Figure 5.26. Basically what is required is to convert a digital value, represented in the controller as a binary number, into an analog value usually in the form of a voltage. The type shown is called an R-2R ladder due to the configuration of the resistances. In practice the mechanical switches shown would be replaced by FETs and the circuit integrated on a chip. In this example the input to the converter from the microprocessor system is in the

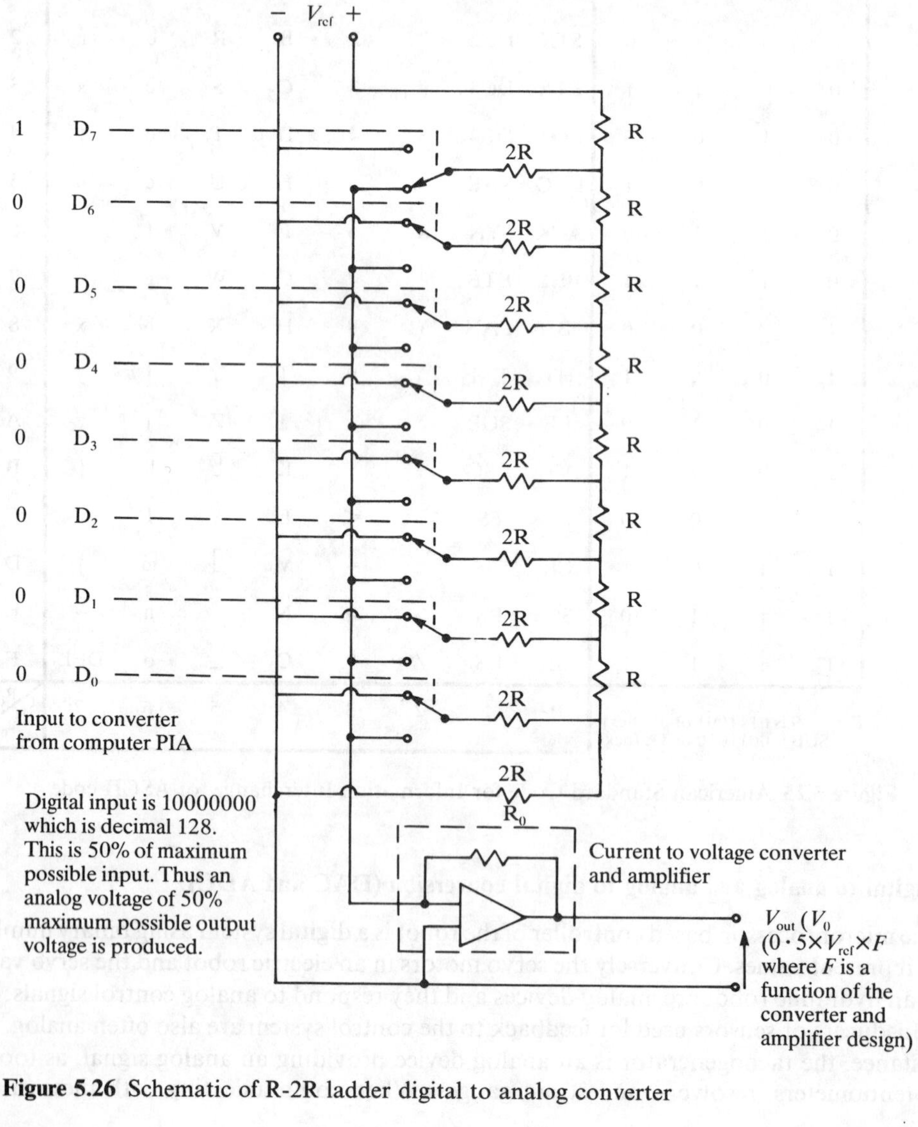

Figure 5.26 Schematic of R-2R ladder digital to analog converter

form of an eight bit word. The most significant bit turns ON or OFF switch D representing a decimal value of 2^7 or 128. The least significant bit only ON will represent 1/256 of the maximum possible value, and all switches on will represent the maximum value. Thus in the configuration shown the output voltage will be proportionate to a reference voltage in a ratio dependent on the binary pattern applied across the FET switches. The analog output will increase or decrease in a series of steps, the total number of steps, or 'resolution', between 0 and maximum depending on the number of bits used in the input.

In order to convert analog input signals to the system into digital, an analog to digital converter such as the one shown in Figure 5.27 can be used. A to D conversion can be implemented in a number of ways, the one shown here is quite common. This successive

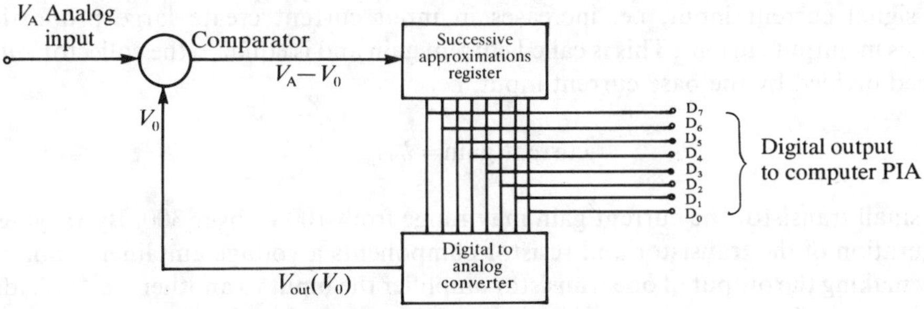

Figure 5.27 Simplified block diagram of analog to digital converter

approximations method operates as follows. The register contains a digital value which it can send as output to the robot system or to the DAC. The DAC converts this to an analog output voltage V which is sent to a comparator. Here the analog equivalant of the value held in the register is compared with the analog input signal. The difference ($V_A - V_0$) is sent to the register to modify the original register value. For example consider that the register is first of all set to 10000000. When the comparison is made, if V_0 is less than V_A then the register value is low in relation to the analog signal and the most significant bit, the 1, is fixed and the next significant bit is set to 1. The test cycle is then repeated and each time the V_0 is less than V_A then the bit just set is fixed. When V_0 is greater than V_A the last bit set is cleared to 0 and the next bit set. This continues until $V_A = V_0$ when the digital value in the register will be equivalent to the analog input voltage. The digital output from the register can then be sent to the control system. Although this 'sampling' occurs hundreds of times per second the analog signal is still being digitized over a finite period. This means that continuously varying signals appear to the control system as changing in discrete steps. This in turn necessitates care in designing the overall analog to digital conversion system to ensure the digital outputs are truly representative of the analog input values.

Analog signal comparison and amplification

In the above section signal amplification was necessary, by using devices called 'operational amplifiers'. Generally in control systems two types of amplifier can be found. The first is the small signal amplifier of which the operational amplifier is typical. This type of amplifier will amplify low current or voltage signals and is ideal for using in the feedback loop of a robot control system, it is commonly available as an integrated circuit chip. The second type is the power amplifier which controls high currents and voltages. This device will be described in the chapter containing electric motor control.

The operational amplifier (op-amp) is an integrated circuit composed of transistors. As mentioned earlier the transistor can be used as a simple ON/OFF switch in digital circuits. However, the fact that a small input base current will allow a much larger collector current to flow means that it is also basically a current amplifier. When using the transistor as a switch we are only interested in it in the state when it is fully OFF and there is negligible current flow, and the state when it is fully ON and there is massive current flow. However, betwen these two states the large current output is proportional to the small signal current input, i.e. increases in input current create large proportional increases in output current. This is called current gain and is equal to the collector current obtained divided by the base current input, i.e.

$$\text{current gain} = I_C/I_B.$$

For a small transistor the current gain may range from 100 to over 300. By varying the configuration of the transistor and resistor components a voltage amplifier is obtained, and by making the output of one transistor amplifier the input to another, i.e. 'cascading', very high amplification ratios can be obtained. The complete circuits are not simple as possible distortion of the signal and instability of the transistors have to be compensated for by careful circuit design.

In the op-amp shown in Figure 5.28(a) either one or both of the input terminals may be used. The input marked '+' is termed the noninverting input, a signal applied here will be amplified and the output will be in phase with the input, see Figure 5.28(b). The input marked '−' is termed the inverting input and a signal applied here will result in an amplified output of opposite phase to the input. As a gain of 10^7 would be possible in an open loop configuration and would make for very unstable conditions, i.e. a very small, possibly unwanted variation in input will cause an undesirably large change in output; negative feedback is built into these circuits.

It is possible to use the amplifier in simple circuits using only one input, the other input being connected to ground. This is shown in Figures 5.28(c) and (d) for inverting and noninverting amplifiers with resistors incorporated into the circuits to provide the negative feedback and so reduce the gain. The amplifier is also often used with both inputs. In this case the total signal experienced by the amplifier is the difference of potential between V_{inv} and V_{noninv}, this is termed the differential input voltage.

Thus these small signal amplifiers can be used in a number of ways. They can be used to carry out the analog equivalent of the digital processes of addition, subtraction, and comparison of signals mentioned earlier. They can also be used in a control circuit as shown in Figure 5.28(c). In this case the feedback voltage signal from, say, a tachogenerator is inverted and amplified before being passed to the inverting input of a second amplifier. Here the reference voltage from, for example, the DAC of the robot microprocessor system is combined with it and the difference amplified. The difference between the two signals is then used as the error signal which could be used for subsequent amplification at the power stage of an industrial robot electric servo motor control.

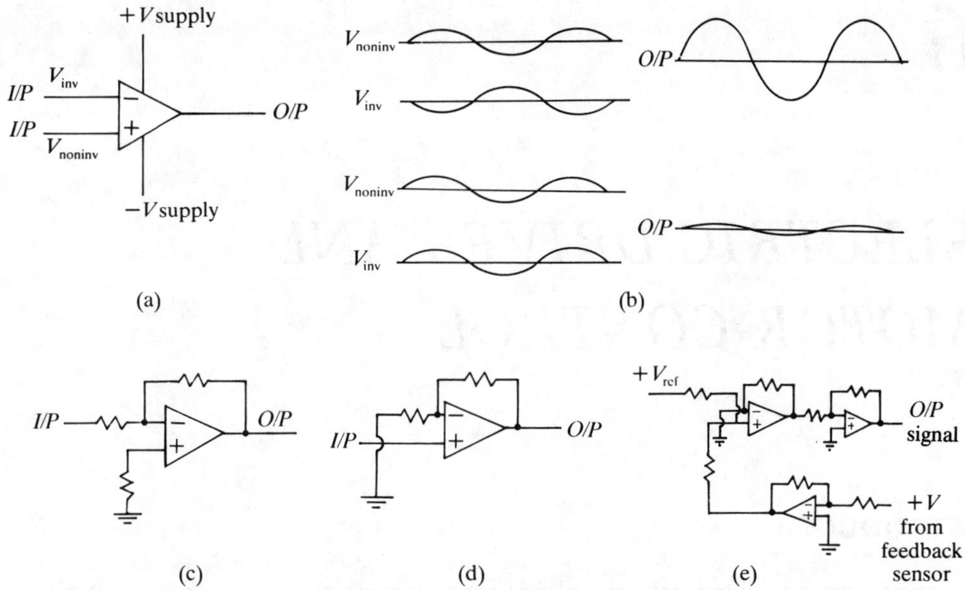

Figure 5.28 Small signal amplification: (a) operational amplifier circuit symbol; (b) phase relationships of input and output signals; (c) inverting amplifier; (d) noninverting amplifier; (e) op-amps in simple feedback loop

SUMMARY

This chapter has described the basic building blocks of a robot control system. Such a 'low-level' knowledge of these concepts and components is not absolutely necessary to be able to *use* a robot. However, if an *understanding* of the robot is the aim, then this section has provided useful background material. The following chapters move on to subjects more specifically related to industrial robots; their treatment assumes the content of this chapter has been understood.

6

ELECTRIC DRIVES AND MOTOR CONTROL

INTRODUCTION

Having examined the general principles of robot control, electric robot motors and their controls are now considered. Fluid power actuators are considered in the following chapter. In the early years of industrial robotics, hydraulic robots were the most common, but recent improvements in electric motor design have meant that most new robots are of all-electric construction. However, there are some advantages and disadvantages for each type, including the pneumatic variety, and these are considered at the beginning of each chapter. Below are some of the attributes of electric robots and drives.

- The basic drive element, in this case the electric motor, is usually lighter than that for hydraulic power.
- The accuracy and repeatability of electric drive robots is normally better than fluid power robots in relation to cost.
- They are very acceptable environmentally being relatively quiet and clean.
- They are easily maintained and repaired.
- Structural components can be lightweight.
- The drive system is well suited to electronic control.
- New motor designs and materials are constantly improving power to weight and power to size ratios.
- One disadvantage is that electric robots often require the incorporation of some sort of mechanical transmission system. This adds mass and unwanted movement, necessitates additional power, and may complicate control. Because of this increase in complexity additional cost is incurred due to the components themselves and also the maintenance costs associated with the degradation of the moving parts. However, this disadvantage is again gradually being overcome with the introduction of direct drive motor systems in which the electric motor is part of the relevant robot arm joint.
- A further disadvantage is that electric robots are not intrinsically safe. They would not therefore be used in, for example, explosive atmospheres. However, the newer

brushless motors do allow electric robots to be used in some fire-risk applications such as spray painting, as the possibility of sparking at the motor brushes is eliminated.

Electric industrial robots generally use one of three types of electric motor: the DC servo motor, the stepper motor, or one of the brushless types of servo motor. These types and their control are examined below.

THE DC MOTOR

Among the more established electric robots the most common drive is the DC servomotor, predominantly of the permanent magnet (PM) type. The principle of operation of the DC motor is first considered.

The very first electrical systems created utilized direct current and therefore DC motors were the first type generally used. When alternating current became widely available the DC motors were largely superseded by the AC type for domestic and light duty appliances. However, DC motors are still widely used in industrial applications due to the ease of control of their torque and speed characteristics. With the emergence of efficient methods of changing the alternating current from national supplies to direct current suitable for DC motors, using for example silicon controlled rectifiers (SCRs), the DC motor has become popular for many applications in manufacturing.

The principle of operation of any electric motor is based upon Ampere's Law which states that a conductor will experience a force if an electric current in that conductor flows at right angles to a magnetic field. Therefore to construct a motor we need two basic components, one to produce the magnetic field usually termed the stator, and one to act as the conductor usually termed the armature or the rotor.

The principle is shown in Figure 6.1 for one element of a DC motor. The magnetic field may be created either by 'field coils' wound on the stator, or by permanent magnets. The field coils, if used, would be provided with an electric current to create magnetic poles on the stator (see Figure 6.2). In the figures shown the rotor current is supplied to a conductor via the 'brushes' and the commutator. The current passing through the field produces a torque on the conductor. The torque is at a maximum when σ approaches 90°. In a full electric motor the stator is made up of an even number of magnetic poles, and the armature is composed of a cylindrical iron core with many conductors embedded in slots and connected to the commutator which is composed of a large number of segments. The stationary brushes in contact with the rotating commutator carry current to and from the conductors in the armature. The commutator segments are organized in such a way that the brushes continually switch the current between the conductors so that the torque angle σ is always 90° thus giving a steady torque.

Some larger robots utilize field control DC motors, i.e. motors in which the torque is controlled by manipulating the current to the field coils. These motors allow high power output at high speeds and can give a good power to weight ratio. It is also possible to use high inertia motors thus removing the necessity for gearing and provide a high torque at low speeds though this is uncommon in robotics. In the more predominant, less heavy applications low inertia motors are required to ensure fast response to control signals.

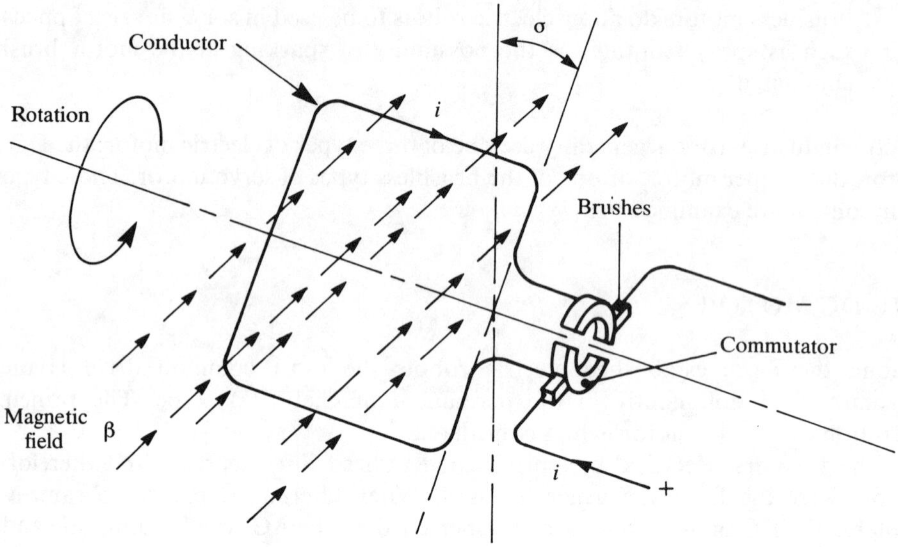

Figure 6.1

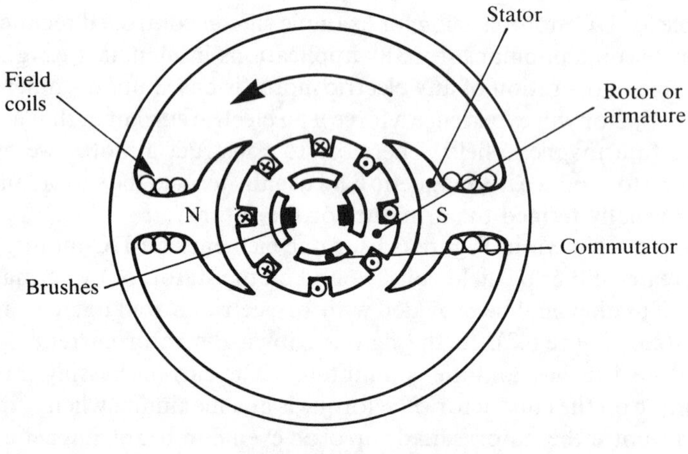

Figure 6.2 A two-pole DC motor with field coils (schematic)

However, for industrial robot systems in general it may be said that current excited field control methods involve too slow a response time and incur losses that make permanent magnet fields and armature control more attractive.

Permanent magnet (PM) DC servomotors

In the permanent magnet motor no field coils are used and the field is produced by the permanent magnets themselves. Some PM motors do have coils wound on the magnet poles but these are simply to recharge the magnets if their strength fails. Due to the field flux being a constant the torque of these motors is directly proportional to the armature current. Some other advantages are; excitation power supplies for the field coils are not required, reliability is improved as there are no field coils to fail, no power loss from field coils means efficiency and cooling are improved. Two types of PM configuration are shown in basic form in Figure 6.3. The permanent magnet segments are given the orientation shown by the arrows and are fixed by adhesive or bolts to the steel housing. The cylindrical motor operates in a similar manner to that already described for other DC motors except that there are no field coils. The disk motor has a large diameter, short length armature of nonmagnetic material. In this type of motor the armature is designed

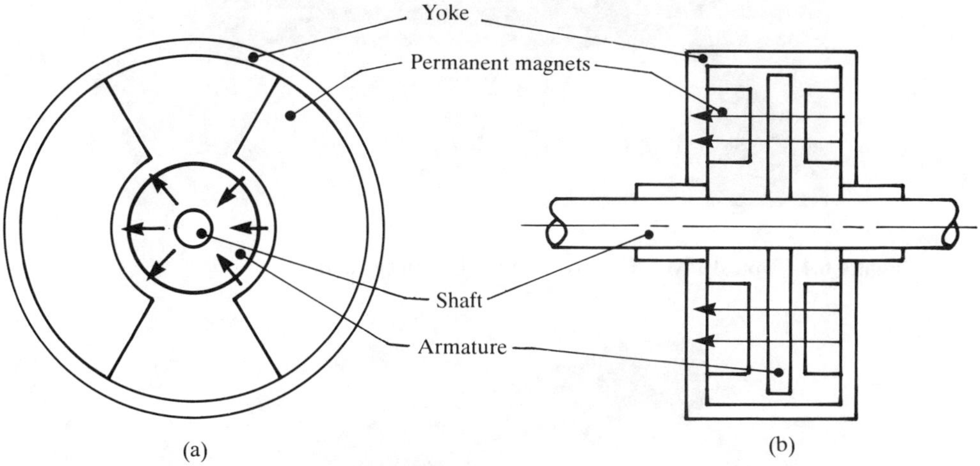

Figure 6.3 Two types of PM DC servo motor configuration: (a) cylindrical; (b) disk

with the conductors lying in layers and stamped from sheet copper. This structure is welded together and the magnetic flux passes from one magnet to the other across the armature and through the yoke thus completing the circuit. However, it is the cylindrical motor that is the more common in industrial robots. Figure 6.4 shows a typical PM DC servo motor, used in many industrial robots, in section.

The popularity of PM DC servo motors has been accentuated by the availability of materials such as Samarium Cobalt for the permanent magnets. These 'rare earth' motors have a much greater flux density per unit of material than the Alnico or Ferrite types thus allowing better torque to weight ratios and the use of more compact lighter motors. It is claimed that rare earth motors are one third lighter than their conventional counterpart.

Figure 6.5 shows the ASEA IRB 6 electric robot in its basic configuration. This robot uses the type of motors discussed above. The two wrist drive motors can be seen protruding on either side of the shoulder axis and one of the three major axis motors can be seen at the rear. These motors each have an integral tachometer and resolver, i.e. feedback components which will be described in Chapter 9.

PM DC servo motor control

The motor control equipment for an electric robot is usually located close to the robot computer control system. It will be noticed that a large number of 'black boxes' are used

Figure 6.4 Typical PM DC servo motor. (Courtesy Inland)

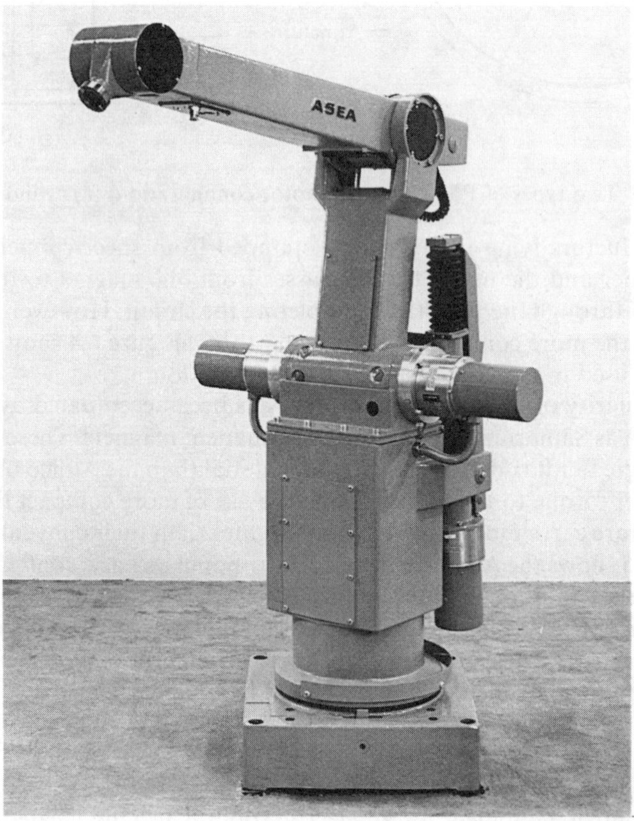

Figure 6.5 ASEA IRB 6 industrial robot with electric drive. (Courtesy ASEA)

here when discussing the control elements. This is because: (1) the basic principles on which their operation is based have been covered in the previous chapter; (2) the full internal structure and theory of operation is too complex to be covered in this text, and (3) the elements are themselves usually available for purchase as 'off the shelf' assemblies or modules.

The sketch in Figure 6.6 shows a much simplified feedback system for robot servo motor control. As mentioned in the previous chapter the command signals are issued from the supervisory system to the individual axis controllers, in this case the *X* axis controller.

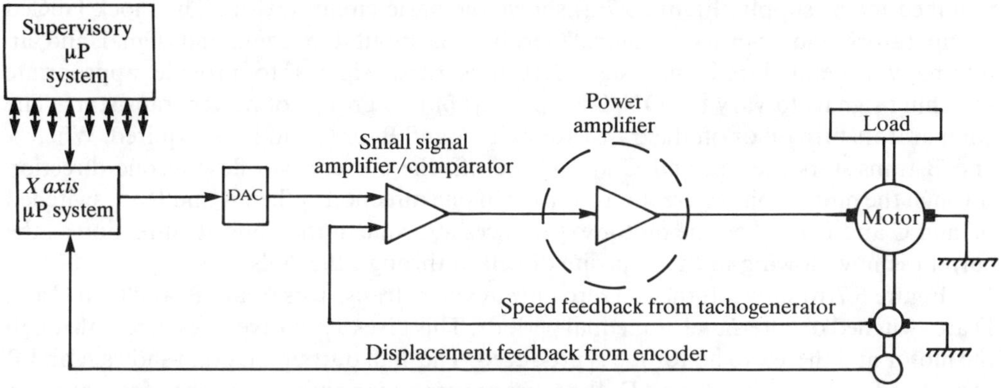

Figure 6.6 Simplified sketch of robot servo motor control

This provides appropriate command signals for the motor. These will be digital signals, and as the motor is an analog device a digital to analog converter is used. This provides a low voltage DC command signal which is fed into a small signal amplifier, this command signal is proportional to the desired speed and direction of rotation required. Generally reversal of direction is signified by a reversal of voltage. This signal is compared with the voltage feedback from the tachogenerator which is providing a voltage signal proportional to the actual speed and direction of rotation. The difference between the two signals is then used as the input to the power amplifier.

In this case the power amplifier will provide a current to the motor armature windings proportional to the input signal, obviously in PM motors field control cannot be used as there are no field coils. The torque produced by the motor is proportional to the current in the windings and is therefore also proportional to the input signal. In this system an increase in torque produces a proportionate increase in speed. The speed achieved is monitored by the tachogenerator attached to the motor shaft which sends a corresponding signal back to the comparator. The angular displacement of the shaft or associated gearing is monitored here by an optical shaft encoder. This is a device providing a digital signal which is fed straight back to the axis microprocessor system. The axis controller will send out appropriate signals to stop the arm as it reaches the required position. The axis controller can also monitor the actual speed by evaluating the rate of change of angular displacement and so adjust the speed commands as necessary.

In the actual robot the supervisory microprocessor system may be on a 'mother' printed circuit board (PCB), the axis microprocessors may have a PCB each, and there may also be a digital control board and an analog control board to handle the input and output signals. The power amplifiers will be separate units designed for handling the higher voltages. These power amplifiers may be of various types and two of the more common are discussed below. The pulse width modulation method using power transistors would be used typically for light and medium duty robots, the thyristor control for applications demanding higher power motors.

Power amplification using pulse width modulation

In PWM the high voltage power supply is DC provided either from a battery or from a modified mains supply. Figure 6.7(a) shows the basic circuit layout. The block labeled 'comparators and transistor control' receives as input the command signal and an internally generated reference signal. It uses these signals to provide appropriate switching signals to vary the ON/OFF periods of two groups of power transistors. The comparators turn on or off the transistor pairs A and B, or C and D as required. When A and B transistors are on, and C and D are off, the current will flow in one direction through the motor coils so creating a torque in one direction. When A and B are switched off and C and D on then the motor experiences a torque in the opposite direction as the current is now flowing in the opposite direction through the coils.

Figure 6.7(b) shows at the top zero output where transistors A and B, and then C and D are switched on alternately for equal periods. This gives a zero average current through the motor and therefore no torque is produced. The 'half output' diagram shows A and B turned on for longer than C and D, the average current is now positive, therefore torque is now experienced and drive obtained. The small time difference between pulses is to protect the transistors by ensuring one pair is fully off before the other is switched on. At full output A and B are fully on and C and D fully off, the motor would be receiving maximum current but speed control would not be possible in this condition. Thus by controlling the relative widths of the pulses for both pairs of transistors the speed of the motor is controlled.

Power amplification using thyristor control

Thyristors can be used to control high voltage to the motor from say two-, three-, or six-phase AC power supplies. The simplified circuit is shown in Figure 6.8(a). A thyristor acts in a similar manner to a diode in that it acts as a one-way valve offering very little resistance or impedance to current flow in one direction but extremely high resistance in the other. The thyristor differs from the diode in that it can be 'switched on' by a 'trigger' pulse to make it conduct, in which case the current increases rapidly to a value which is determined by the supply voltage and other factors. The thyristor remains in this conducting mode until the forward current is reduced below a certain level when it will then return to the current blocking state. A schematic of the thyristor is shown in Figure 6.8(b) and the effect of operation is shown in (c). The thyristor will not, of course, conduct when the voltage from the AC source drops below zero in its cycle. A number of these thyristors in the one unit thus provide an effective method of speed control as by initiating the trigger pulse at the appropriate part of the AC cycle the thyristor can be made to conduct at any required forward voltage.

In the amplifier shown there are two groups of thyristors connected to the AC supply, one group controlling the forward voltage and one group the negative voltage. The

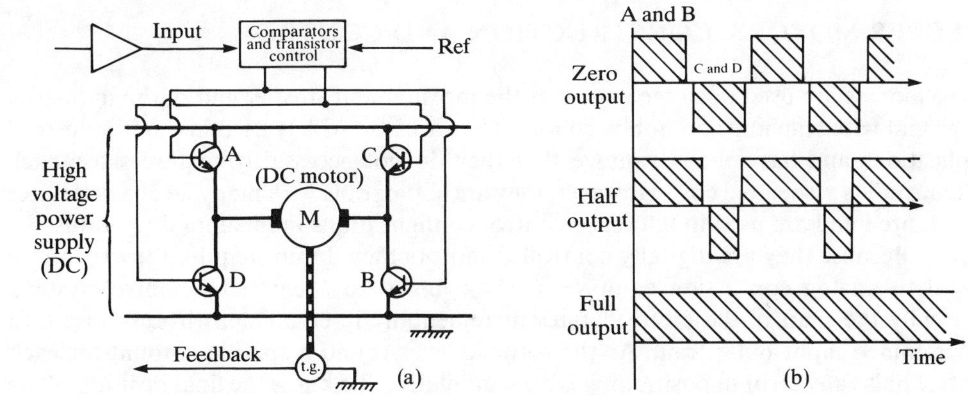

Figure 6.7 Power amplifier using pulse width modulation (PWM)

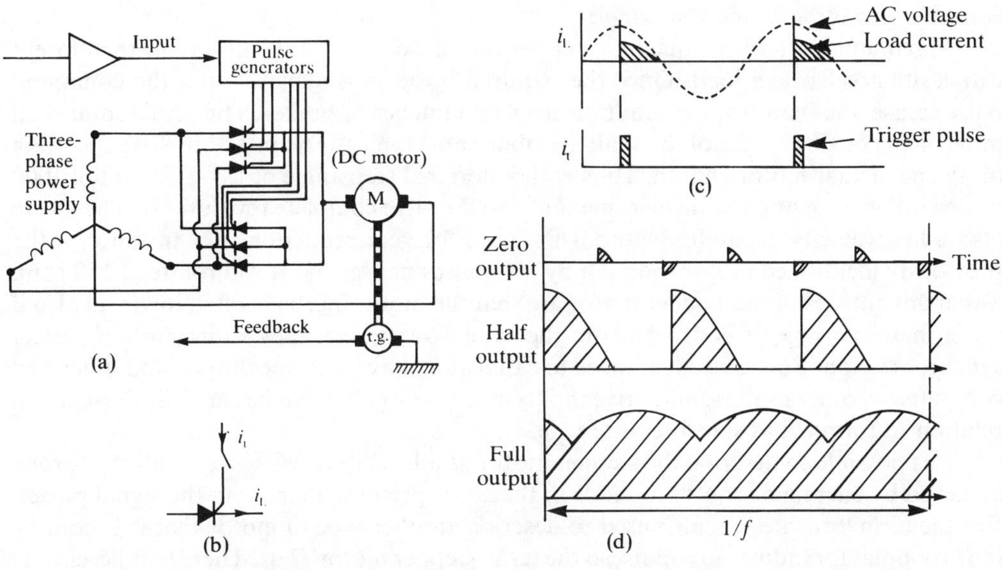

Figure 6.8 Power amplifier using thyristor control (three-phase AC power supply)

thyristors are turned on and off by trigger pulses from the pulse generator in response to the control signal input. In Figure 6.11(d) the zero output figure shows what happens when the positive and negative thyristors are turned on for equal periods, i.e. the mean output current is zero. For half output the positive thyristors are on for one quarter (90°) of each phase cycle and the negative thrysitors are left off. The motor will therefore experience a torque and drive is obtained. At full output the positive thyristors are on for the full time that they are forward biased allowing current to flow for 180° of each cycle in the positive direction. Under these conditions the motor experiences maximum torque and speed cannot be controlled. Thus by controlling the timing of the trigger pulses to the thyristors the motor speed is controlled.

STEPPER MOTORS: CONSTRUCTION AND CONTROL

These motors are used on some robots at the medium and smaller end of the industrial range and with teaching and hobby robots. They are also widely used in other industrial applications and have the advantage that they do not necessarily require a feedback system and its associated cost. However, they are compatible with many feedback devices if so desired and are used in full servo control configurations in medium duty industrial robots. Because they are digitally controlled motors they do not require the expense of digital to analog conversion equipment when linked to a computer control system. Normally the shaft of the motor rotates incrementally in equal steps in response to a programmed input pulse train. As the rotor indexes round a specific amount for each control pulse any error in positioning is noncumulative. To know the final position of the rotor all that is required is to count the number of pulses fed into the motor stator phase winding. The number of pulses per time unit determines the motor speed. The rotor can be made to index slowly, coming to rest after each increment, or it can move rapidly in a continuous motion termed 'slewing'.

Maximum dynamic torque in a stepper motor occurs at low pulse rates therefore it can easily accelerate a load. Once the required position is achieved and the command pulses cease, the shaft stops without the need for clutches or brakes. The actual rotational movements or step angles of the shaft are obtainable typically from 1.8° to 90° depending on the particular motor chosen. Thus with a nominal step angle of 1.8° a stream of 1000 pulses will give an angular displacement of 1800° or five complete revolutions. They have also a low velocity capability without the need for gear reduction. For instance, if the previously mentioned motor is driven by 500 pulses per second it will rotate at 150 rpm. Other advantages of the stepper motor are that the motor inertia is often low, and also if more than one stepper is driven from the same source then they will maintain perfect synchronization. Some disadvantages are that they have a lower output and efficiency compared to other motors, and drive inputs and circuitry have to be carefully designed in relation to torque and speed required.

It is possible to describe the stepper motor as a brushless DC motor with the torque achieved by energizing the stator coils in the appropriate sequence via the signal pulses. But the term 'brushless' is also used to describe another type of motor that is becoming more popular for industrial robots, so the term 'stepper motor' is used here to differentiate this type from the other brushless motors described later. There are various types of stepper motor, i.e. permanent magnet, variable reluctance, and hybrid. The intention here is not to examine these in detail but simply to consider the general concept of stepping motors. To this end the basic method of operation of a permanent magnet type is illustrated schematically in Figure 6.9.

In the motor shown there are two coils A and B each of them producing four poles but displaced from each other by half a pole pitch. The rotor is of permanent magnet construction and has four poles as illustrated. It can be seen in Figure 6.9(a) that the motor is at rest with the poles of the permanent magnet rotor held between the residual poles of the stator (the field excitation coils, or windings, have been omitted for clarity). In this 'detent' position the rotor is locked unless a turning force is applied. If the coils are energized and, in the first pulse the magnetic polarity of the poles of coil A is reversed, the rotor will experience a torque and will rotate counterclockwise as shown in (b), the

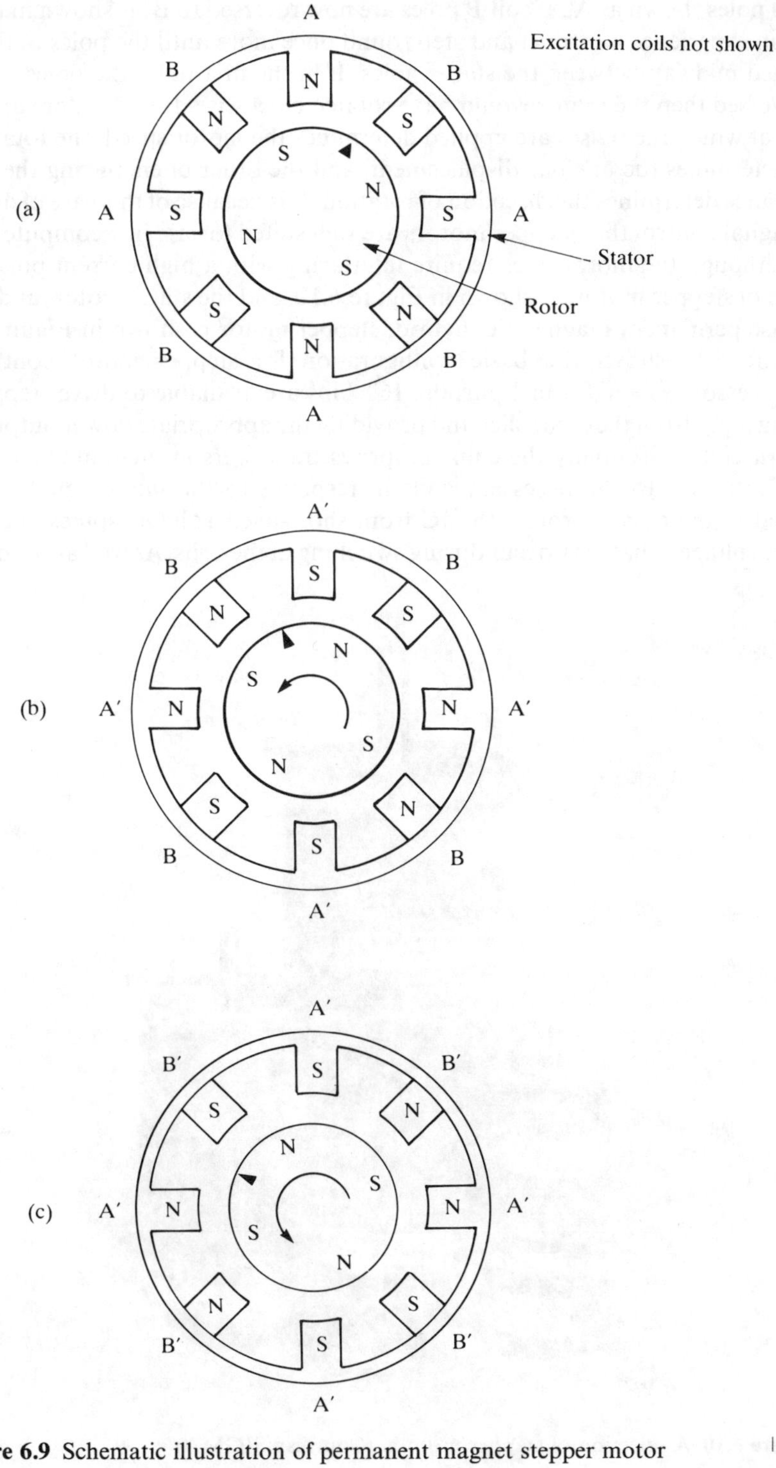

Figure 6.9 Schematic illustration of permanent magnet stepper motor

reversed poles shown as A′. If coil B poles are now reversed to B′ as shown in (c) the rotor will again experience a torque, and step round once more until the poles of the rotor are positioned midway between the stator poles. If in the first pulse the poles of coil B had been reversed then the motor would have rotated clockwise. It is therefore apparent that the rate at which the pulses are applied determines the motor speed, the total number of pulses determines the angular displacement, and the order of energizing the coils in the first instance determines the direction of rotation. It is because of this ease of driving using direct digital control that stepper motors are well suited for use in a computer controlled robot, although the motor does require interfacing with a high current pulse source. A selection of stepper motors is shown in Figure 6.10, and the stator, rotor, and wiring of a two-phase permanent magnet, i.e. hybrid, stepper motor is shown in Figure 6.11.

Figure 6.12 shows the basic configuration for stepper motor control from a microprocessor system. Special purpose IC chips are available to drive stepper motors, accepting input from the controller and providing the appropriate power output signals to the motor coils. Essentially the chip comprises transistors in integrated circuit form to switch the power to the necessary coils in response to the input signals. The diodes contained in the circuit protect the IC from short-lived voltage 'spikes', i.e. unwanted transient voltages, that may occur during switching of the coils. As well as the drive pulses,

Figure 6.10 A selection of stepper motors. (Courtesy SIGMA)

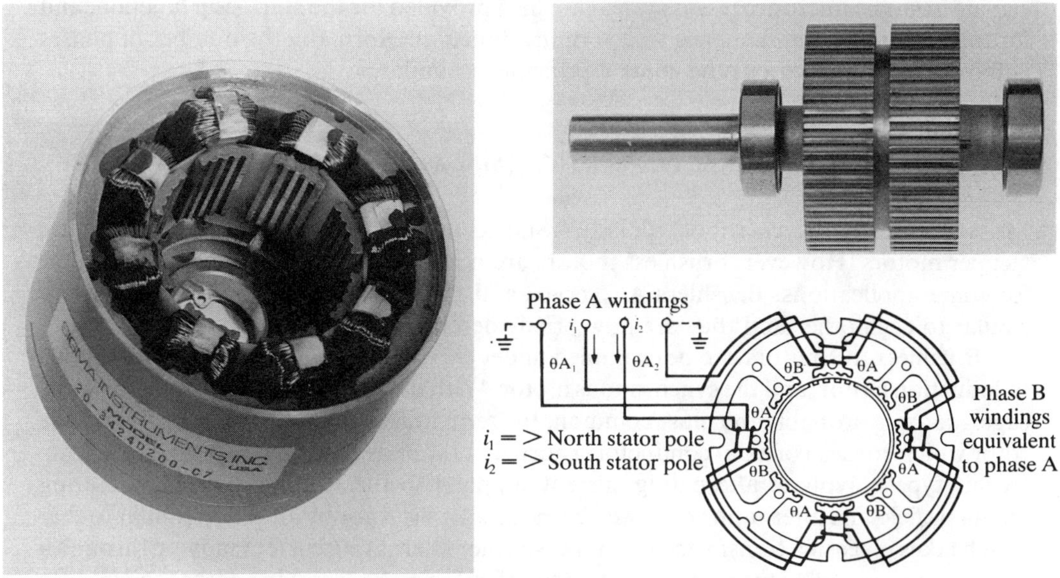

The rotor has a large number of poles (in this case 50) in the form of teeth. The stator has eight poles, each slotted to give five smaller poles. This configuration gives a step angle of 1.8° or 200 steps per revolution.

Figure 6.11 A two-phase stepper motor. (Photographs courtesy of SIGMA)

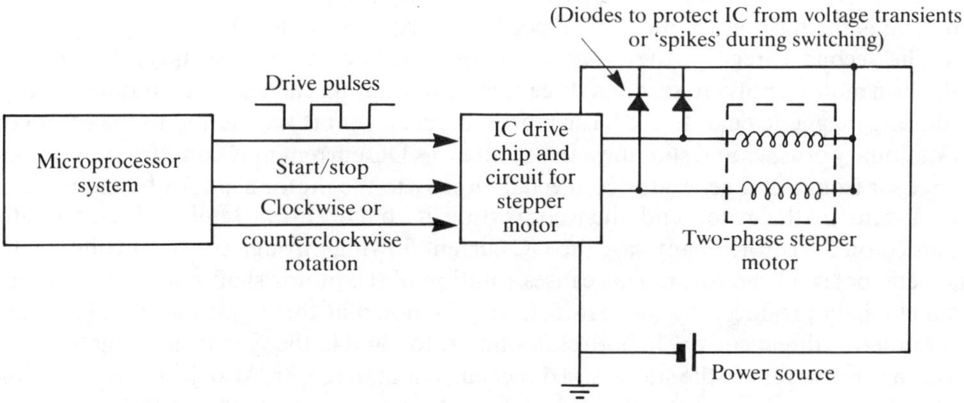

Figure 6.12 Open loop control of stepper motor

signals from the microprocessor system are also provided for the start/stop function, and for the coil energizing sequence to determine direction. Normally the number of phases will be more than two but the control principle is similar.

BRUSHLESS MOTORS: CONSTRUCTION AND CONTROL

Most electrically driven robots already installed use DC servo motors and a few use stepper motors. However, brushless motors are now beginning to become very popular for some applications. Brushless AC servos and brushless DC servos are in fact very similar to each other and they are now briefly described.

Brushless AC motors can be examined under two main categories: the asynchronous or induction motor, and the synchronous motor. With the asynchronous type the 'squirrel cage' motor is probably the most common, the term arising from the construction of the rotor which consists of axial conductors shorted at the ends to form a cage-like structure. In this type polyphase alternating current supplied to the stator produces a rotating magnetic field and alternating currents are induced in the rotor by an action similar to that which takes place in a transformer. In a transformer alternating current moving through a conducting coil, called the primary, is capable of inducing a current in another coil in close proximity called the secondary. In the motor the combined electromagnetic effect between the induced rotor currents and the magnetic field produced by the stator produces torque. The fact that a component of the current induced in the rotor goes to producing a magnetic field, and a component goes towards producing torque, makes it difficult to use this motor for speed control. However, although recent developments have led to these two components being able to be separated for control purposes, the AC induction motor is not yet used significantly on servo controlled industrial robots. It is used widely elsewhere for servo control although the induction motor does tend to 'cog' at low speeds, (cogging is the term used to describe 'negative' torque pulsations).

The second category, the synchronous motor, is of more relevance. As with the induction motor a polyphase, often three-phase, alternating current in the stator produces a rotating magnetic field. In the basic motor direct current is transferred to the rotor coil by stationary brushes and slip rings. In practice this DC current may come from a separate source, or from a DC generator connected directly to the motor shaft, or by inducing an AC current in the rotor and then converting it to DC using semiconductor diodes mounted on the rotor. In any case the DC current flowing through the rotor coils creates magnetic poles on the rotor. This causes rotation of the motor shaft due to the rotating magnetic field produced by the stator. It may be noted at this point that in a basic DC motor there is direct current in both stator and rotor, and in the synchronous motor there is alternating current in the stator and direct current in the rotor. Also, just as it is possible to use permanent magnets on the stator to create the magnetic field in a DC motor, it is possible to use permanent magnets on the rotor to provide the poles in a synchronous motor. In fact the PM synchronous motor is the previously mentioned brushless AC servomotor now being applied in some robot drive systems. Figure 6.13 shows a typical brushless AC motor as used on the ASEA IRB 2000 robot.

Figure 6.14 shows schematically a three-phase synchronous motor with permanent magnets mounted on the rotor. Synchronous motors are so-called because they act in

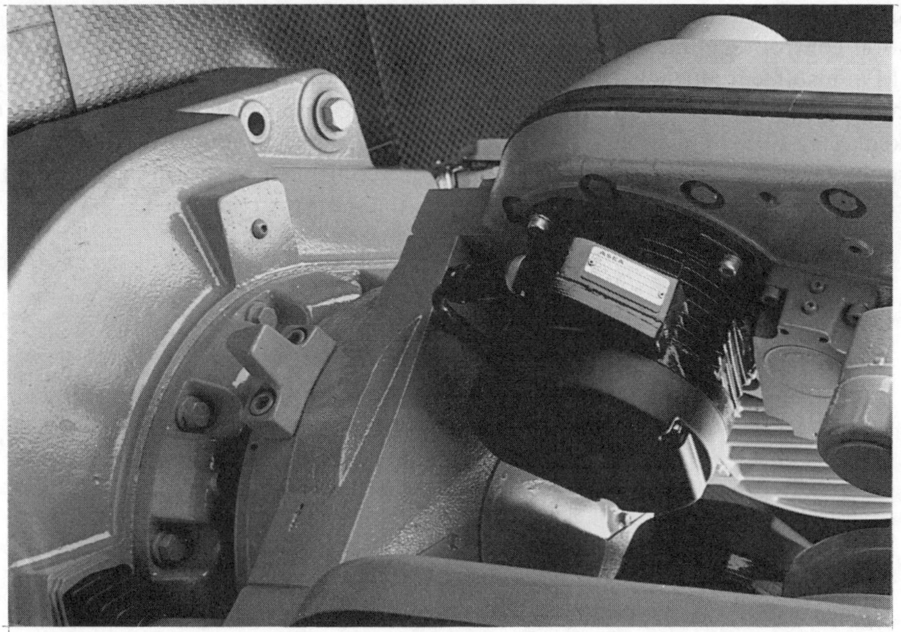

Figure 6.13 Brushless AC servomotor used on ASEA IRB 2000 robot. (Courtesy ASEA)

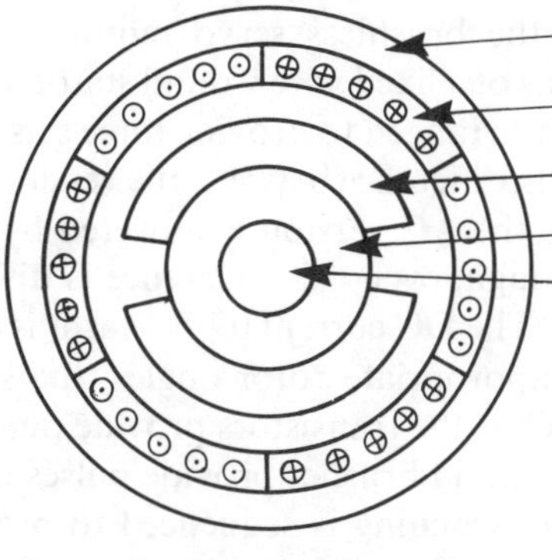

Figure 6.14 Schematic of brushless motor

absolute synchronism with the frequency of the supplied alternating current. The rotor speed is determined by the number of pairs of poles and is a ratio of the supplied AC frequency. Thus by electronically controlling frequency and voltage, full control is gained of the motor speed under load. The term 'commutation' will be remembered from the DC servo motor, i.e. the switching of current from one coil to another. In this case the field windings in the stator are electronically commutated to turn the rotor. As with DC servo motors pulse width modulation is used although with the AC system this control is more complex. The AC controller has to control three sets of windings in the motor stator. Commutation has to be controlled by monitoring rotor shaft position, this is achieved by using sensors such as very accurate optical shaft encoders. For robot arm speed and position control tachometers and encoders are used as with the DC servomotors. The PWM adjustable frequency drive operates as follows. Firstly the three-phase AC supply from the mains is rectified to produce a constant DC voltage. This DC voltage is then broken up by a PWM inverter, i.e. an inverse rectifier, which produces alternating voltage pulses of varying controlled widths. These can be controlled as shown in Figure 6.15 to give an approximation of a sine wave as far as the motor is concerned.

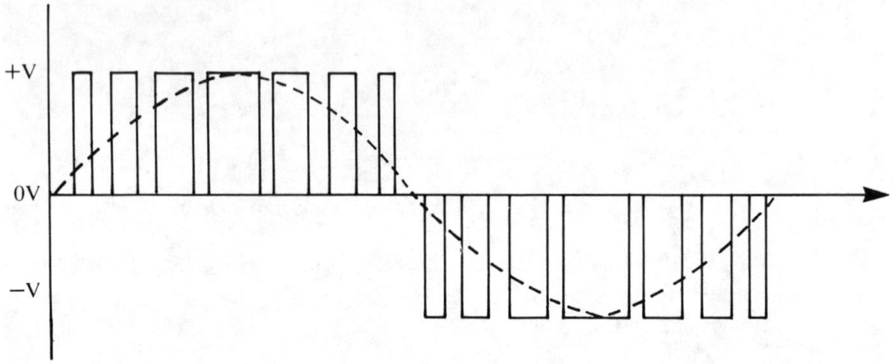

Figure 6.15 A pulse width modulated waveform for a brushless motor

The corresponding DC version of the brushless servo motor is very similar in construction to the AC type. It too consists of a fixed wire wound stator and a permanent magnet rotor. Also the stator windings are arranged to provide three sets of poles equally distributed around the periphery. The main difference between the versions is that the DC motor is powered by DC pulses whereas the AC version is powered by polyphase sine wave alternating current. In practice, though, even this difference is difficult to discern depending upon the exact waveform used. The DC current to the stator is commutated by switching transistors on and off at the appropriate rotor angles, the switching signal coming from sensors on the rotor shaft. Thus the transistors provide pulses of current to the stator windings just as the commutator and brushes provide pulses of current to the rotor coils in a conventional motor. The switching is sequenced to provide a rotating magnetic flux in the airgap that remains at the optimum angle to the magnetic flux produced by the permanent magnets on the rotor. Figure 6.16 shows an example of a control circuit for a brushless DC servomotor.

The 'brushless' motors have many advantages over conventional DC servo motors:

- They have better heat dissipation, heat being more easily lost from the stator than the rotor.
- There is reduced rotor inertia.
- The motors in themselves are less expensive.
- They are more durable.
- They weigh less.
- They are of smaller dimensions for comparable power.

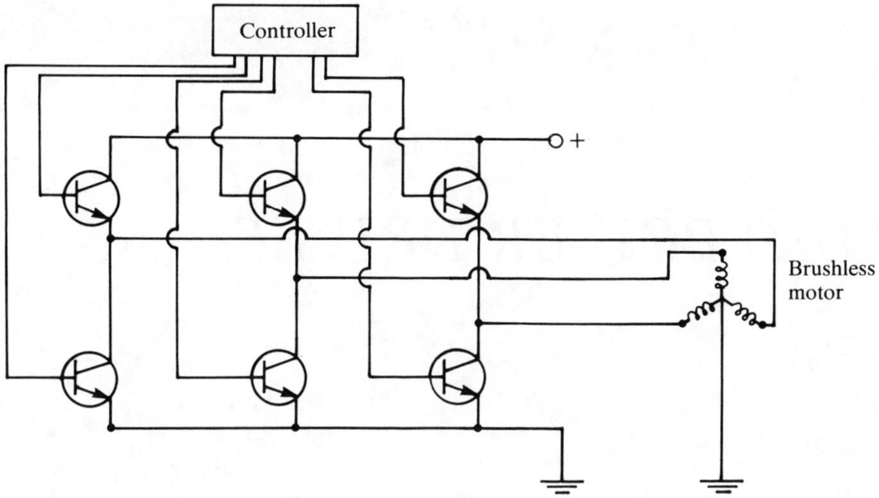

Figure 6.16 One servo amplifier configuration using transistors to obtain commutation in a brushless motor

- The absence of brushes reduces maintenance costs due to brush and commutator wear, and also allows electric robots now to be used in some hazardous areas with flammable atmospheres such as are found in spray painting applications.

One disadvantage is that the control systems for brushless motors are relatively expensive at the moment. However, the constant reduction in the cost of electronic components and improvements in motor design mean that the combined cost of motor and controller is now commercially attractive. This indicates that these brushless motors, combining the best qualities of AC and DC motors, are going to continue to increase in popularity for industrial robot applications.

7

FLUID POWER DRIVES

INTRODUCTION

Fluid power drives for industrial robots are mainly used today in the form of full servo controlled hydraulic systems at the heavy duty end of the market, and non-servo-controlled pneumatic systems for high speed, light duty operations. Hydraulic power systems utilize a high pressure fluid such as oil to transmit forces to the point of application desired, while pneumatic systems use high pressure, or 'compressed', air, though at much lower pressures than those used in hydraulics. This chapter considers the components necessary to provide the fluid power to drive the robot arm. The components required for servo control and sequencing of movements are considered in later chapters. The actuators used in both hydraulic and pneumatic systems are very similar in appearance, though the hydraulic equipment is designed to operate at much higher pressures. Listed below are some of the characteristics of fluid power drive systems.

Hydraulic systems

Advantages

- A high power to size ratio is obtained with hydraulic actuators and motors. Actuators are components that convert the fluid power into a movement of the arm, e.g. a piston is an actuator.
- Few backlash problems occur due to the stiffness and incompressibility of the fluid especially when the actuator acts as the joint itself such as in the rotary vane actuators described later. Backlash is the term used to describe the unwanted 'give' in transmission components from their intended position, when put under load conditions, due to wear or clearances between surfaces.
- Large forces can be applied directly at the required location.
- They can be used in hazardous areas with high fire risks as only small amounts of electrical energy are required for the control components. This assumes of course that the power supply for the hydraulic pressure is located remotely from the work area, and also that nonflammable hydraulic fluid is used.
- They generally have a greater load carrying capacity than electric and pneumatic robots.

- No mechanical linkage is required, i.e. a direct drive is obtained with mechanical simplicity.
- Because an accumulator, which acts as a 'storage' device, is used the system can meet sudden demands in power.
- Hydraulic robots are more capable of withstanding shock loads than electric robots.

Disadvantages

- They can be less reliable than electric robots.
- Leakages can occur causing loss in performance and general contamination of the work area.
- The power pack can be noisy, typically about 70 dBA (70 decibel average which is about the noise level of heavy traffic) or louder if not protected by an acoustic muffler.
- Changes in temperature alter the viscosity of the hydraulic fluid. Thus at low temperatures fluid viscosity will increase, possibly causing sluggish movement of the robot.
- For smaller robots hydraulic power is usually not economically feasible as the cost of hydraulic components does not decrease in proportion to size.
- Servo control of hydraulic systems is complex and is not as widely understood as electric servo control.

Pneumatic systems

Advantages

- It is the cheapest form of drive, components are readily available and compressed air is normally an already existing facility in factories.
- They have few moving parts making them inherently reliable and reducing maintenance costs.
- Pneumatic systems are common throughout industry therefore relevant personnel are often very familiar with the technology.
- Very quick in action and response time thus allowing fast work cycles.
- No mechanical linkage is usually required.
- Pneumatics can be intrinsically safe in explosive areas as no electrical control is required. Also in wet conditions there is no danger of electrocution.
- The systems are usually compact.
- Control is simple, e.g. mechanical stops are often used.
- Individual components can be easily interconnected.

Disadvantages

- Sophisticated movements and speed control are not easily obtainable unless more complex electromechanical devices are incorporated into the system. This means that only a limited sequence operating at a fixed speed is often available.

- If mechanical stops are used resetting the system can be slow.
- Pneumatics are not suitable for moving heavy loads under precise control due to the compressibility of air. This compressibility necessitates the application of more force than would normally be necessary to ensure the actuator is held firmly in position against its stop when under load conditions.
- If moisture penetrates the units, and ferrous metals have been used, then damage to individual components can result.

THE HYDRAULIC SYSTEM

The basic principle of hydraulic power was introduced in Chapter 2 on robot history, this section now considers how this principle is applied to industrial robot drives. The hydraulic system for a robot is basically similar to that in many other applications although fast response times, accuracy, repeatibility and control are parameters especially important. The four basic hydraulic elements required are: (1) a reservoir to hold the fluid; (2) a pump to pressurize or force the fluid through the system; (3) appropriate valves to control the fluid pressure, flow rate and direction of flow, and (4) the actuators to convert the fluid pressure or flow into a mechanical action. Both rotary and linear actuators are used in robots. Typical working pressures are between 70 and 170 bar, that is roughly 70 and 170 times ordinary atmospheric pressure.

A very much simplified diagram of the basic hydraulic system is shown in Figure 7.1. The actual system in a robot is more complex and utilizes servo valves which will be considered in the chapter on control components, however the figure does illustrate the essential elements. Not shown is the electric motor which drives the pump. Normally in hydraulic systems the separate hydraulic components shown inside the dotted line are contained within the one unit. i.e. a power unit or power pack. This consists of a reservoir, pump, drive motor, filter, heat exchanger or cooler, and basic system valving. An example of a power pack is shown in Figure 7.2. This particular unit contains the fluid reservoir, a gear pump, a 5.5 hp electric motor, and sensors to enable the fluid level and temperature to be continuously monitored by the control system. As well as the actuators, valves and power pack, there are other components required. For example, an accumulator is normally used to allow storage of the hydraulic fluid under pressure and thus provide a backup facility when sudden demands are made on the system. Hydraulic motors with continuous rotation are not considered here as they are not usually employed in robots. This is due to the necessity for transmission components which would incur the loss of many of the previously mentioned advantages of hydraulic power.

The pump

The pump, which is in effect the heart of the hydraulic system, is driven by an electric motor. This motor will vary in size depending on the robot type, for paint spraying robots a 5 kW motor would be typical in a pump producing about 70 bar, and in heavy duty robots a 22 kW motor in a pump producing around 170 bar. Two types of pump commonly used are the variable displacement vane pump and the gear pump. The principle of operation of these pumps is shown in Figures 7.3 and 7.4.

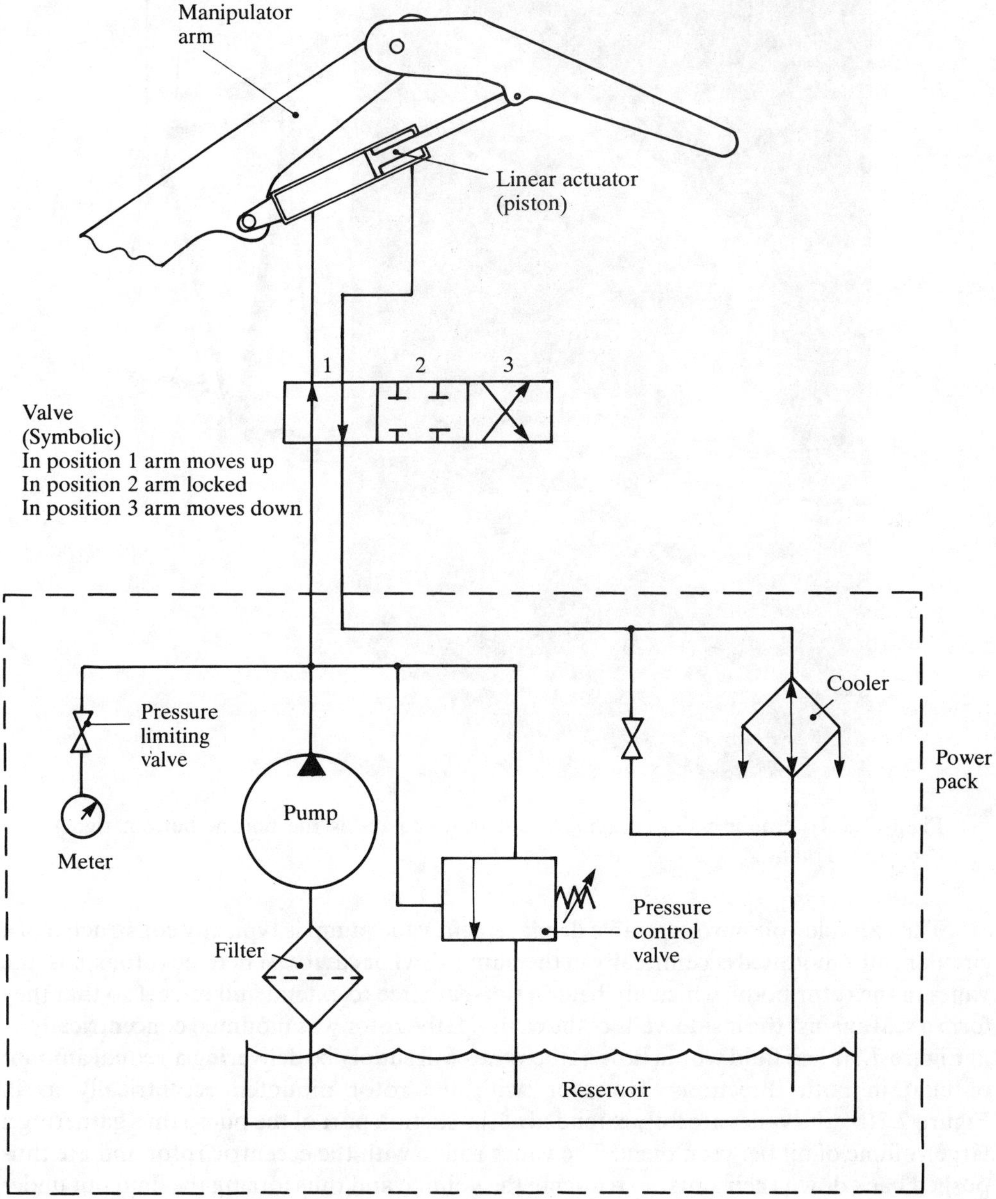

Figure 7.1 A basic hydraulic system

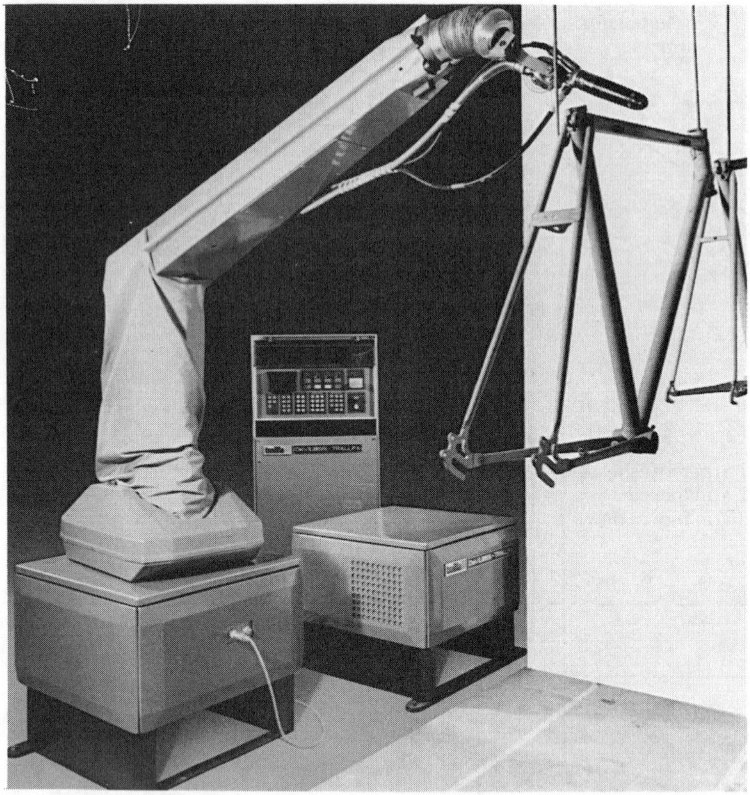

Figure 7.2 Hydraulic paint spraying robot (power pack is the unit at bottom right). (Courtesy Trallfa)

The variable volume or variable displacement vane pump is typically constructed of a circular rotor mounted eccentrically in the pump bowl or cavity. When the rotor spins, the vanes in the rotor body, which are held in slots, are free to extend and retract so that they form a seal against the inside walls of the cavity. If the rotor was mounted concentrically as in Figure 7.3(a) no fluid would flow as the vanes will simply be delivering an equal amount of fluid in both directions. However, with the rotor mounted eccentrically as in Figure 7.3(b), the vanes are fully extended at the suction port of the pump thus gathering a large volume of oil between them. The vanes rotate with the eccentric rotor and are thus pushed back down their slots, so reducing the volume and thus forcing the fluid out under pressure as they pass the discharge port. A common configuration is to have the rotor fixed and rotating inside a 'cam ring'. This cam ring is then moved to achieve the required amount of eccentricity. Altering the eccentricity alters the fluid volume pumped and determines the direction of flow. These pumps are used especially in the larger heavy duty robots.

The gear pump is used on the lighter hydraulic robots and operates as shown schematically in Figure 7.4. Various configurations of gear pump are available. The one illustrated comprises two gears of equal size meshing with each other inside the pump

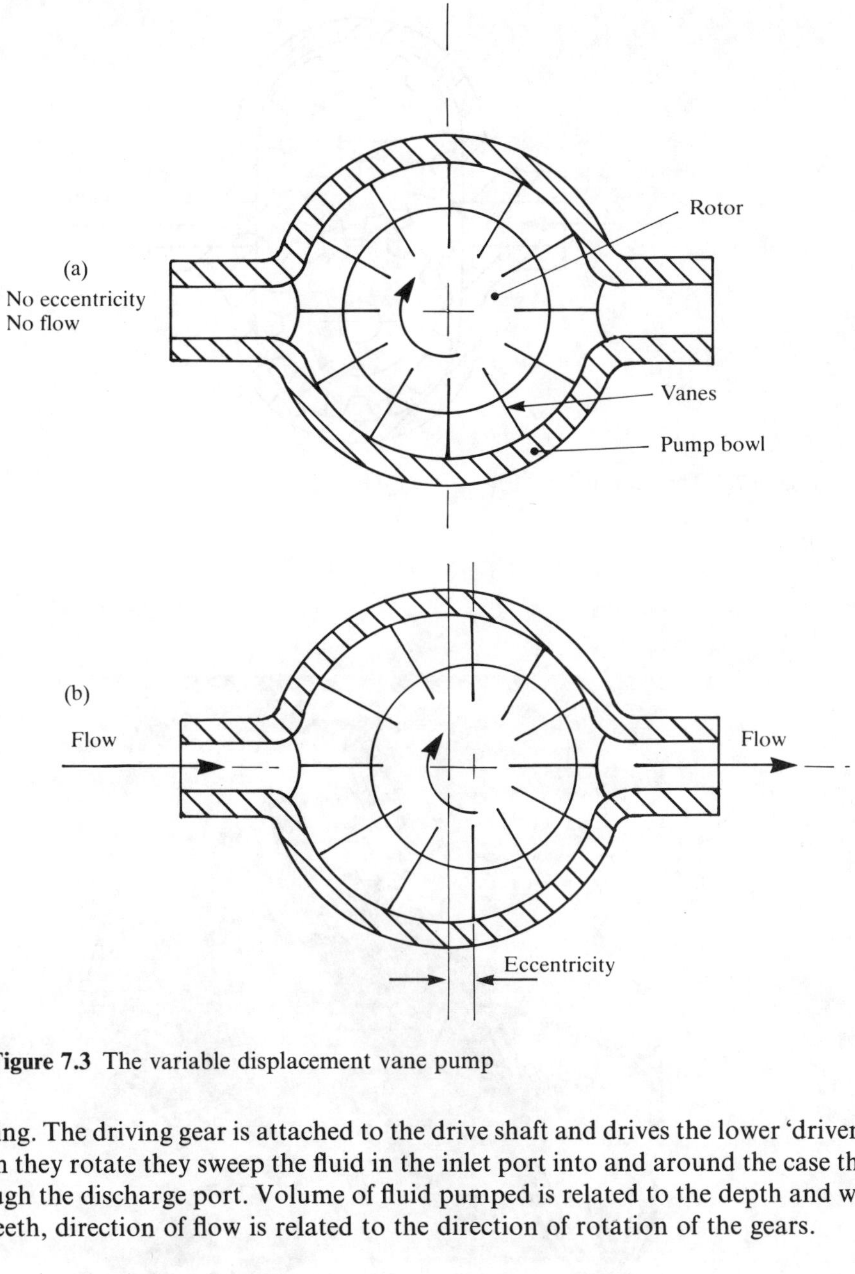

Figure 7.3 The variable displacement vane pump

housing. The driving gear is attached to the drive shaft and drives the lower 'driven' gear. When they rotate they sweep the fluid in the inlet port into and around the case then out through the discharge port. Volume of fluid pumped is related to the depth and width of the teeth, direction of flow is related to the direction of rotation of the gears.

Hydraulic actuators

Figure 7.5 shows the arm of a hydraulic robot in which can be seen a linear hydraulic cylinder and a single vane rotary actuator, both of these actuators are now described.

Considering firstly the linear actuators, or cylinders, it should be noted that there are many types available. For example, there are telescoping cylinders, multiposition

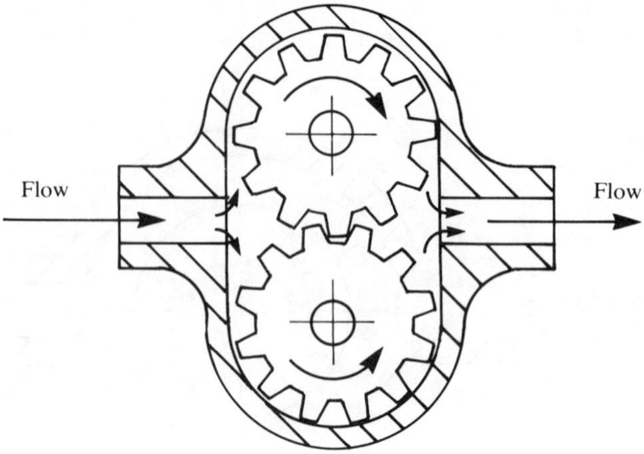

Figure 7.4 The gear pump

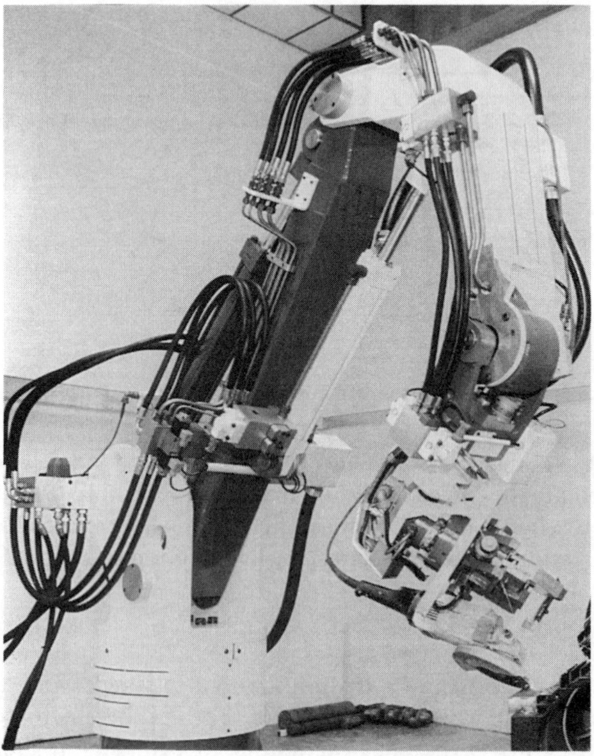

Figure 7.5 'Ajax Workmaster' hydraulic robot arm. (Courtesy Thorn-EMI)

cylinders, and additional refinements such as shock absorbers or cushions. The simplest type is the single acting cylinder which can deliver force in only one direction. The opposite action to return the piston must come from the load or a return spring inside the cylinder. However, the most common is the double acting cylinder which can produce force in both directions. The piston is acting as a mechanical interface between the hydraulic pressure and the mechanical action. The pressure acts on the high pressure side of the cylinder (see Figure 7.6) producing a force that is equal to the pressure differential between the chambers and the total area of the piston face on which it is acting. This force pushes the piston and the attached piston rod forward to enable mechanical work to be done. Extremely high forces can be obtained by using these cylinders.

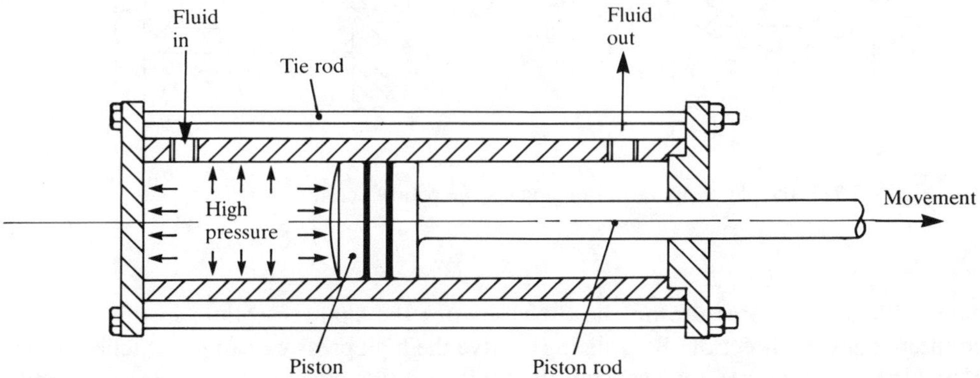

Figure 7.6 A double acting hydraulic cylinder of the tie rod configuration

Rotary actuators are commonly used in hydraulic robots and in the case of the vane type often constitute an actual joint of the arm. They are also commonly used in other applications such as in construction equipment and other heavy plant. They produce reciprocating rotary motion by turning an output shaft through a fixed arc. They are usually simple, compact, and of high efficiency. The direct rotary output of the actuator permits a simple and versatile design and removes the need for complex linkages. They produce excellent starting and stopping torques and are ideal for the acceleration of large masses as found in heavy robot applications.

The vane type rotary actuators are very compact devices converting hydraulic pressure into torque via a vane and shaft which rotates through an arc of up to 280° total travel. The fact that they can be used as a direct drive means no backlash problems are experienced. All of the working components of the actuator are enclosed and sealed and the wear and contamination problems associated with piston action are largely eliminated. These actuators are available in various sizes and can have a double vane which increases the torque but reduces the angle of arc to about 100°. The principle of operation is shown in Figure 7.7. When high pressure fluid enters port X from the hydraulic system it will cause chamber 1 to increase in volume. Chamber 2 which is connected to the low pressure side of the system through port Y will now decrease in volume. The fluid flow caused by the action of the high pressure liquid on one side of the

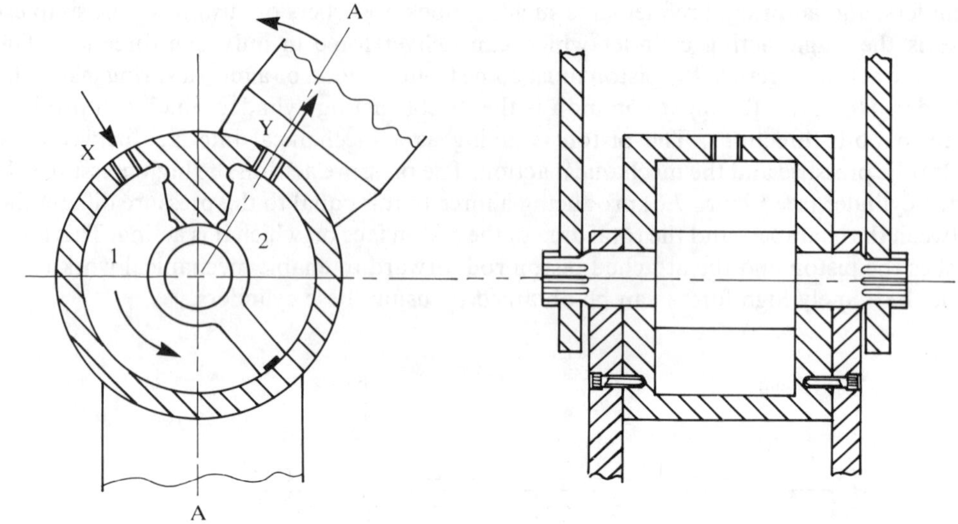

Figure 7.7 Principle of operation of single vane rotary actuator

vane with the low pressure on the other, causes the shaft to rotate, in this case in a counterclockwise direction. By utilizing a valve the high pressure can be switched to port Y and the low to port X thus causing the shaft to rotate clockwise. Torque developed is proportional to the area of the vane and the fluid pressure differential, speed is related to the rate of flow and pressure capabilities of the overall system.

The actuator in the figure is shown connected in a typical robot arm configuration. The lower arm is bolted to the actuator body, thus acting as a relatively fixed platform. The upper arm is attached to the actuator shaft by passing it over the splined shaft ends and enabling it to move in response to the application of hydraulic pressure as outlined previously. It should be noted that the stationary fluid barrier is not used as a stop. There are external mechanical stops fitted to the arm to prevent the vane hitting the internal fluid barrier, and even these stops are only there as a safety measure as the actual amount of rotation is controlled by the computerized control system. A typical rotary vane actuator as used in existing industrial robots is shown in Figures 7.8 and 7.9.

In the rack and pinion type the hydraulic pressure drives a piston which is connected to a gear rack, this in turn rotates a pinion to provide the turning force. It is possible to obtain rotation of 360° with this type of actuator and in the form illustrated in Figure 7.10 very high output torques can be achieved. The principle of operation is as shown in the figure and is similar to that of the previous actuator except that in this case there are two input ports. High pressure applied through the ports X increases the volume in chambers 1a and 1b and forces the pistons inward, thus causing their racks to rotate the pinion and shaft counterclockwise. The low pressure fluid in chamber 2 flows over the pinion shaft and through the port Y. Conversely high pressure applied through port Y forces the pistons apart allowing the low pressure fluid to be pushed out through ports X. An example of this type of actuator is shown in Figure 7.11.

Figure 7.8 Section through rotary vane actuator. (Courtesy Bird-Johnson)

Figure 7.9 Rotary vane actuator components. (Courtesy Bird-Johnson)

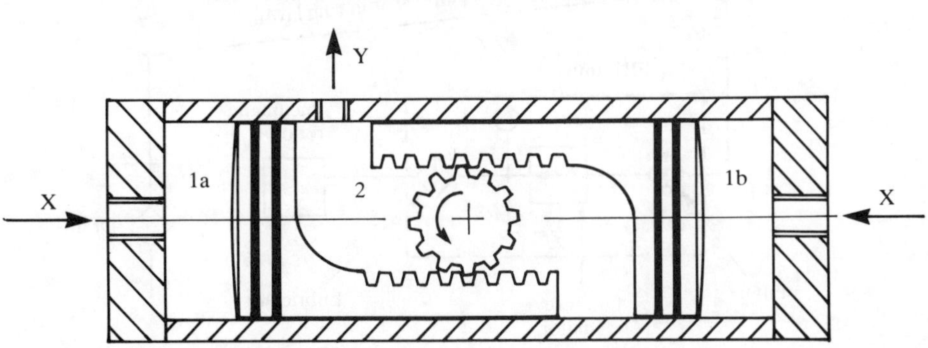

Figure 7.10 Rack and pinion rotary actuator

Figure 7.11 A double unit, rack and pinion type, hydraulic rotary actuator. (Courtesy Flo-Tork)

THE PNEUMATIC SYSTEM

Hydraulic and pneumatic power systems both utilize the same principles, i.e. pressurizing a fluid and delivering it under pressure to where it can do work, as in the hydraulic components described above.

In pneumatic power an air compressor is used in place of an hydraulic pump and is usually located in a remote position within the factory. This location is usually away from the main shopfloor and in this case the robot installation. The compressor will normally be of such a size as to be able to provide a compressed air supply to the whole factory for various pneumatically operated and controlled equipment. High pressure air for power in robots is usually about 1 to 7 bar and low pressure for control purposes about 1 bar.

As a result of the air being compressed, heat, water and oil are introduced into the air supply. Well designed coolers at the compression stage will remove the heat and much of the water. Most of the remaining water will require draining off at regular points in the system, and as water is heavier than air the falling ring system shown in Figure 7.12

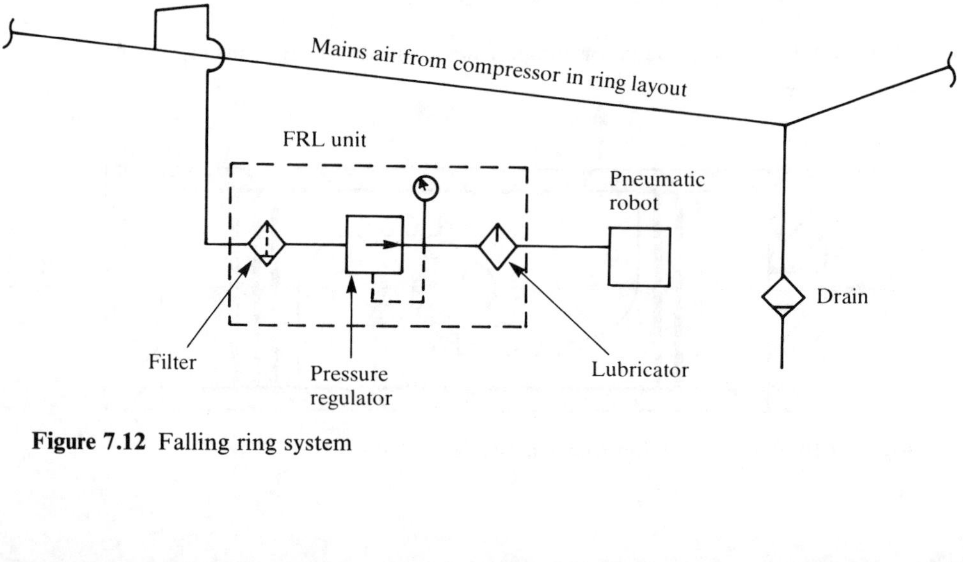

Figure 7.12 Falling ring system

facilitates this. Additional filtering of the air before the robot removes residual moisture and also any foreign particles and oil which when combined, can form a sludge capable of 'gumming up' the pneumatic components. Filtering, pressure regulation, and lubrication of the air is carried out in a service unit (FRL) close to the robot as shown. An FRL unit is shown in Figure 7.13.

If the compressor in a pneumatic system is regarded as analogous to the pump in an hydraulic system then an 'air receiver' may be taken as analogous to the hydraulic accumulator. Air receivers are used to store the air under pressure until required, and to smooth out the air flow from the compressor.

The air system is usually constructed in the form of a ring around the factory to reduce the pressure drop throughout the supply.

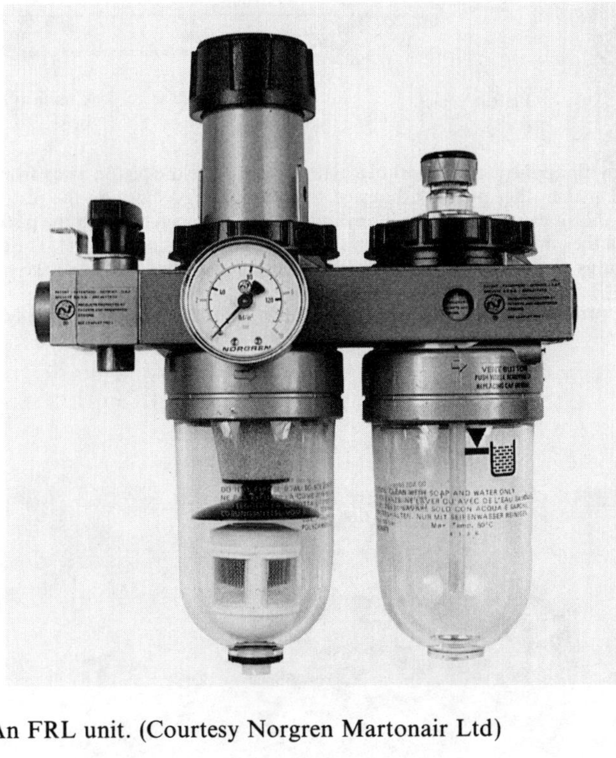

Figure 7.13 An FRL unit. (Courtesy Norgren Martonair Ltd)

Pneumatic actuators

These again bear many similarities to their hydraulic counterparts. Figures 7.14(a) and (b) show a double-acting cylinder and sectional view illustrating the principle of cushioning. Here too a variety of cylinders is available, e.g. single acting cylinders, midget cylinders and cylinders with position sensing devices. Cycle times are the fastest of all the drive systems and control can easily be achieved electrically by means of solenoid operated control valuves which are discussed in Chapter 9.

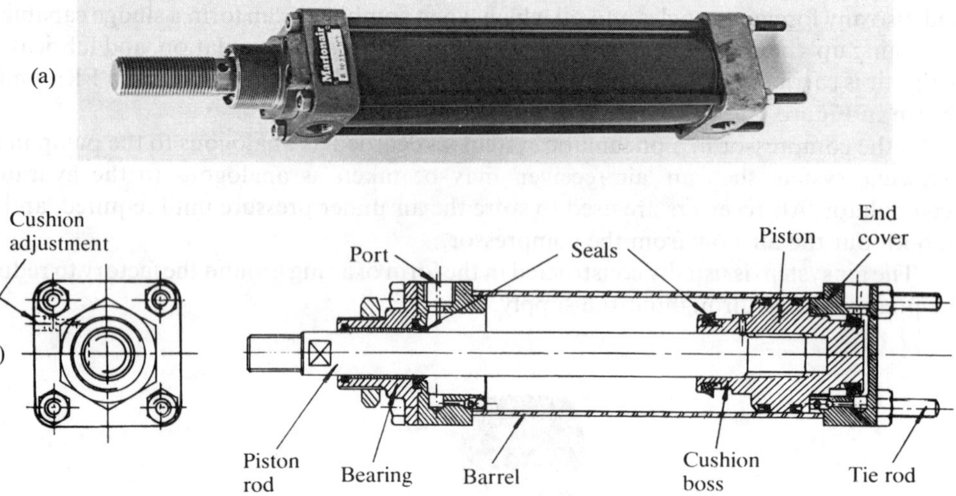

Compressed air applied to one port of a cylinder will drive the piston away from the port, provided that the other port is exhausted to atmosphere. A boss on the piston enters a recess, and the air exhausting from the cylinder is now trapped between the piston and the end cover in the annulus formed by the piston boss and the cylinder barrel. The air trapped in this annulus can only escape by a small orifice, effectively cushioning the cylinder.

Figure 7.14 A pneumatic double-acting cylinder. (Courtesy of Norgren Martonair Ltd)

Figure 7.15 A single unit, rack and pinion type, pneumatic rotary actuator. (Courtesy of Norgren Martonair Ltd)

Rotating movement or torque can be achieved by a rack and pinion system similar to that used in hydraulics, see Figure 7.15. However, in robotics a regulated velocity with a smooth rotation is often required and this can be achieved with an oil/pneumatic system. In this configuration the compressed air system is linked with an oil system such that movement of the pneumatic cylinder forces oil through a restriction thus regulating the speed of movement of the actuator.

This concludes the introduction to drive components. This chapter and the previous one have illustrated the variety of methods available for providing power and motion to robots, the next chapter shows the components used to transmit the movement of the prime movers to the required point of application. Ideally no transmission components should be used at all, this is normally the case in the fluid power systems mentioned above and in some of the more recent electric robots; however, the vast majority of electric robots in use at present do still employ transmission systems.

8

TRANSMISSION SYSTEMS

INTRODUCTION

The purpose of transmission systems is to transfer movement from a prime mover to one or more members of the robot, these may be robot arm joints, wrist joints, or gripper fingers. In fluid power systems force is applied directly via the actuator usually without the need of mechanical linkages. However, in electric robots, due to the speeds, torques and masses of the electric motors, some sort of mechanical drive transmission is usually necessary. Direct drives with the electric motor integral with the joint are becoming more popular but are mainly found only on the lighter duty assembly robots.

Transmission systems can allow rotary movement to be changed to linear or vice versa, they can give reductions or increases in speed between the prime mover and arm member, and they can provide transfer of force or torque. The one major advantage of a transmission system is that the mass of the prime mover can be located remotely from the point of application of force. This reduces inertia and gravitational effects and thus promotes a lighter faster arm construction. The disadvantages are the extra cost of the linkages and the opportunity for 'play' created by backlash and wear to occur in the system. Indeed play is probably the major problem in transmission systems and some adjustment facility usually needs to be built in to the system to allow compensation to be made. However, as play develops in a random manner compensation can be difficult.

The components and configurations of the transmission system used take many forms and the choice of configuration is governed by the purpose of the robot and other design constraints. In a robot transmission system the strength to weight ratio must be at an optimum, backlash and wear must be minimized or compensated for, inertia should be kept as low as possible by good design and material choice, rigidity should be maximized, and every attempt made to reduce friction. This will all help to improve control, resolution, accuracy, and repeatability, reduce vibration and power requirements, and give better performance in terms of shorter cycle times.

This section introduces some of the transmission components used in industrial robots then goes on to show how they can be composed into functional systems.

TRANSMISSION COMPONENTS

The transmission system will comprise a number of individual components, configured in such a way as to give optimum operating characteristics for the anticipated applications of the particular robot being designed. Only some of these 'building blocks' can be considered here but this should be adequate to convey the general concepts employed in transmission system construction.

Gears

Of all mechanical power transmissions gears are the most long-lasting and reliable. High quality gears are expensive but will be found in some form or other in nearly all electrically driven industrial robots. High quality is required in robotic applications due to the precision, speed and low noise requirements. One factor that is considered in gear design, and was mentioned previously, is backlash. This is the shortest distance between noncontacting surfaces of adjacent teeth and is evidenced by nonmovement of the gears for a proportionate amount of time when the direction of rotation is changed by the motor. The amount of nonmovement is directly related to amount of backlash. For high precision, backlash should obviously be kept to a minimum; however, if not enough clearance is given excessive noise, overheating and overloading can occur. Some gear types are shown in Figure 8.1.

Spur gears. These are commonly found in robots, often for rotation of the waist axis. The gear teeth are parallel to the gear shaft axis. They are relatively inexpensive and easy to maintain but do not have the load carrying capacity of the other types and can be more noisy.

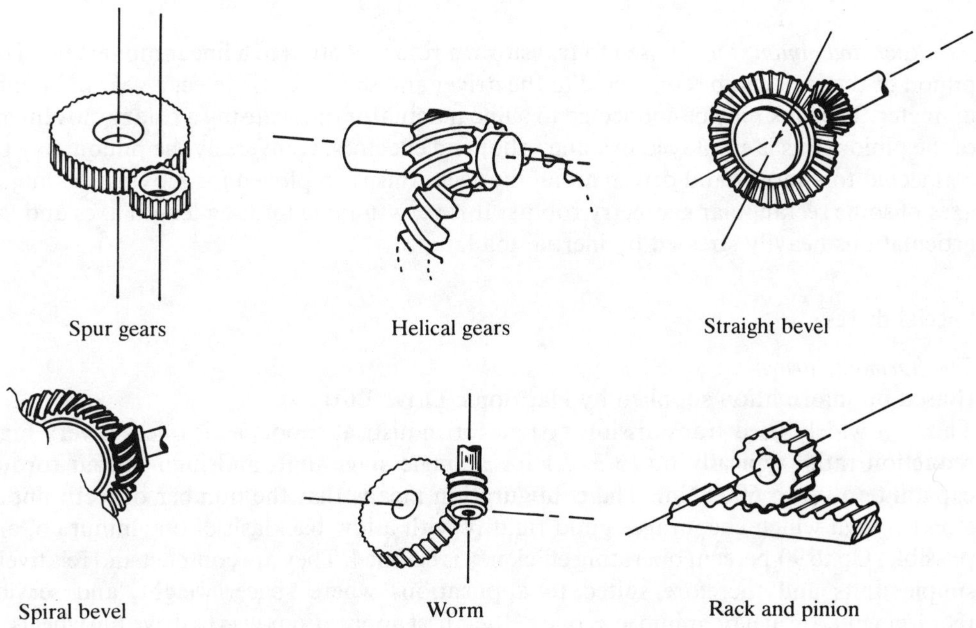

Figure 8.1 Some gear types

Helical gears. These are more expensive than spur gears but can carry higher loads, are quieter and smoother running. Due to their construction they develop end thrust on their shaft and require thrust bearings. To overcome this double helical and herringbone gears are produced, though the latter are more expensive to manufacture.

Crossed helical. These are used with nonparallel shafts and although visibly similar to parallel shaft helical gears, are often called spiral gears. They have a lower load capacity.

Straight bevel. Bevel gears are widely used in robot transmission systems with nonparallel shafts. The straight bevel is the simplest type of gear for intersecting shafts. Their supporting shafts and bearings must be rigid to ensure proper tooth contact. They are often used on shafts intersecting at right angles.

Spiral bevel. A similar effect as when spur gears are exchanged for helical gears occurs when spiral bevel gears are used instead of the straight bevel type. The load is distributed over two or more teeth at any moment and therefore higher loads can be handled. They are also quieter than the straight bevel but are more expensive and develop thrust load.

Hypoid. Used for higher loads than the spiral bevel gear. This is because the shafts do not intersect and therefore bearings can be mounted at both ends of each shaft. They are smoother and quieter than spiral bevel gears and allow high drive ratios.

Worm gears. These are used for speed reduction and the shafts are usually at right angles. They provide the highest reduction ratios for a given center distance and have a high load capacity.

Rack and pinion. This is used to transform a rotary motion to a linear movement. The pinion or the rack can be connected to the driver and the rack is in effect a gear of infinite diameter. The rack may be connected to a linear actuator thus causing a rotary movement of the pinion, this is employed in some robot end effectors. Conversely the pinion may be connected to a motor and driven along the rack, this is employed for moving the major axes of some rectangular geometry robots. It is very suitable for long length axes and for articulations heavily stressed by inertial load.

Special drives

The harmonic drive

(Based on information supplied by Harmonic Drive Ltd)

This is a widely used transmission system for industrial robots as it offers a very high reduction ratio, typically up to 320:1 for a single stage unit, and high output torque capabilities, up to 6000 Nm. The configuration means that the number of teeth under stress is high which encourages good rigidity with a low backlash of one minute of arc possible. Up to 90 percent operating efficiency is claimed. They are compact and relatively simple units and therefore suited to applications where space, weight, and service requirements are at a premium, e.g. one of their first applications was to drive the wheels of the Lunar Rover during the Apollo missions.

The principle of operation of the device is shown in Figure 8.2. There are three basic parts, i.e. the wave generator (input), the flexspline (output), and the circular spline (fixed). The wave generator comprises an elliptical former on which is fitted a thin walled ballrace. The former is keyed onto the input shaft. The flexspline is in the form of a flexible thin walled steel cylinder one end of which is attached to the output shaft. It has external teeth and fits closely over the wave generator. The teeth engage with the circular spline across the major axis of the ellipse of the former. The circular spline is a fixed solid steel ring with internal teeth of the same pitch as those on the flexspline. The flexspline has less teeth than the circular spline and the difference gives the reduction ratio. That is, the reduction ratio $= Nf - Nc/Nf$ and as Nf is less than Nc the two splines contrarotate. Figure 8.3 shows the components.

Although the above conforms to the most common configuration a number of others can be used, e.g. the wave generator can be fixed, the flexspline used as input, and the circular spline used as output to give a different ratio and input and output rotation in the same direction.

In industrial robots the systems are often driven by a closely coupled servo motor giving accurate and compact servo control. The drive is claimed to allow good dynamic response with minimal overshoot and hunting problems, and also good absolute positional accuracy.

The cycloidal speed reducer

(Based on information supplied by CENTA Transmissions Ltd)

This is becoming popular for robot drives where higher transmission forces are involved.

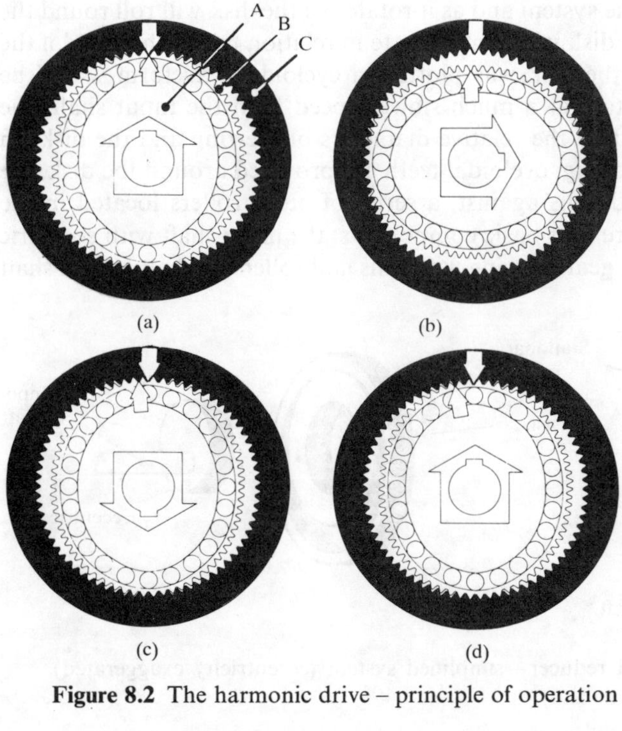

(a) 'A' Wave generator (input)
'B' Flexspline (output)
'C' Circular spline (fixed)

(b) Immediately the wave generator begins to rotate clockwise the zone of tooth engagement travels with the major elliptical axis.

(c) When the wave generator has turned through 180° clockwise the flexspline has regressed by one tooth relative to the circular spline, when the tooth ratios are as shown in the sketches.

(d) Each turn of the wave generator moves the flexspline two teeth backwards on the circular spline.

Figure 8.2 The harmonic drive – principle of operation

Output torque to 6000 Nm
Ratios from 78:1 to 320:1

Figure 8.3 Harmonic drive HDUC component set. (Courtesy Harmonic Drive Ltd)

A simplified system is shown in Figure 8.4, a schematic diagram in 8.5, and a sectional photograph in Figure 8.6. These drives also offer compact size, low weight, low inertia and low backlash. Single stage reduction ratios from 6:1 to 87:1 and multi-reduction ratios in excess of 10 000 000:1 are possible. Output torques up to 65 000 Nm can be obtained.

The basic concept and principle of operation is as follows. Consider a disk mounted on an eccentric shaft with its surface in contact with the inside wall of a stationary ring. The shaft provides the input to the system and as it rotates so the disk will roll round the inside of the stationary ring. The disk will contrarotate in relation to the shaft and if the path of a point on the edge of the disk was plotted, a cycloidal waveform would be produced. The disk will also rotate at a much slower speed than the input shaft, the reduction ratio being determined by the relative diameters of the ring and the disk. In practice, to prevent slip during rolling, cycloidal teeth are provided around the disk, the edge of which engages with, and rolls against, a series of outer rollers located in the stationary housing. Thus there are four basic components: the input shaft with eccentric and bearing; cycloid disks; a ring gear housing with pins and rollers; and an output shaft with drive pins and rollers.

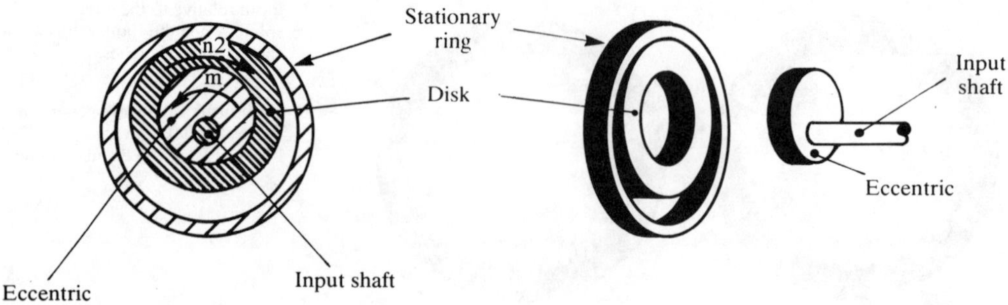

Figure 8.4 The cycloidal speed reducer – simplified system (eccentricity exaggerated). (Courtesy CENTA Transmissions Ltd)

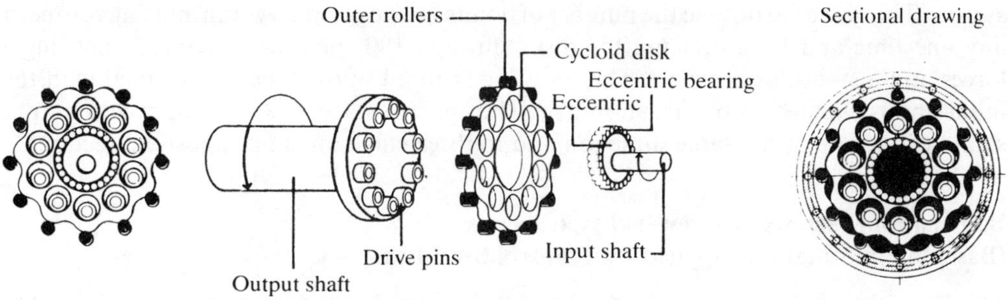

Figure 8.5 The cycloidal speed reducer – schematic diagram. (Courtesy CENTA Transmissions Ltd)

Figure 8.6 Sectional photograph of 'Cyclo' speed reducer. (Courtesy CENTA Transmissions Ltd)

The cycloid disc is mounted on the eccentric bearing which rotates at the speed of the input shaft causing the cycloid disk to roll around the inside of the ring gear housing. As there is one less cycloid tooth on the disk than the number of rollers in the ring gear housing, the cycloid disk rotates in the opposite direction to the input shaft by one tooth pitch for each revolution of the input eccentric. Thus output rotation is created in the opposite direction to input rotation and at a reduced speed. The reduction ratio is determined by the number of teeth on the cycloid disk, e.g. 87 teeth on the cycloid disk and 88 rollers in the ring gear housing gives a reduction of 87:1.

The rotation of the cycloid disk is transmitted to the output shaft by a number of pins and rollers projecting through holes in the disk. Most cyclo gearboxes employ a two-disk system. This serves to double the number of points on the periphery transmitting torque at any one time and by displacing the disks through 180° provides balanced centrifugal forces and a smooth operation. The resulting reduced output speed is coaxial with the input and is claimed to be vibration free. Thus in two stage units the input and output shafts will rotate in the same direction and in three stage units in opposite directions.

Ball and roller screws or screw–nut systems
(Based on information supplied by Transrol SKF Ltd)

This is another widely used method of drive transmission in industrial robots. It is used to combine the speed reduction function with the conversion of rotation to linear movement. Figure 8.7 shows principles of operation and examples.

In robotic applications the screwed shaft is normally rotated by an electric servo motor and possibly via a speed reduction system such as the harmonic drive or cycloidal gear. The nut mounted on the shaft is usually of the ball or roller screw type. As the shaft rotates so the nut is driven by the screw thread and moves linearly along the axis of the shaft. In operation the shaft is attached to one member of the robot, and the nut to the following member, the two members being linked by a joint, see Figure 8.13(c). Good rigidity, resistance to vibration, smooth action, excellent repeatability, and high reliability are some of the features of this type of transmission.

Figure 8.7(a) shows the Acme sliding screw which typically consists of a steel shaft with trapezoidal form thread and accompanying nut. The resulting sliding action with associated friction gives an efficiency of about 30 per cent. However, by placing balls or rollers between the nut and the screw shaft, see Figure 8.7(b), a screw of high efficiency of about 90 per cent is obtained. The load carrying capacity of each ball in the screw is limited by its diameter, see Figure 8.7(c), which is in itself limited to 60 to 70 percent of the lead of the screw. However, the load carrying capacity of the screw also depends on the number of contact points. To increase the number of contact points roller screws of threaded (see Figure 8.7(d)) or grooved (see Figure 8.7(e)) form are used.

Figure 8.7(f) shows a section through an internal recirculating ball screw. In this type the balls return by a tube after 1.5, 2.5, or 3.5 turns of load carrying. Although the system has the disadvantage of a relatively large nut, it is the least expensive recirculation method; it is also reliable and is used in a wide variety of industries.

An internal recirculation system is shown in Figure 8.7(g). Here a stainless steel insert is used in which the balls jump back over the screw thread after one turn of load carrying. This makes it more expensive than the previous system but a compact nut is obtained.

For higher loads roller screws are used and Figure 8.7(h) shows a planetary roller screw. This has threaded rollers with a helix angle exactly the same as the nut thread, thus ensuring the roller does not move axially relative to the nut as it rolls. It is robust, capable of high rotational speeds, and it can run smoothly even in dirty or poorly lubricated conditions.

Grooved rollers are used in recirculating roller screws as shown in Figure 8.7(i). As these rollers rotate inside the nut they move axially thus necessitating recirculation as in ball screws. This type of screw is useful for high loads and small leads down to 1 mm, with system resolutions as small as 0.025 μ having been achieved. This arrangement provides high rigidity and facilitates simplification of the complete system.

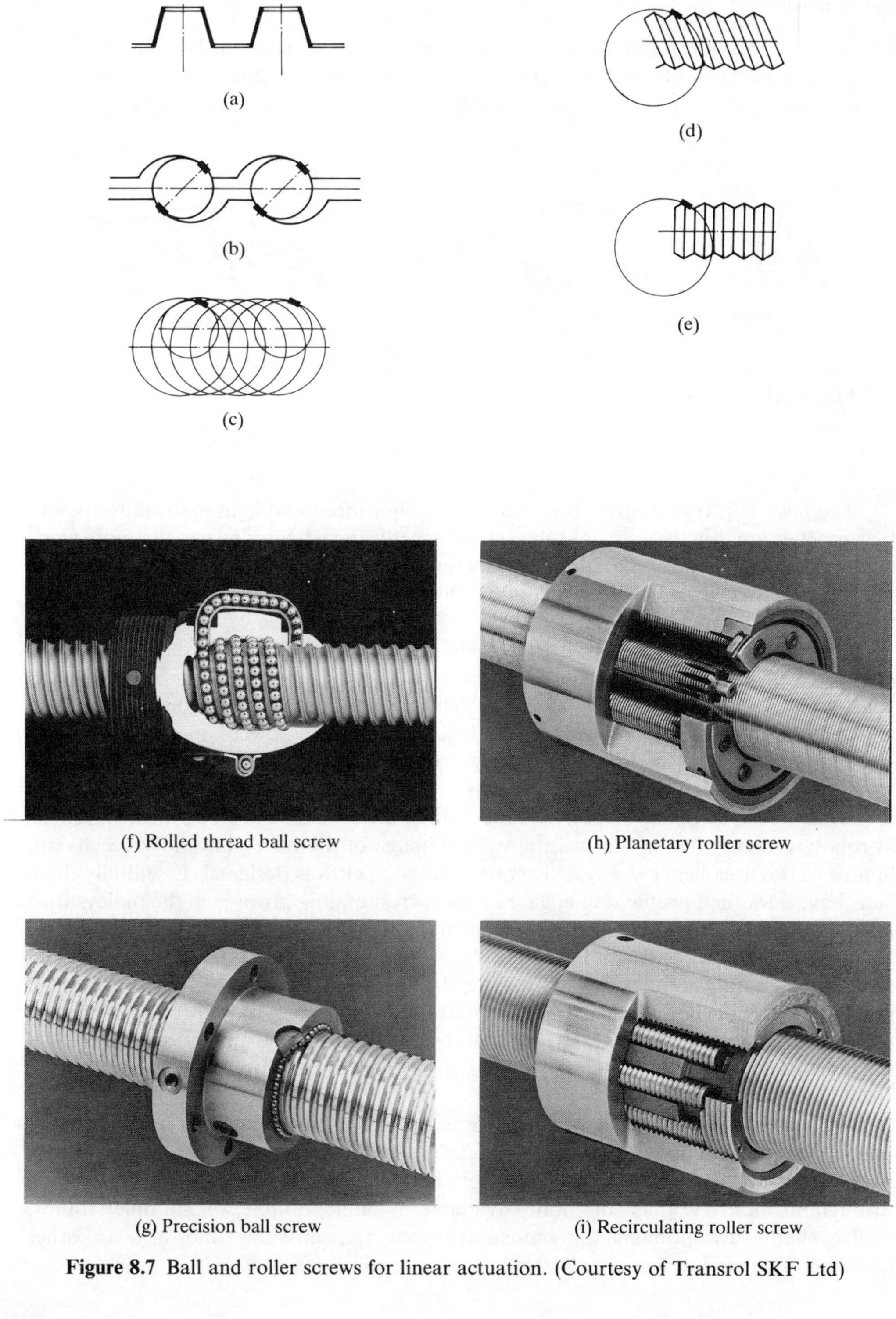

Figure 8.7 Ball and roller screws for linear actuation. (Courtesy of Transrol SKF Ltd)

Pulley drives

These drives are widely used in robotics, particularly the synchronous belt and the roller chain. They have low inertia, high rigidity, and efficiencies often greater than 95 percent. Some pulley-type drives are discussed briefly below and three are shown in Figure 8.8.

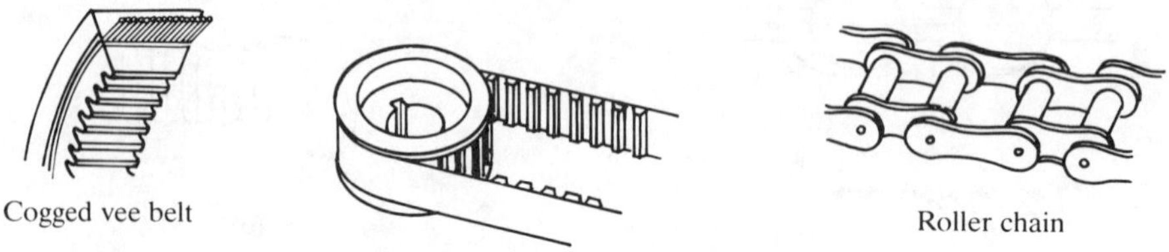

Figure 8.8 Some pulley drives

Flat belt. This type of drive is suited to very high speed and light load transmission. Because they are thin they are not subject to high centrifugal loads and can therefore work well even over small pulleys at high speeds. However, they rely on belt tension to produce grip over their pulleys which can cause a reduction in bearing life.

Cogged vee belt. Vee belts utilize the wedging action that occurs between the belt and pulley thus providing large driving forces with much less belt tension than required by flat belts. A recent type is the cogged vee belt which allows the use of smaller diameter pulleys and hence more space efficient design. Vee belts provide high power handling ability in a small volume and are suited to heavy duty work, shock loads, and high starting forces.

Synchronous or timing belt. Where input and output shafts must be synchronized, as in robotics, there must be no possibility of slippage of the belt. Flat and vee belts are limited in this area hence the synchronous or timing belt was designed. Essentially these belts have a toothed profile that engages with corresponding grooves in the pulleys thus giving them the same sort of positive engagement that is achieved with gears or chains. They are an extremely efficient form of drive, producing relatively little bearing wear and capable of handling high loads. They are therefore popular even where synchronization of input and output is not necessary. Modern developments in belt tooth profiles mean that these belts are becoming quieter, stronger and smoother running, and indeed can be used in applications where previously chain drives were the only solution.

Roller chain. Chains are generally cheaper than gears but have lower load capacities and service lives. In comparison to belt drives they have greater load capacities and service lives but may be more expensive. The roller chain is the basic type of power chain used throughout industry. It is commonly available in single, double, or multiple strands. Roller chains are popular for industrial robot transmissions and also in other manufacturing engineering applications such as in conveyor systems.

Linking methods

Only the more commonly used linking methods on industrial robots are considered here although there is a large variety of different types available.

Concentric transmission rods. These are in effect concentrically arranged hollow tubes capable of transmitting torque from the motor and gear drive units located at one end of an arm to the opposite extremity. These tubes are designed to resist twisting thus providing a compact, positive and accurate transmission device. At the driven end the tubes are presented to the drive gears in step fashion as shown in Figure 8.9(a). At the opposite end the tubes are connected in an appropriate manner to the driven members. This is illustrated in the example of the Cincinnati Milacron wrist shown in Figure 8.15.

One piece helical coupling. Shown in Figure 8.9(b) the one-piece helical coupling is flexible and is therefore in operation a curved beam. It can transmit motion from one shaft to another with zero backlash and with constant velocity even when misaligned. The one piece construction compensates for parallel and angular misalignment, torsional vibration, and axial motion. The design requires no lubrication and is of low mass, making it suitable for high speed applications such as are found in robotics. They are used in the Unimation Puma™ robots as illustrated in Figure 8.14.

The 'rod end'. The rod end is a spherical bearing with the ball enclosed at the end of a link as shown in Figure 8.9(c). The outside diameter of the bearing is spherical thus allowing a ball and socket action to compensate for shaft and mount misalignment. They are commonly used in oscillating linkages, e.g. the rod ends in Figure 8.13 on the ASEA robot.

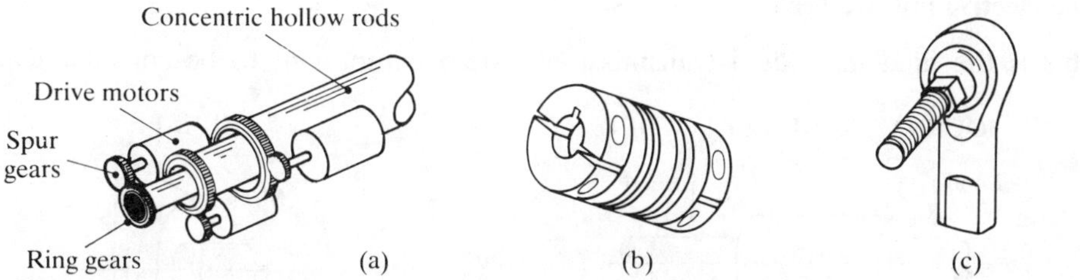

Figure 8.9 Examples of linking methods: (a) concentric transmission rods, principle of operation; (b) one-piece helical joint; (c) spherical bearing or rod end

Bearings

The variety of bearings, their application and construction is such that only a general indication of their importance can be given here. Bearings allow smooth, low friction movement between two surfaces. This movement may be rotary as is found in a cylindrical robot joint, or linear as is found in a prismatic joint. In each case the bearings will use either a sliding or a rolling action and an attempt will be made to keep the surfaces separated by using a film of oil or other lubricant. Providing contact of the moving metal parts can be avoided then satisfactory service lives can be obtained. The bearings that utilize a sliding action are called 'plain' bearings, those that use a rolling action are termed 'rolling element' bearings. Although plain bearings can be appropriate for use in many situations and are usually low cost, the rolling element bearings are considered to be more dependable and of higher quality.

Bearings can therefore carry loads along their axis of rotation or movement and/or at right angles to their rotation or movement. Plain and rolling element bearings that carry loads along their axis of rotation are termed 'thrust bearings'. Plain bearings carrying loads perpendicular to their rotational axis are termed 'journal bearings'. Rolling element bearings carrying similar loads are called 'radially loaded' bearings. These bearings are all illustrated schematically in Figure 8.10.

Factors that have to be considered when choosing and using bearings are as follows. The maximum load that can be carried by the bearing at particular speeds. This can be determined from past statistics of tests in operation of the bearings. Load variations and impact loads should also be considered. Lubrication is obviously important and can be of several types. Hydrodynamic lubrication is a natural consequence of lubricant flow between the bearings and the shaft, and is probably the most desirable type for plain bearings. Hydrostatic lubrication is provided by an external pumping of the lubricant. This type allows minimum deflection, precise shaft positioning and smooth operation. Another type of lubrication arises in the case of self-lubricating bearings. These bearings are usually constructed of sintered metals with oil impregnated into the pores of the metal structure. As the bearing or shaft rotates so the oil is gradually exuded and lubrication occurs often in the hydrodynamic manner. Other factors to consider are the deflection of the bearing, the power required to overcome starting friction, the size constraints, the noise created, and total costs including hardware, maintenance, and failure costs, etc.

The electric failsafe brake

This is the final individual transmission system component to be considered and its

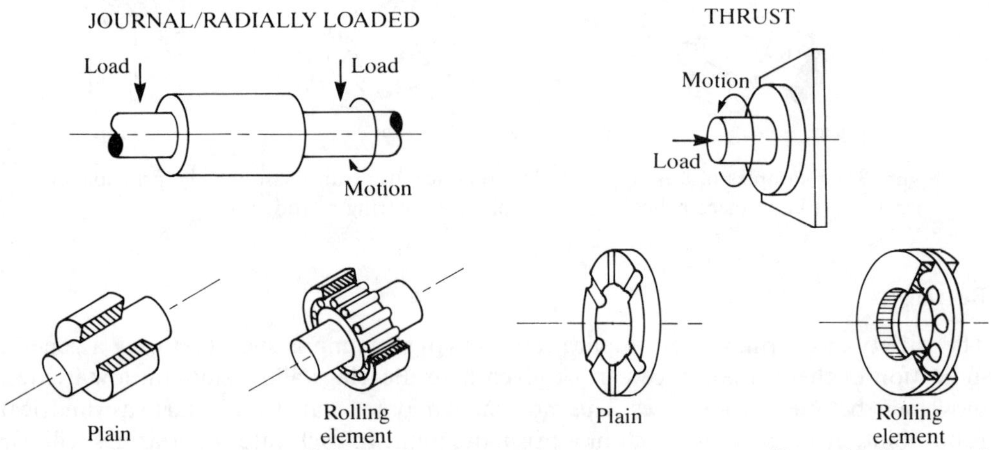

Figure 8.10 Bearings

principle of operation is shown in Figure 8.11. When power is removed from the drive system in an electric robot there will be a tendency for the robot articulations to move due to gravity. This will be exacerbated if a heavy load is suspended from an outstretched end effector. Some method is therefore required to lock the system when power is removed to prevent drastic loss of position and possible damage to the robot, material being handled, or surrounding work area.

The method usually employed is that of the electric failsafe brake. The particular type will probably be of the plate type disk brake which allows a high braking force in a small unit. If cycled frequently it can tend to overheat but this should not be a concern in robotic applications as power removal would be occurring at fairly long intervals. In operation when power is applied to the robot the brake is held in the disengaged position against the force of a spring. The friction surface is pulled back against the spring by a powerful electromagnet, loss of power de-energizes the magnet and allows the spring to engage the brake. The brake is thus failsafe and in this configuration is very fast operating.

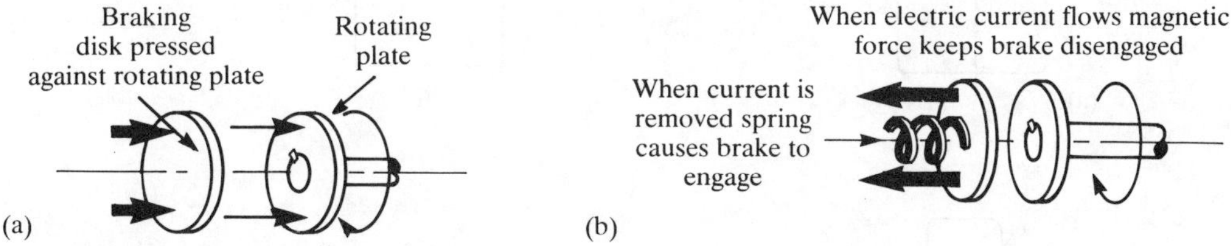

Figure 8.11 Electric failsafe brake: (a) disk brake of full plate type; (b) spring actuated failsafe disk brake

TRANSMISSION SYSTEM EXAMPLES

The following are examples of how individual transmission components are configured in some commercial robots. All sketches are based on drawings kindly supplied by the manufacturers concerned.

FANUC 'M' series robots: some transmission methods

Figure 8.12 shows four different configurations used by FANUC in their 'M' series robots. Figure 8.12(a) shows the linear axis transmission for the model 0 robots. The motor drives a screw via bevel gears. The screw nut is fixed to the bracket which runs on parallel guides. A failsafe brake is incorporated. The rotational axis of the model 0 shown in Figure 8.12(b) utilizes a timing belt from the motor shaft to a worm shaft and wheel. The vertical Z axis of the model 1 robot is shown in Figure 8.12(c). This has an overhead mounted motor, operating through a failsafe brake, and driving a ball screw. The nut of the ball screw is attached to the horizontal body of the robot thus allowing vertical up and down movement on two guide pillars. The model 3 rotational axis configuration is shown in Figure 8.12(d). The horizontally mounted motor drives a worm and worm wheel arrangement which drives a pinion. This pinion rotates a gear which rotates the body of the robot.

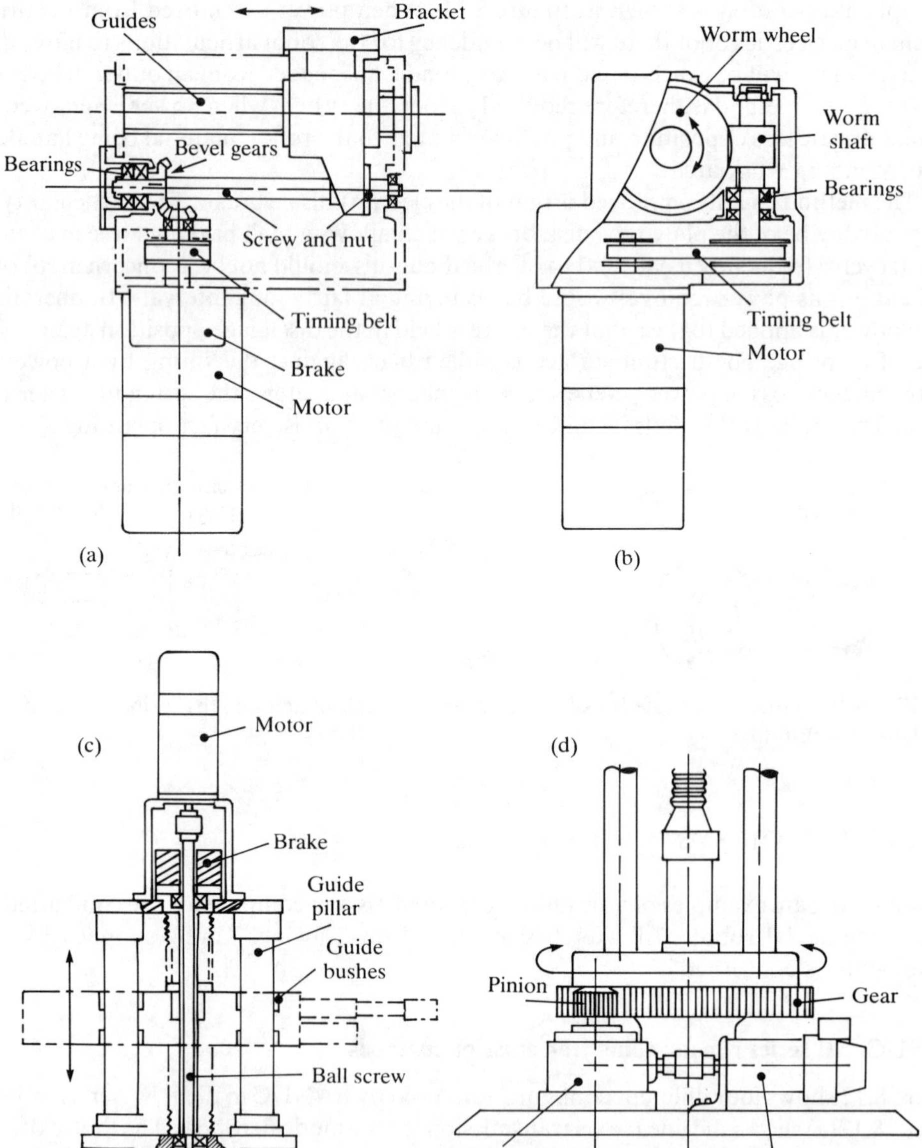

Figure 8.12 Some FANUC transmission configurations: (a) FANUC robot M, model 0 – *XYZ* axes transmission configuration; (b) FANUC robot M, model 0 – 'B' axis transmission configuration; (c) FANUC robot M, model 1 – *Z* axis transmission configuration; (d) FANUC robot M, model 3 – θ axis transmission configuration. (Courtesy FANUC)

ASEA IRB 6 transmission system

The transmission system for this robot is shown in Figure 8.13. Figure 8.13(a) shows the pedestal and body arrangement. Rotary movement is obtained by rotating the body which is mounted on bearings on the robot pedestal. This rotary movement is effected by a motor unit which rotates the body via an harmonic drive with a reduction ratio of 158:1. The upper and lower arm movements shown in Figures 8.13(b) and (c) are effected by means of ball screws. The motors to drive these screws are suspended from the ends and the bases of the screws and motor casings are fixed in a joint to the robot body as shown. The ball screw nuts are secured via bearings to levers on the upper and lower arms, a parallelogram structure being adopted for the upper arm. Thus rotation of the motors causes rotation of the screws, this gives a linear movement of the nuts which causes rotation of the arms about their respective pivot points.

Figure 8.13(d) shows the transmission for the wrist turn motion. Linkage rods are also used for the wrist bend motion and are mounted on the other side of the arm. Point X shows the axis on which the drive motors for the wrist movements are mounted, one horizontally on either side of the lower linkage center for the lower arm. The rotary movement of the motors is transmitted via harmonic drives to the linkage disks located inside the arm. Two further disks are located in the joint of the upper and lower arms and a further two inside the wrist. The linkage rods are connected to the disks by spherical bearings, or 'rod ends'. Figure 8.13(e) shows the wrist construction for the turn and tilt motions, the turn motion being obtained by the use of bevel gears.

Forearm and wrist transmission in Unimation Puma™ 500 series robot

This is shown in Figure 8.14 in schematic form. The motor to rotate the forearm about its axis is mounted in the upper arm. The drive from the motor is transmitted via a bevel pinion to a bevel gear on one end of an idler shaft. A spur pinion at the other end of the shaft engages a spur gear which is fixed to the forearm. This effects rotation of the forearm about its axis or elbow. The motors to drive the wrist axes are located at the elbow end of the forearm, this minimizes the weight at the end of the arm and also acts as a counterbalance. The shafts from these motors are connected to the wrist spur and bevel reduction gears by one piece helical joints.

The Cincinnati Milacron 'three roll wrist'

This is an interesting design initially used on the electric robots of Cincinnati Milacron. The major components of the wrist are driven by concentric hollow shafts or rods. These shafts are driven through gears by electric motors mounted at the elbow end of the robot forearm, the principle of which was shown earlier in Figure 8.9(a). The wrist configuration is as illustrated in Figure 8.15 and three degrees of freedom are obtained as shown. Rotation of the outer concentric shaft produces rotation of the whole assembly in the direction R_1 as shown. Rotation of the second shaft causes rotation of the upper hemispherical shell in the direction R_2. Finally rotation of the central shaft causes rotation of the faceplate in the direction R_3. Each element is controlled independently and their movements are coordinated to provide the roll, pitch, and yaw wrist articulations

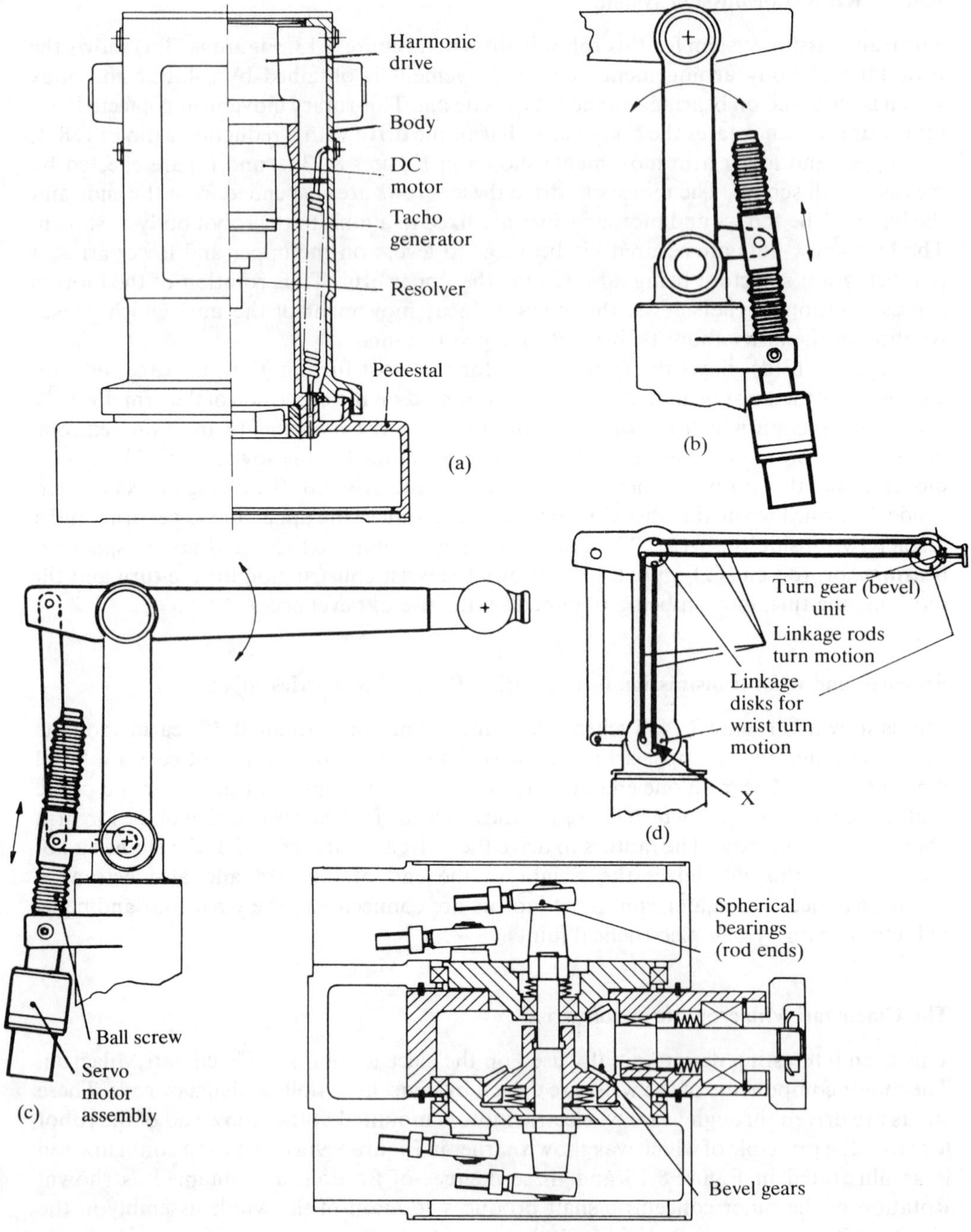

Figure 8.13 Transmission system of ASEA IRB 6 industrial robot: (a) pedestal and body; (b) lower arm; (c) upper arm; (d) wrist (one axis only shown); (e) wrist with turn and tilt motions. (Courtesy ASEA)

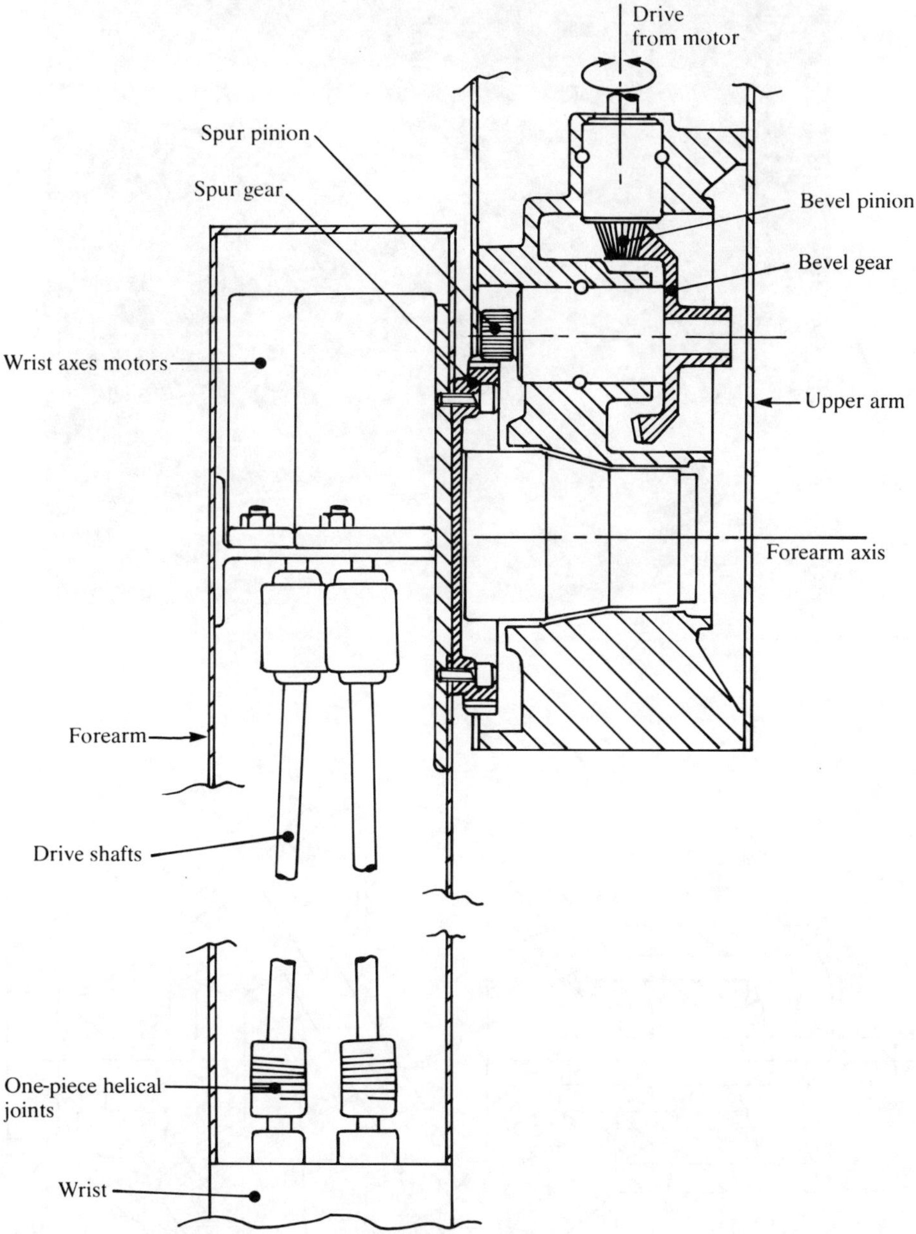

Figure 8.14 Transmission system for a forearm and wrist on Unimate Puma™ 500 series robot. (Courtesy Unimation (Europe) Ltd)

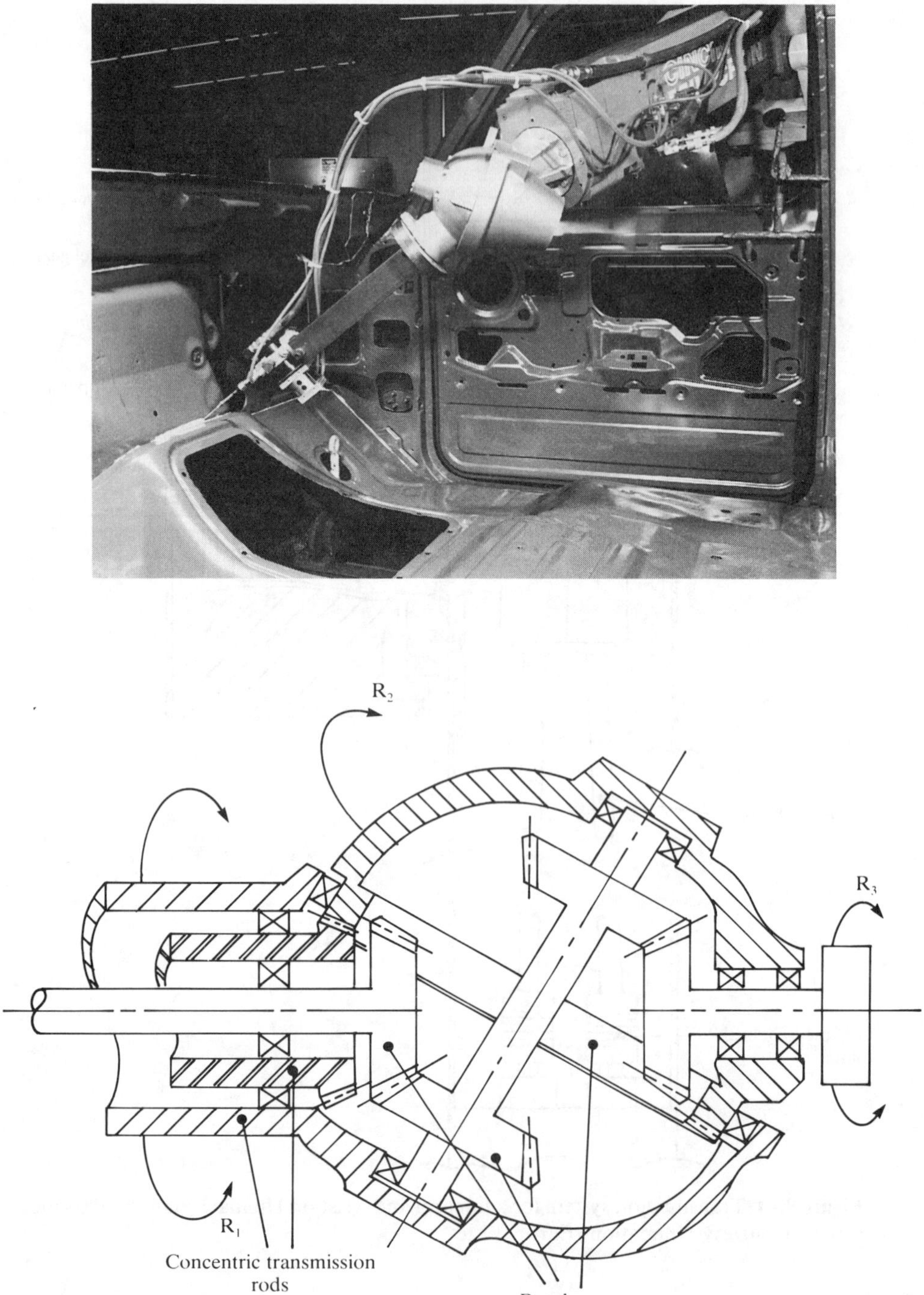

Figure 8.15 Cincinnati Milacron 'three roll wrist': (a) in operation; (b) schematic. (Courtesy Cincinnati Milacron)

described later in Chapter 12. The design offers several advantages over simpler but more cumbersome types of wrist construction, such as the one shown in Figure 12.13. As the motor drive is remote and the three axes of rotation intersect at the one point, a compact design results. Control is simplified and continuous reversible rotation is available at all times on each axis. Using lightweight materials a low inertia, high stiffness wrist is obtained that will perform well in most applications even where space is restrictive.

The Spine robot transmission

The transmission technique applied in the Spine robot is not common and is included here to highlight the almost limitless transmission configuration possibilities. It is shown in Figure 8.16. The structure allows great flexibility, it is constructed in two arm units each unit capable of moving throughout a work envelope of hemispherical shape. The robot is unusual in that it is an hydraulic robot with a complex transmission system. The under and upper arms comprise a number of disks with hemispherical contact surfaces. Four wires are connected through the underarm and a similar number to the upper arm. The tension corresponds to a force of several tonnes per wire pair. The under arm wires are attached to hydraulic cylinders in the lower section and the wires in the upper arm are connected to the hydraulic cylinders in the arm joint. The cylinders are under constant pressure from the hydraulic system and this pressure creates the wire tension. Movement of the cylinders causes movement of the wires which results in movement of the arm. This movement of the cylinders is effected by a redistribution of hydraulic pressure via the robot's control system.

Direct drive systems

As mentioned at the beginning of this chapter, direct drive robots, i.e. robots in which the drive motors are an integral part of the joint, are becoming more popular. At present they are almost always limited to the SCARA type configuration assembly robots. This configuration is particularly suitable as the vertical axis construction is good for carrying the extra weight of the motors while maintaining a light overall structure. The ability to incorporate direct drives in an electric robot design offers many advantages, the main one being the elimination of transmission systems and their associated problems and costs. The simplicity of direct drive construction should reduce component and maintenance costs. A well designed robot of this type will achieve higher speeds, precision, and stiffness, as well as a lighter total arm construction.

In most practical engineering designs optimum performance and cost are usually obtained from a combination of concepts. This is also true of direct drive robots which are often of hybrid construction using direct drive at the most suitable joints, and transmissions for other joints where appropriate. When using direct drive on revolute type robots some sort of counterbalancing structure, such as a parallelogram configuration, will still be required. As well as rotary motors for direct drive it is also possible to use linear motors to achieve high speed in rectangular and cylindrical type robots, although this is as yet uncommon.

The AdeptOne was one of the first robots incorporating direct drive on some of its

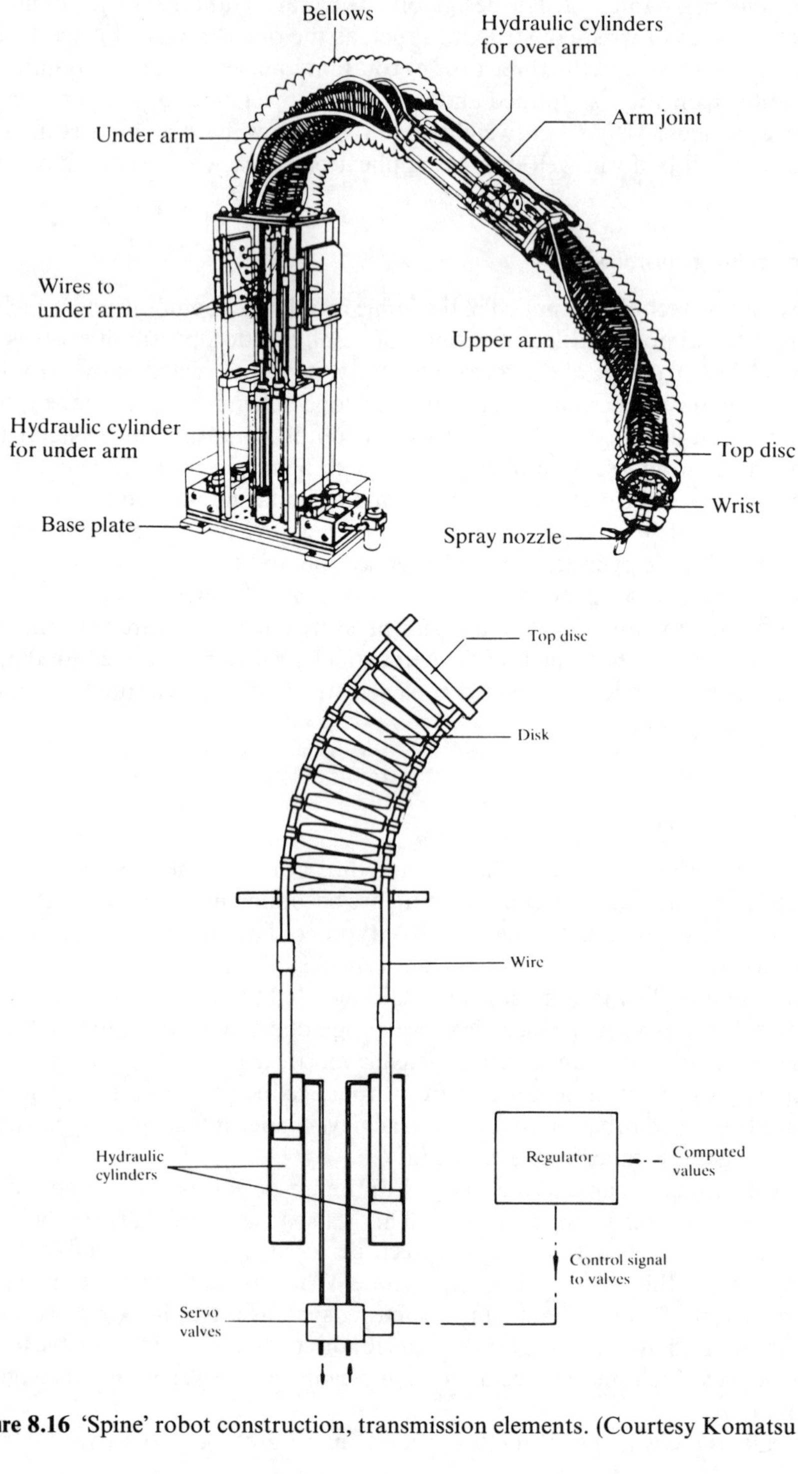

Figure 8.16 'Spine' robot construction, transmission elements. (Courtesy Komatsu Ltd)

axes, other axes being driven via transmission components, i.e. prestressed steel bands for the elbow joint and a toothed belt drive for the tool head. Figure 8.17 shows the AdeptOne drive construction.

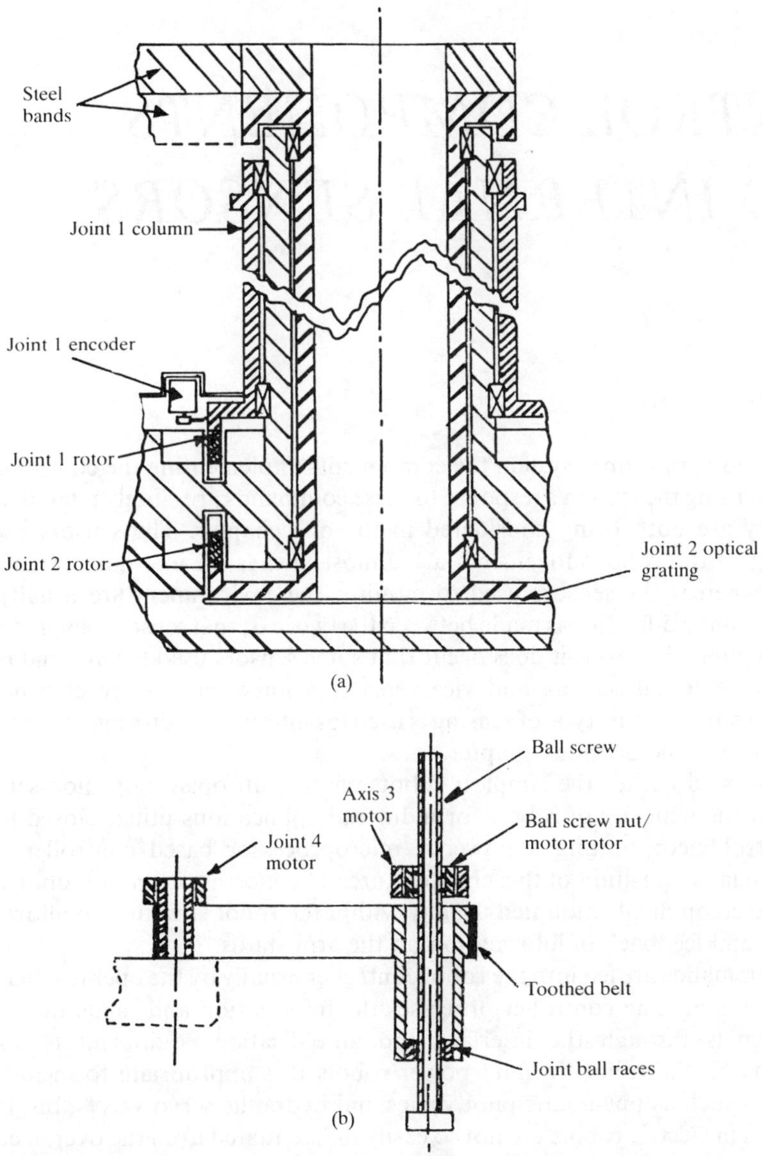

Figure 8.17 AdeptOne direct drive arrangement: (a) main column and drives of axes 1 and 2; (b) drives of axes 3 and 4. (Used with permission of Adept Technology Inc.)

9

CONTROL COMPONENTS AND INTERNAL SENSORS

INTRODUCTION

Within an industrial robot system the components implementing the control commands and those sensing the system's response to these commands are closely related. It is for this reason they are both being considered in the one chapter. The sensors used for the monitoring of the robot movements are almost always an internal part of the robot structure, whereas the sensors used to monitor the environment are usually mounted externally. Thus a difference is made between the two, external sensors being considered in another chapter. However, it does occur that some sensors used for internal monitoring are used for external sensing and vice versa, e.g. force sensors are considered under 'external sensors' but this type of sensing is used for internal monitoring in the Spine robot system shown in the previous chapter.

As discussed earlier the simplest robots employ an open loop, non-servo-control system, but the majority of robots for industrial applications utilize closed loop or full servo control incorporating a powerful microprocessor based controller. Figure 4.1 shows the relative position of this chapter's area of concern within the robot system, i.e. the discrete components mounted usually within the robot structure to allow control of movement and feedback of information on the arm status.

The commands are fed into the robot controller usually by the operator via a terminal or teach pendant. The controller processes the information and sends the appropriate control signals through the interface and amplification equipment to the control components on the robot. In fluid power robots it is appropriate to include here the components such as pneumatic pilot valves and hydraulic servo valves, but the control components in electric robots are not so easily differentiated from the overall control and drive system. It was for this reason that the control components associated with electric motor control, such as thyristors and PWM units, were included in the chapter on electric drive systems. The control components respond to the command signals thus enabling for example an electric motor to turn or a pneumatic cylinder to be moved. In an open loop system, where there is no feedback, either the control signal will be applied for an appropriate period of time to allow the actuator to attain the required position, or the movement of the actuator will be physically restricted by a mechanical stop. In the closed loop system, as the actuators and robot limbs are moving, signals will be sent back to the computer control by the feedback components, containing information that can be used to determine position, velocity, and direction of travel.

These control and feedback components may be analog or digital and these concepts have already been covered in Chapter 5 together with conversion techniques.

CONTROL COMPONENTS

Mechanical stops

These are the simplest method of control, they are employed on non-servo axes, and are considered in Chapter 12 on Programming and Languages.

Mechanical control by cams

The cam is one of the earliest control devices; Figure 9.1 illustrates some types. It can be used to trigger on/off responses in robots by activating pneumatic control valves or electrical microswitches. Or it can position or move a machine component in response to

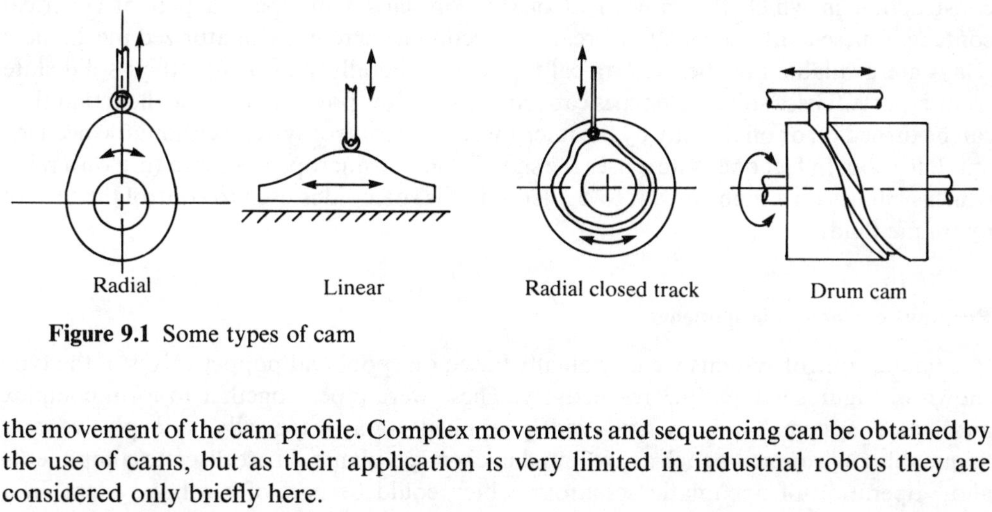

Figure 9.1 Some types of cam

the movement of the cam profile. Complex movements and sequencing can be obtained by the use of cams, but as their application is very limited in industrial robots they are considered only briefly here.

The most common type of cam is the flat plate cam which is typically cut from steel plate. This type of cam may be radial, i.e. using rotary motion as an input, or linear, i.e. using straight line motions as input. Here the control signal is represented by the position of the cam surface relative to a set datum. This is 'read off' the cam by the cam follower. The follower is a device one end of which rests on the cam surface and is constrained to respond to the cam movement. Many other types of cam are available, e.g. closed track cams, and drum cams with the cam track cut into the surface so that when the drum rotates the cam follower responds with a linear movement. Apart from their use in very basic non-servo-control robots for sequencing, cams are seldom used as, due to their purely mechanical nature, reprogramming is manual, tedious and time consuming.

Solenoids

These are digital components in either the on or off state; they are widely used to control pneumatic and hydraulic systems. Essentially they comprise a cylindrical iron core free to move within a conducting coil. When a current is passed through the coil the core, being within the magnetic field, experiences a force that moves it linearly along its own axis. Thus the flow of current can be turned on or off by the robot controller to provide a mechanical action. The signal from the microprocessor system will normally require to be amplified to provide sufficient power to the coil. Although a mechanical action is obtained in the solenoid this is not used directly for actuation but is rather used to control a secondary power source. An example of such an application is shown in Figure 9.6 where a solenoid is used in a pneumatic pilot valve. They are very easily interfaced to other control equipment due to their simple on/off operation.

Relays

These are also basically digital components widely used in many types of control system. They may be mechanical or solid state, both reacting to a simple on/off signal. Essentially they allow a low power signal in one circuit to turn on a higher power supply in another circuit. For example, in the mechanical type, one configuration has a solenoid type construction in which the movement of the core closes or opens a pair of electrical contacts. These contacts make or break the secondary circuit. Miniaturized mechanical relays are available but the mechanical types are generally slower than their solid state counterparts. These solid state types can require such low power to operate them that they can be turned on or off directly by the microprocessor control system without the need for amplification. In fact one system uses the signal from the microprocessor to turn on a relay which switches on a high current to a solenoid, this solenoid is used to control the flow of hydraulic fluid.

Pneumatic control components

Pneumatic control systems were originally based on spool and poppet valves of the type shown in Figures 9.2 and 9.3 respectively. These were piped together to form complex automated systems and are still widely applied today especially in explosive and flammable environments such as are found in chemical plants. These discrete components allow operation of pneumatic actuators, which could be part of a robot system, in a controlled sequence.

The clearest way of understanding their application is to consider the simple circuit shown in Figure 9.4 in which the valves and actuator are represented by symbols. The symbol convention adopted here for representing the various components and their status is a standard one. Both types of valve are represented by two boxes, one box shows the flow of air through the valve in one state, the second shows the air flow with the valve spool in its alternative position. In the circuit shown the piston A will move out and in repeatedly following the sequence A+ A− A+ etc. This system consists of the double acting cylinder A, a four-way four-port control valve, and two spring return poppet valves.

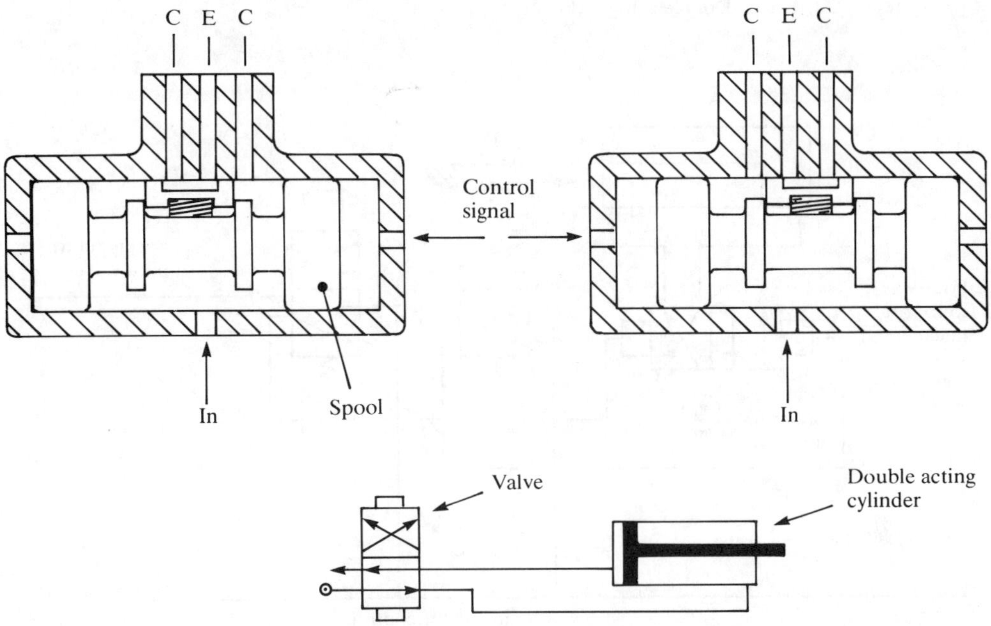

These valves can be used to control double acting cylinders. The valve has two cylinder ports, an exhaust port and a supply port.

Figure 9.2 Four-way, four port linear slide valve

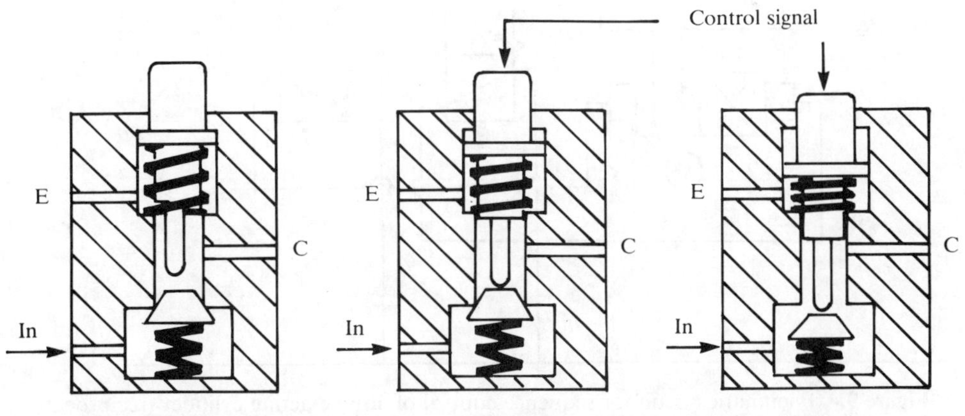

Pilot valves, like the poppet valve above, are usually used to provide an air signal to operate air operated control valves such as the four-way four-port valve in Figure 9.2. When the plunger is depressed the exhaust port closes first and then the inlet port opens. Thus air cannot be passed to atmosphere if the valve stops in mid position.

Figure 9.3 Spring return poppet valve

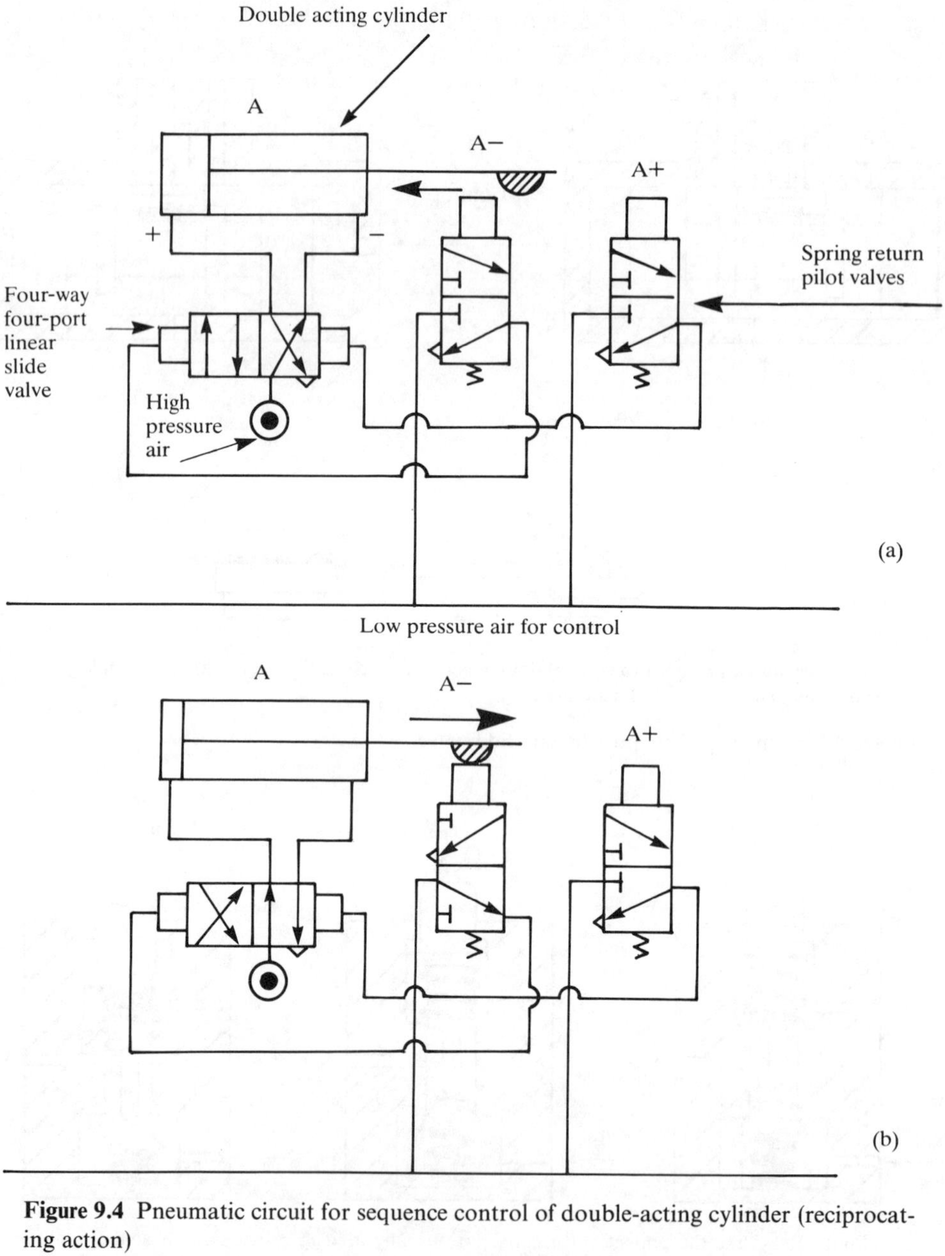

Figure 9.4 Pneumatic circuit for sequence control of double-acting cylinder (reciprocating action)

In Figure 9.4(a) the cylinder with the cam, or trip dog, mounted on the piston shaft is moving backward in the A− direction. A pulse of air from the A+ pilot valve has pushed the spool in the slide valve to allow main high pressure air through to the negative side of the double-acting cylinder. This has caused the piston to move backwards in the A− direction. Immediately the cam left the A+ pilot valve, the action of the spring returned the valve spool to allow air to exhaust from the slide valve when necessary, otherwise the system would lock. The cam then pushes the plunger down on the A− pilot valve as shown in (b). This sends a signal pulse through the pilot valve to the slide valve and pushes the spool in the opposite direction. Thus the air from the negative chamber of the cylinder can exhaust to atmosphere and mains air passes to the positive side of the cylinder so moving it forward. A simple reciprocating action is obtained by this arrangement but the same principles are applied to much more complex sequences with large numbers of cylinders.

Many other pneumatic control components are also commonly used and two examples are shown in Figures 9.5 and 9.6. The shuttle valve is very useful in control circuits when it is desired to be able to select a particular air signal in preference to another

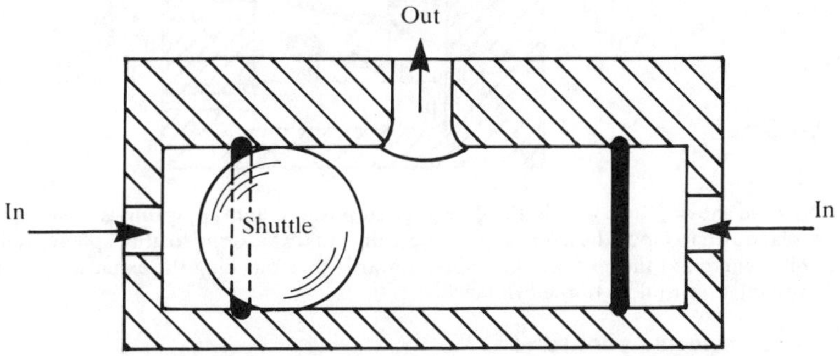

This is used to direct flow when two alternative air signals are to be used in a circuit. Incoming air from the right pushes the shuttle to the left thus preventing flow of air to the left. Air can therefore only flow through the side that last received a signal. A subsequent signal at the left will change the state of the valve by pushing the shuttle to the right.

Figure 9.5 The shuttle valve

at a suitable point in time. The solenoid pilot valve is very common in automation and robotics where electrical control signals can be used. These solenoid valves can be used to control air supply to, for example, a pneumatic gripper. A signal from the robot controller operates a relay, which allows the coil to be energized and sends air to the gripper cylinder, so closing the gripper. After the appropriate sequence of moves has been made by the robot another signal will allow the coils to be de-energized so allowing the gripper air to exhaust to atmosphere and the gripper to open, possibly under the action of a return spring. Similar solenoid operated valves are also common for simple control functions on hydraulic robots.

Miniaturization of pneumatic components has taken place in recent years. Miniaturized, completely pneumatic, sequential controllers have been popular but now very cost effective electronic control equipment is married to pneumatic control components making hybrid electropneumatic control systems common. For example, electronic programmable logic controllers, considered in Chapter 17, are often used to control pneumatic systems.

The hydraulic servo valve

This is the first control component to be considered that is used on full servo controlled

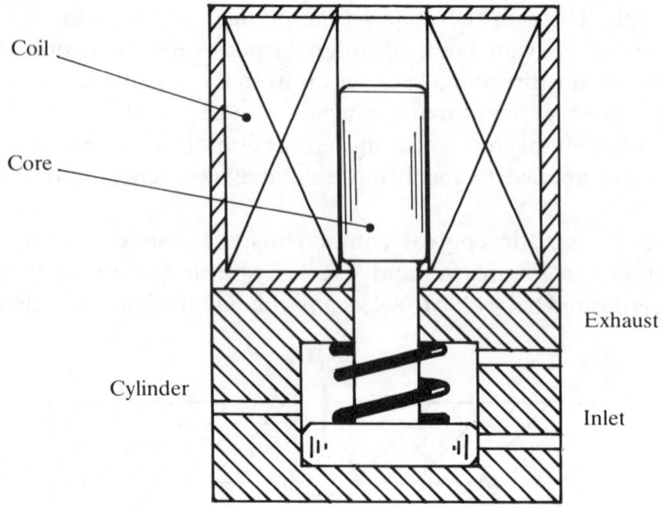

In the valve shown, the coil is in the de-energized position and the spring is pushing the valve seat down to block the inlet supply. The cylinder is exhausting to atmosphere. When the coil is energized the core will be pulled upwards thus blocking the exhaust port and allowing inlet air to flow to the cylinder.

Figure 9.6 The solenoid pilot valve

fluid power robots. All hydraulic robots with servo control use servo valves and the valve shown in Figures 9.7(a), (b) and (c), the Moog valve, is one of the most widely used. It is worth noting at this stage that in an hydraulic system the fluid pressure controls the force or torque, the flow rate controls speed, and the flow direction controls direction of movement.

There are various configurations of hydraulic servo valves; however, the one shown illustrates the principle well and embodies some interesting techniques. There are usually two stages to the valve. The first is the pilot stage in which an electrical signal from the servo amplifier is changed into a mechanical motion. This mechanical motion is used to direct the control pressure to the second, or power, stage of the control valve. The second stage directs the high pressure hydraulic fluid to the actuator. Types of first stage include double flapper, single flapper, spool, deflector jet and jet pipe amplifiers. The servo valve shown incorporates a double flapper servo pilot stage. The second stage incorporates a piston or spool which is free to slide to and fro within the cylinder. The spool has a body diameter which is smaller than the large diameter portions called lands which fit flush against the cylinder wall. Control pressure pushes the spool in the required direction so allowing the high pressure to be directed between the chambers created by the lands and so out to the actuator.

Before examining the operation further it is appropriate to consider the basic methods of operation of an hydraulic servo system. This is shown schematically in Figure 9.8. All hydraulic servo systems have essential elements in common. Firstly there must be a command signal, then a comparator and amplifier, a control valve, an actuator, and a sensor to determine the state, or effect, of the actuator and provide a feedback signal to the

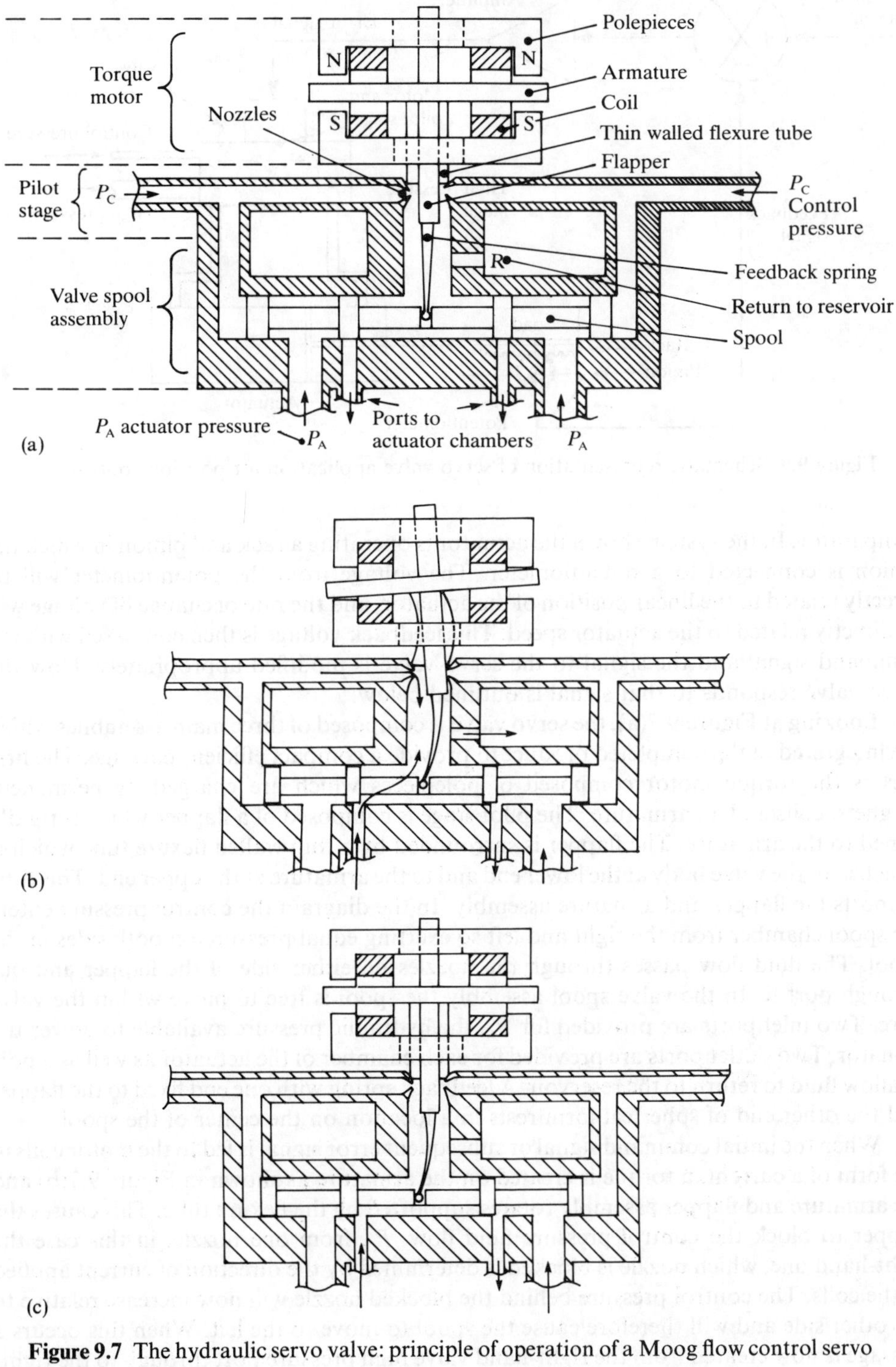

Figure 9.7 The hydraulic servo valve: principle of operation of a Moog flow control servo valve

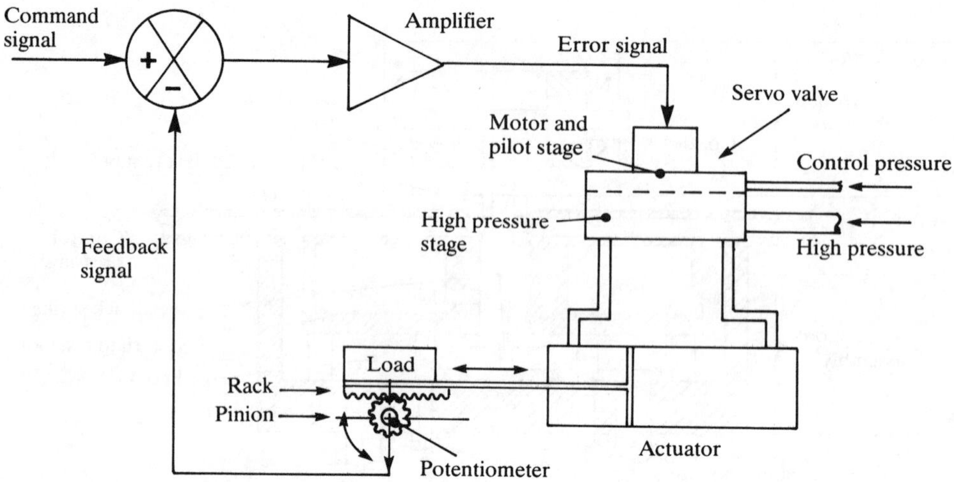

Figure 9.8 Schematic representation of servo valve application for position control

comparator. In the system shown the actuator is operating a rack and pinion in which the pinion is connected to a potentiometer. The voltage from the potentiometer will be directly related to the linear position of the actuator, and the rate of change of voltage will be directly related to the actuator speed. This feedback voltage is then compared with the command signal and the signal to the servo valve is modified appropriately. How the servo valve responds to that signal is outlined below.

Looking at Figure 9.7(a), the servo valve is composed of three main assemblies which are integrated in the completed product to provide a compact efficient package. The first part is the torque motor composed of polepieces which are charged by permanent magnets, coils and an armature. The pilot stage is composed of a flapper which is rigidly joined to the armature. The flapper is surrounded by a thin walled flexure tube which is attached to the valve body at the lower end and to the armature at the upper end. This tube supports the flapper and armature assembly. In the diagram the control pressure enters the spool chamber from the right and left so exerting equal pressure on both sides of the spool. The fluid flow passes through the nozzles on either side of the flapper and out through port R. In the valve spool assembly the spool is free to move within the valve bore. Two inlet ports are provided for P_A, the hydraulic pressure available to power the actuator. Two outlet ports are provided for each chamber of the actuator as well as a port to allow fluid to return to the reservoir. A feedback spring with one end fixed to the flapper and the other end of spherical form rests in a location on the center of the spool.

When the initial command signal or subsequent error signal is fed to the motor coils in the form of a current, a torque is created on the armature as shown in Figure 9.7(b) and the armature and flapper assembly rotates supported by the flexure tube. This causes the flapper to block the control pressure fluid flow, P_C, from one nozzle, in this case the right-hand one, which nozzle is blocked is determined by the direction of current applied to the coils. The control pressure behind the blocked nozzle will now increase relative to the other side and will therefore cause the spool to move to the left. When this occurs a passage is now created from the right-hand valve high pressure port through to the right-hand actuator chamber port causing the load, e.g. a robot arm, to move. As the spool moved to the left it also pulled the feedback spring left. This caused a torque on the flapper and armature assembly, opposing the torque produced by the signal current. When the two torques are equal the flapper assembly comes to rest in the center position again, and the spool comes to rest in a position proportional to the signal current applied. Thus if pressure is kept constant the flow to the actuator will be proportional to the spool position. When the load reaches the desired position the signal to the motor will cease, removing the applied torque on the flapper and allowing the spool to return to center under the action of the feedback spring. With the spool stationary in this position the system will be locked and no undesired movement of the actuator can occur due to the load or gravity.

The servo valve can be used to control position, as in the configuration shown, or velocity, or force, by using appropriate sensors and control circuitry.

INTERNAL SENSORS

These sensors are mounted on the shafts of the drive motors, the joints of the robot limbs, or on any other suitable location where position, movement, or force can be measured. There are many types available and some of the more common are described here. Force is more often used for external sensing and is therefore considered in the next chapter.

The limit switch

This is probably the simplest type of sensor for robots, it is used on non servo pick and place types and also on full servo control robots. As internal sensors they are often used to sense the end of travel of a limb movement in simple robots, or to act as a safety measure to prevent overtravel by triggering the halt movement of a limb before it reaches a mechanical stop. They can also be used to sense cam surfaces for sequence control purposes, or as external sensors to detect, for example, presence or absence of a component in a workspace or in the robot gripper. They are also used widely elsewhere in industry for tasks such as stopping tables and slides on machine tools, identifying loads on conveyors, and on safety equipment such as gates and guards. Thus they will be found to be quite ubiquitous components being present not only as internal and external robot sensors but also as process sensors in the work area surrounding the robot.

The limit switch usually incorporates a 'micro switch', i.e. a very small movement is required to open or close the electrical connection, usually by a snap action. These limit switches are available in a variety of configurations with, for example, plunger, lever, roller lever, and 'cat's whisker' actuation. They can be used in a normally open (NO) or normally closed (NC) mode. In the former mode touching of the switch will close the contacts, thus if the switch was mounted on the robot arm a signal would be sent to the robot controller saying that an object had been encountered. Figure 9.9(a) shows the plunger and roller types normally used as internal sensors to prevent overtravel. Figure 9.9(b) shows the whisker type which can be mounted on the robot gripper as an external sensor.

(a)

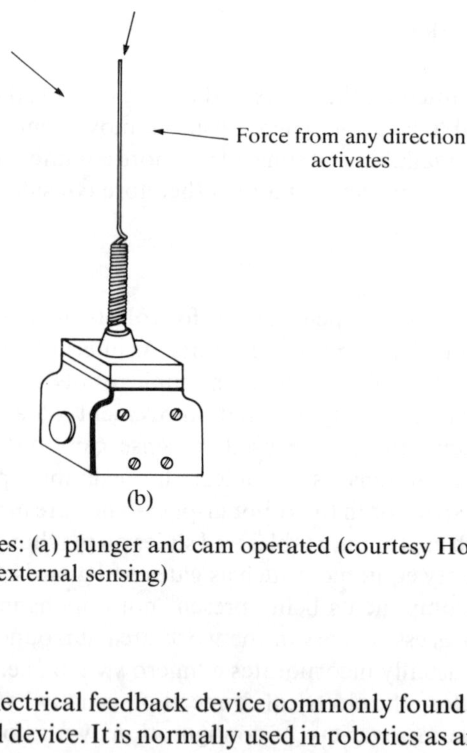

(b)

Figure 9.9 Limit switches: (a) plunger and cam operated (courtesy Honeywell); (b) 'cat's whisker' type (used for external sensing)

The potentiometer

The potentiometer is an electrical feedback device commonly found on robots and it can also be utilized as a control device. It is normally used in robotics as an angular position or motion sensor but it can also be used to measure linear travel by means of a rack and pinion drive to convert the straight line motion to a rotary one, as was shown in Figure 9.8. The principle of operation of the potentiometer is shown in Figure 9.10. Figure (a) shows the basic circuit and (b) the actual layout. Considering the first sketch, if a DC supply of V volts is applied across a resistance R and the output measured by tapping the resistance shown over a distance r, the value $v = V * r/R$. This v will vary between 0 when $r = 0$ and V when $r = R$. By taking the resistance element and making it

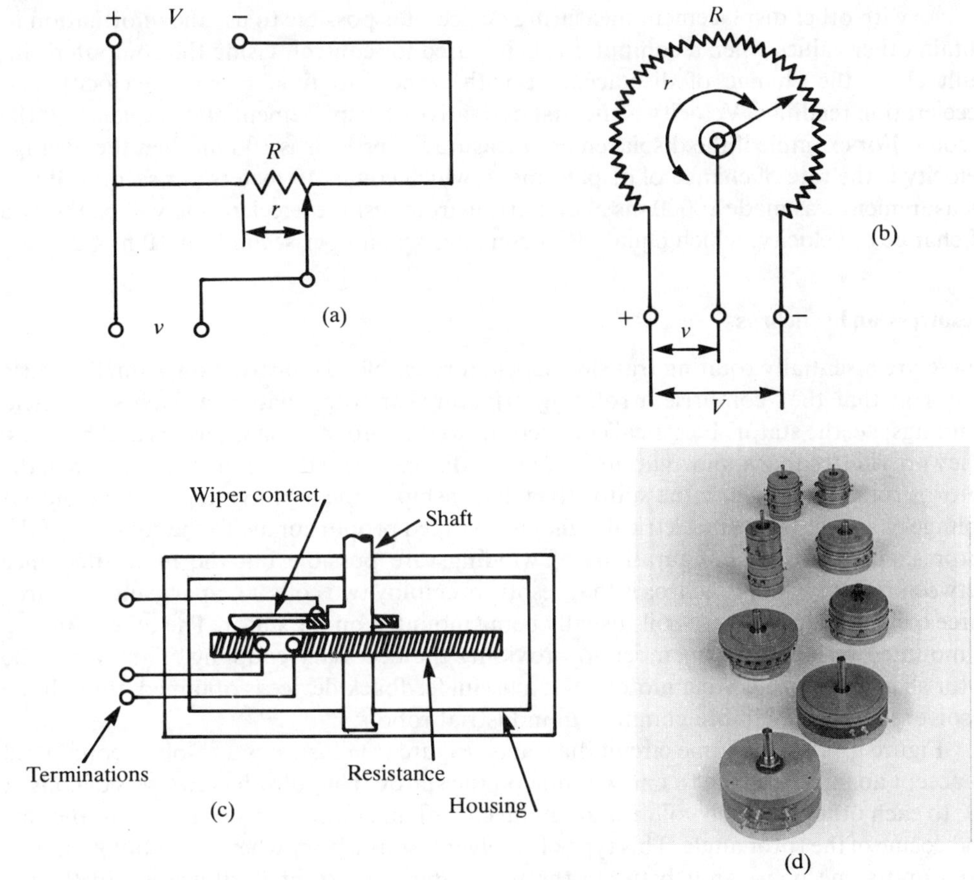

Figure 9.10 The potentiometer: (a) basic circuit; (b) layout; (c) configuration; (d) wirewound potentiometers (courtesy of Penny & Giles Potentiometers Ltd)

circular and tapping the element by using a wiper arm pivoted at the center and connected directly to a shaft, a device is obtained for measuring angular displacement. As the shaft is rotated the wiper moves over the element thus reducing or increasing the effective resistance in proportion to the angular displacement. Thus the value of voltage measured indicates the displacement, and the direction of motion is determined by whether the voltage is increasing or decreasing. Typical construction and examples are shown in Figures 9.10(c) and (d) respectively.

In operation on the robot the shaft of the potentiometer, or 'pot' as it is often called, can be connected directly to the joint axis, the pinion shaft, or linked to the rotating axis through a timing belt arrangement. The potentiometer may be used to obtain readings for use directly, or to provide rough preliminary readings before using a shaft encoder to make fine measurements. For example in some robots a movement will occur in response to a command, the potentiometer will provide an absolute reading in relation to a start or home point, and then an incremental encoder will provide incremental data of a more precise nature to allow movements to be made with the necessary accuracy.

As with other displacement measuring devices it is possible to use the information to obtain other values when a computer is being used for control. Using the computer's in-built clock the change of displacement with respect to time provides velocity and acceleration readings. Velocity is the first derivative of displacement and acceleration the second. For example if the displacement measured over 1 ms is 10 mm then the average velocity is the rate of change of displacement, which equals 10 meters per second. If this measurement was made 10000 ms after starting from rest the acceleration will be the rate of change of velocity, which equals 10 meters per second per second, or 10 m/s^2.

Resolvers and synchros

These are essentially rotating transformers and resemble in construction a small electric motor in that they comprise a rotating armature, or rotor, within a fixed set of field windings, i.e. the stator. Electrical connections to the rotor are via slip rings and brushes. They provide feedback information on angular displacement due to the fact that when the rotor is rotated relative to the stator the relationship of the input voltages to the output voltage changes. Thus an electrical signal is obtained proportionate to the rotation of the rotor shaft. Various configurations of windings are possible but the main difference between resolvers and synchros is that resolvers employ two coils 90° apart, and synchros three coils 120° apart, these coils usually being mounted on the stator. The other winding is mounted on the other member so providing the necessary coupling. Thus when the rotor shaft is connected to a motor shaft a useful feedback device is obtained, though the resolver is by far the more common in industrial robots.

Figure 9.11 shows some circuit diagrams. Figure (a) illustrates a resolver configured to accept an input voltage to one winding on the stator. The rotor has two sets of coils at 90° to each other in which voltages are induced with magnitudes proportional to the sine and cosine of the rotor angle. This type of resolver is satisfactory where the input voltage is fixed and some phase shift between the input and the output is allowable. In (b) the

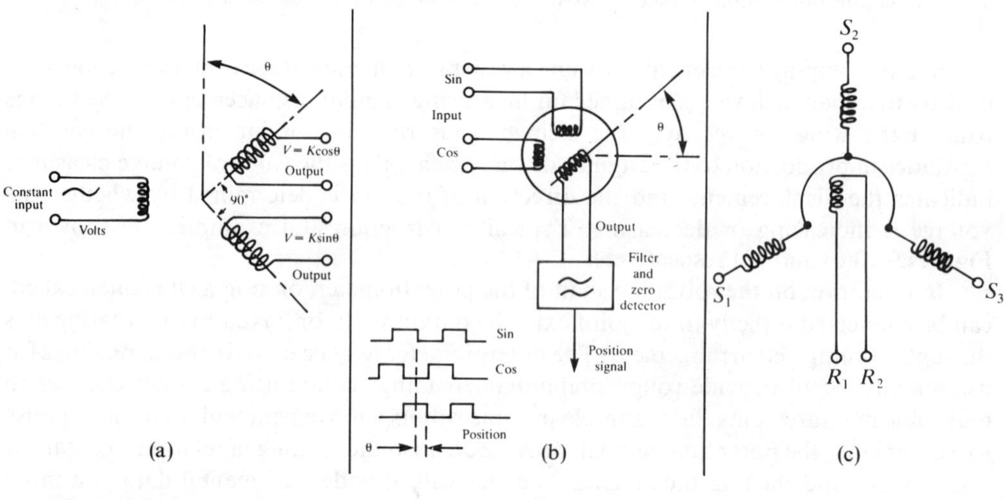

Figure 9.11 (a) and (b) Resolver; (c) synchro

configuration adopted by ASEA is shown. In this case the stator is wound with two coils for the input voltages, and the resolver has one winding in which the output voltage is induced. The stator coils are supplied with two square voltages 'sin' and 'cos'. The frequency of both voltages is 2 kHz, their amplitudes are the same, but they are 90° out of phase with each other. The signal induced in the rotor winding is filtered and converted into a square wave signal whose phase position relative to the 'sin' voltage is a measure of the angular position of the resolver axis. In the ASEA robot one rotation of the resolver corresponds to 200 robot increments, one robot increment being the smallest programmable motion. The axis units on the robot joints utilize counters to record the number of revolutions of the resolver from its previous position. In Figure (c) a synchro circuit is shown. When the rotor is energized voltages are induced in the stator windings. These induced voltages have an amplitude and phase governed by the relative position of the rotor winding and stator coils. The rotor shaft would be connected to the shaft of the motor or gearing of the axis being monitored.

The tachogenerator

The tachogenerator is an analog device, just as are the potentiometer, resolver and synchro. The construction of an electric generator and an electric motor is almost identical, in fact it is possible to design a machine that can be rotated to generate electricity, or be supplied with electricity and rotate like a motor. As electric motor principles have been covered in an earlier chapter the operation of the tachogenerator can be described briefly. The current flow in motors and generators is shown in Figure 9.12(a) and (b). In a permanent magnet generator the poles are permanent magnets thus the field flux is basically constant. Therefore if the speed of rotation is altered a change in output voltage will result. The 'tacho' circuit is shown in Figure (c) and the voltage graph in (d). Thus a tachogenerator as used in robot drive configurations is really a small machine that has been designed to produce a voltage proportional to the speed at which it is being driven. It is thus acting as a velocity feedback device. They are often mounted in a manner integral with the DC servo motor as shown schematically in Figure 9.13, the motor and tachogenerator being mounted on a common shaft but magnetically separate.

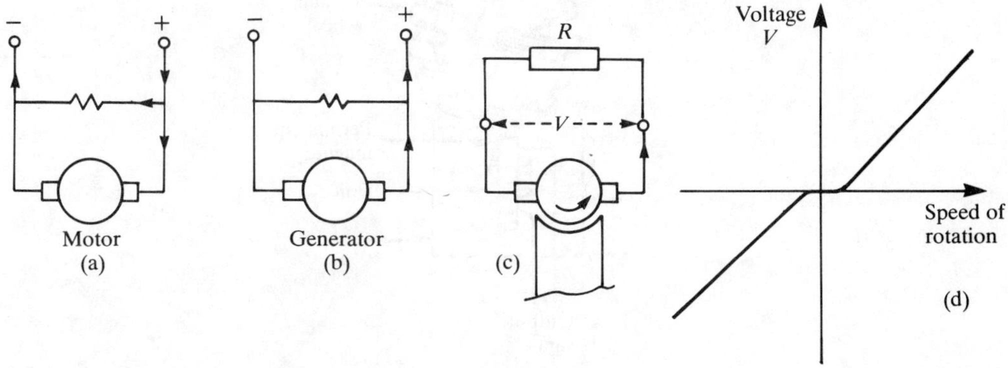

Figure 9.12 Tachogenerator

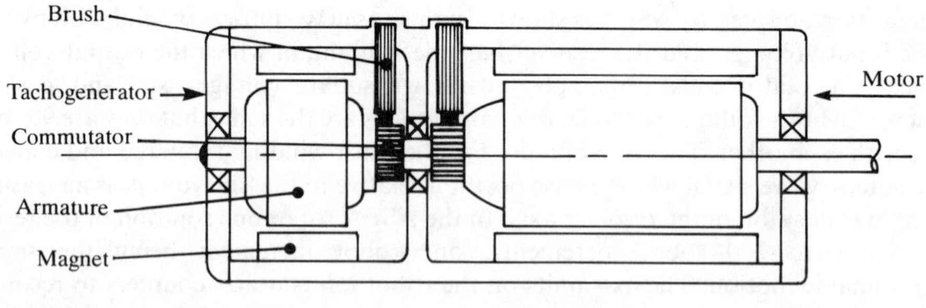

Figure 9.13 DC servo motor with integral tachogenerator

Digital tachometers

Digital tachometers can be magnetic using a metallic toothed wheel with a coil and permanent magnet, or optical using a rotary disk and light sensor.

The magnetic device in its most simple form consists of a wire coiled round a permanent magnet. When a ferrous object approaches the magnet and coil the magnetic flux through the coil is changed thus generating a voltage at the coil terminals. In a practical sensor assembly the ferrous 'object' is usually a 'tooth' on a wheel as shown in Figure 9.14. This wheel is connected either directly or through gears to the shaft whose angular velocity is being measured. A voltage pulse is induced in the coil each time a tooth passes through the field. Thus the pulse frequency generated will be equal to the rate of rotation of the shaft in revolutions per second multiplied by the number of teeth on the shaft. Conversely, by using a frequency counter the robot controller can calculate the speed of rotation by dividing the pulse frequency by the number of teeth. These magnetic devices require no external power source and have a high resolution. Because of their simplicity they can operate at extremes of temperature and pressure and can withstand high shock loads. They can also measure accurately high rotational speeds with pulse frequencies of many kilohertz. However, due to their method of operation they are not suitable for measuring speeds near zero.

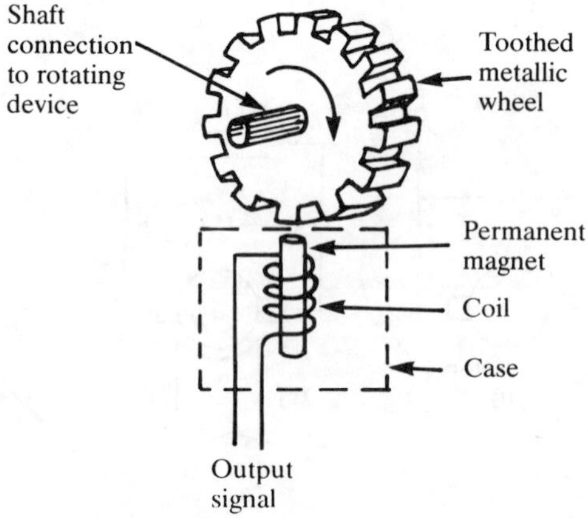

Figure 9.14 Magnetic tachometer

The simplified construction of an optical tachometer is shown in Figure 9.15. The light source projects a beam of light onto a rotating disk. The opaque disk has a number of equispaced apertures in it. As an aperture passes the light source the beam shines through the hole onto the light sensor. This sensor may be one of a number of types (photaic sensors are covered in the next chapter) but basically as a result of the light impinging on the sensor it produces a measurable voltage. If the disk is rotated at a steady rate the sensor will produce a voltage pulse every time the aperture passes the light source. Thus the rotational speed can be found in a similar manner to that used in the magnetic device the pulse frequency being divided, in this case, by the number of apertures.

In both the magnetic and the optical devices accuracy is increased by increasing the number of voltage pulses obtainable at each revolution. These digital devices have the advantage that the sensing element has no direct contact with the rotating component.

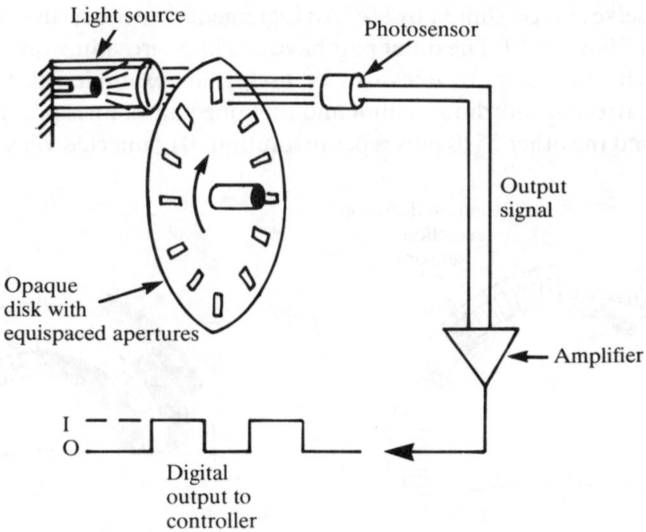

Figure 9.15 Optical tachometer

Rotary optical shaft encoders

These feedback components operate in a similar manner to the optical tachometers described above. Essentially they are very accurately manufactured rotary pulse generators used for the precise measurement of angular displacement and therefore also velocity and acceleration. Basically the encoder is comprised of a transparent or opaque glass disk, in which opaque segments or clear apertures respectively have been formed. The glass disk is rotated by a shaft connected to the robot motor, actuator, or transmission system, and a light beam is passed through the disk onto a photosensor, or pattern of photosensors. The opaque segments, or clear apertures, make and break the beam and the resulting signal from the sensors is amplified and conditioned to give a pulse output, usually in the form of a square wave, to the robot or motor controller. The light source is usually an LED (light emitting diode). Sketches of disks are shown in

Figure 9.16(a) and (b) and a sectional sketch of an encoder is shown in Figure 9.17. Figure 9.18 shows examples of some complete encoders.

Encoders are usually classified as either incremental or absolute. The incremental encoders generally have glass disks with uniform patterns of equispaced segments. These, when used in the above mentioned configuration, produce detector pulses that are used to determine the total angular displacement of the shaft relative to some original reference or 'home' position. A single additional segment suitably placed on the disk can act as an additional reference giving a zero pulse once per revolution. To provide information as to direction of rotation two sensors can be used as shown in Figure 9.16(a). These are focused on the row of segments such that they are exactly one-half segment width apart. This means that as the disk rotates the sensor signals produced will be 90° out of phase with each other. Thus by checking which sensor is activated first the direction of rotation is known. An alternative to this method is to manufacture the disk with two sets of lines which are themselves phase shifted by 90°. An incremental encoder disk with two pairs of lines is shown in Figure 9.19. The outer pair have 360 lines providing one pulse per degree of revolution. The two pairs of lines can be useful, for example say the encoder was connected to a cartesian coordinate robot and that one pair of lines provided 1000 pulses per revolution and the other 2540 pulses per revolution. If connected via a pinion to a rack

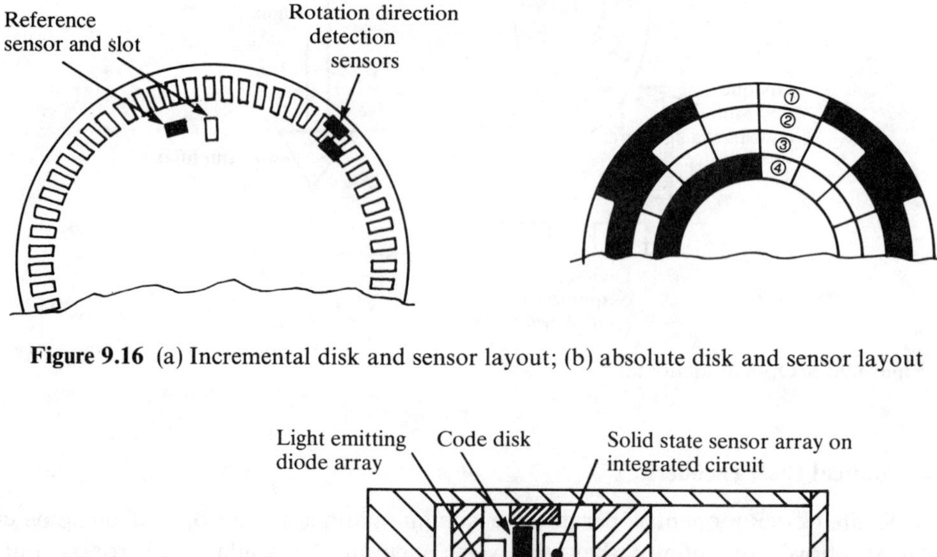

Figure 9.16 (a) Incremental disk and sensor layout; (b) absolute disk and sensor layout

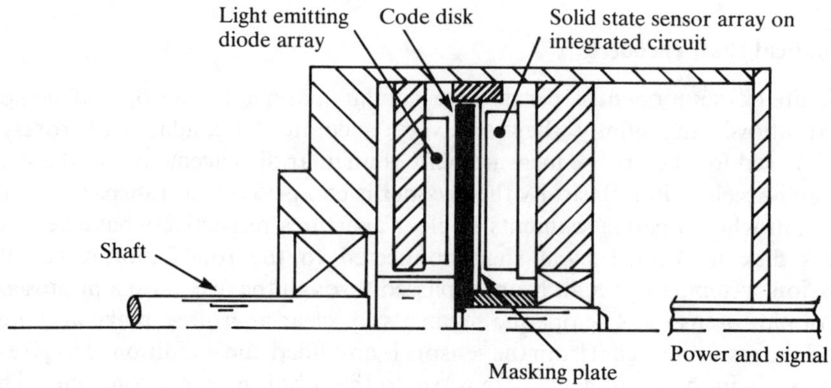

Figure 9.17 Sectional view of optical shaft encoder

Figure 9.18 A selection of rotary optical shaft encoders. (Courtesy Penny & Giles Ltd)

Figure 9.19 Incremental shaft encoder disk. (Courtesy Gaebridge Ltd)

that provides one rotation of the pinion for one inch of linear travel of the rack, then one set of lines would provide a resolution of 0.001 inches, and the other a resolution of 0.01 mm. Incremental encoders are more widely used than absolute encoders mainly because they are cheaper.

Absolute encoders allow the position of the shaft to be uniquely defined. They utilize disks that simultaneously allow activation of a number of sensors, see Figure 9.16(b), whose outputs form a digital word which represents the absolute position of the encoder shaft. Although these encoders are more expensive than the incremental type a major advantage is that if power is removed from the encoder and the robot arm is then moved from its last recorded position, the encoder when powered up will record the new true position. It is possible to utilize a simple binary coded disk. There can, however, be problems with this as ambiguity can occur. In Figure 9.20(a) it can be seen that between sectors 7 and 8, and 15 and 0, all four bits change simultaneously. This is very difficult for the counting and recording software to handle and can result in wrongly recorded information. In some designs an anti ambiguity track is added to indicate to the system when it is allowable to sample the encoder status. A much more commonly adopted approach is to use a Gray code pattern. This is a pattern in which only one bit of information changes in adjacent sectors. Figure 9.21 shows the Gray code in comparison

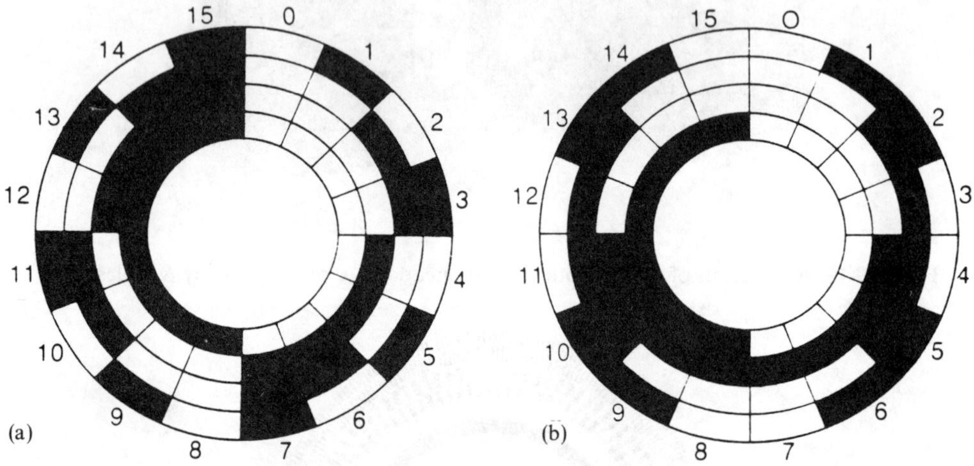

Figure 9.20 (a) Absolute encoder using binary code; (b) absolute using Gray code

Decimal	Natural binary	Hex code	Gray code
0	0000	0	0000
1	0001	1	0001
2	0010	2	0011
3	0011	3	0010
4	0100	4	0110
5	0101	5	0111
6	0110	6	0101
7	0111	7	0100
8	1000	8	1100
9	1001	9	1101
10	1010	A	1111
11	1011	B	1110
12	1100	C	1010
13	1101	D	1011
14	1110	E	1001
15	1111	F	1000

Figure 9.21 Gray code in comparison with decimal, binary and hex

with the decimal, natural binary, and hex codes. A Gray coded disk for four sensors, providing a four bit word to define each sector, is shown in Figure 9.20. However, to achieve a good resolution a large number of sensors will be required and this is indicated by the disk example shown in Figure 9.22.

The control components and feedback sensors discussed here allow the robot to respond in a controlled, but blind, manner to the programmed instructions provided by the robot computer system. However, should the robot be required to make decisions based on the state of its own environment, then additional sensory abilities are required. These are the subject of the next chapter.

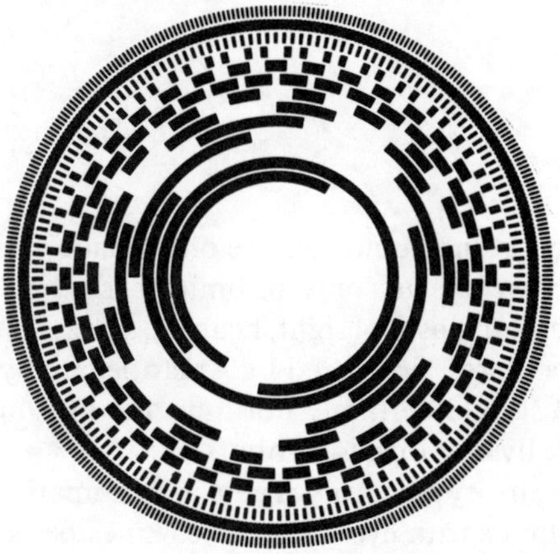

Figure 9.22 Absolute shaft encoder disk. (Courtesy Gaebridge Ltd)

10

EXTERNAL SENSING

INTRODUCTION

When considering robotic sensing, analogies are often made with the sensory abilities of living organisms; these, however, are only of limited use. For instance, it is often the practice to compare the human senses of sight, hearing, smell, taste and touch, with senses given to robots. These five senses, first listed by Aristotle, are by no means all the senses available to humans, and should certainly not limit the imagination when considering what senses could be usefully employed in robots.

Consider the need for sensory receptors. There is, in humans, the ability to be aware of the position of one's own limbs with eyes closed, i.e. 'kinesthesis'. This is achieved by the use of receptors within the body which contribute to the 'enteroceptive' system. The analogous receptors within the robot are items such as optical shaft encoders, tachometers, and resolvers as described in the previous chapter. Humans also have a 'vestibular' sense, i.e. the sense of equilibrium and the ability to sense internally the effect on the whole body of gravity and acceleration. It is possible that these senses may be appropriate for robots in some situations in the future, e.g. in space or undersea work, and could be included in the overall robot system.

Today the average commercial servo controlled robot comes complete with a reasonably adequate enteroceptive system, i.e. they are aware of where their end effector is, or should be, at any point in time even though they are totally unaware of their surrounding environment. This they can do because of their internal sensors and a knowledge of their own limb geometry. However, compared with robots, humans have the ability to work much more efficiently with greater flexibility, and tackle jobs of extremely high complexity, because we can sense our external environment. Humans use the previously mentioned five senses coupled with the ability to sense, for example, temperature, pressure and pain. For the body to experience these senses it uses receptors located between the body and the outside world, these receptors are part of the exteroceptive system. It is therefore possible that robots, if provided with an exteroceptive system of their own, could work more efficiently, be more flexible, and tackle more complex tasks. This of course assumes that the computing ability to gather, analyze, and constructively use the data obtained from the receptors is available.

It is possible to argue that industrial robots should, in the perfect factory, have no need of external sensors. After all they add cost and increase the complexity of the control system and initial programming. It can be said that a factory should be a highly structured and controlled environment. Within the factory walls all events are capable of being controlled, monitored, and programmed with a high degree of certainty. This would be especially true if a totally computer integrated manufacturing system with no human presence was involved. Thus all material, parts, and products are controlled in such a way as to be presented to the robot in a suitable location and orientation for the robot to carry out its required task. The problem with this concept of perfection is that it is impossible to achieve at realistic cost levels. The random piece of material that arrives will always be out of tolerance, the odd component that arrives in a pallet will always be slightly out of position, and the occasional product will always have some defect that will prohibit further processing. In an automated system this necessitates some method to sense when one of these deviations occurs, and to initiate the remedial action to be taken. As far as robotics is concerned this sensing device may be on the robot, as in a tactile sensor on a gripper, or part of the overall manufacturing unit, as in a vision system interfaced to the robot and used to inspect, accept, or reject, components moving along a conveyor belt.

The need to present material, components and products to robots in a controlled position and orientation in itself incurs cost and decreases flexibility. This is because of the need to employ specially designed transport and orientation devices such as bowl feeders, specially designed pallets, or special packaging of materials from the vendor. These in themselves involve additional costs and also limit flexibility as even more cost is involved if a new product is initiated, and bowl feeder mechanisms, pallets, and packaging have to be redesigned. A robot with appropriate external sensors, however, will be able to handle a certain amount of disorder amongst what is presented to it, and will enable the manufacturing unit to be more simply reprogrammed when component or product changes takes place.

Another reason for considering giving external sensors to robots is that they would open up opportunities for robots to work in highly unstructured, disorganized environments. For example, undersea work such as pipeline maintenance or sea farming, the construction industry, outer space, or domestic situations. Here the ability of the robot to perceive its environment and react appropriately to situations will require some degree of artificial intelligence, high level language programming, and good sensory abilities.

Finally if the cost of sensors and associated software can be reduced, then it may be possible to build faster, lighter robots that may also be cheaper than their earlier generation counterparts. For the average robot without external sensors to know the exact position of its end effector it must rely on its internal sensors and the assumption that the links or arms have not been subject to bending, stretching, or other movement. This demands strong construction to ensure rigidity and this in turn means a relatively heavy robot. However, a lightweight construction could be adopted that could move very quickly to a rough location, then move from there with high precision to its required final position by using senses such as vision and touch.

If it was assumed that no improvements in the data processing capabilities of robots were to be expected in the future, then the addition of external sensors would mean a slower overall processing time which would cause lower work cycle times. However, microprocessors, their attendant components, and their system architectures, continue to become faster, cheaper and more efficient. This means that multiprocessor systems are now commonly available and sensors can have their own dedicated microprocessors to reduce the quantity of data to be handled by the robot's central processor. Thus the addition of an exteroceptive system need not be expected to reduce the overall operating and response time of the robot.

Before looking in some detail at the external sensors available it is worth attempting to classify robot sensing abilities. This is not an exhaustive classification but it does provide a framework within which to discuss the topic and it is shown in Figure 10.1. Internal and external sensors have been separated for convenience of analysis. The function and role of internal sensors is best understood in conjunction with the robot control components and system and this is why they were discussed in the previous chapter. However, the external sensors, although used essentially to gather data on the outside environment may also in some cases be part of the internal structure of the robot.

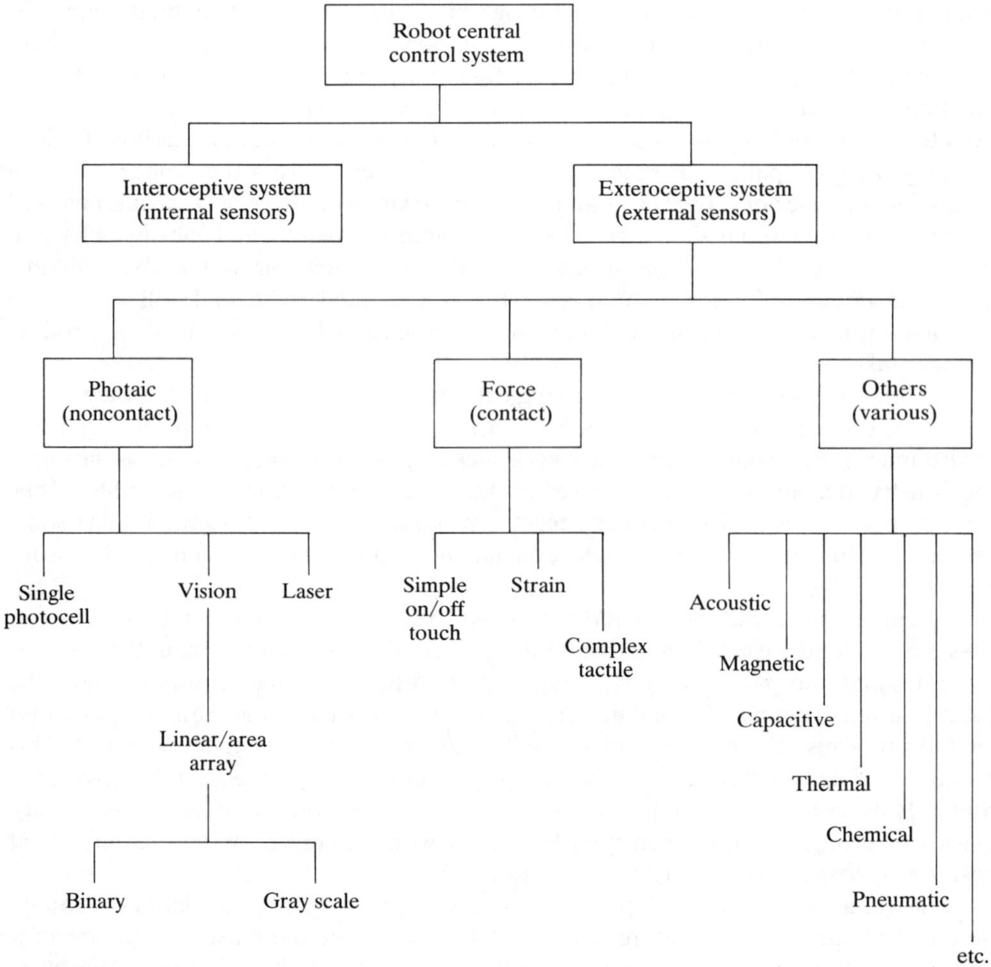

Figure 10.1 Robot sensing

For example, strain gauges could be mounted within the robot arm to measure deflection in response to a load. Conversely it is possible to mount shaft encoders, classed here as internal sensors, on the motor of an external powered conveyor supervised by the robot control system. The rationale adopted here is that internal sensors relate to sensors necessary for full servo control of the robot without regard to the environment beyond the surface of the robot arm. External sensors provide information on the environment surrounding the robot, either passively as in a vision system, or actively as in a tactile sensor. The external sensors mentioned here are also used to measure the influence of the environment on the robot or the robot on the environment, as in the case of strain gauges mounted on the robot gripper. The simplified classification adopted is adequate for the present state of commercial robotics, but it could be easily elaborated to cope with future developments. The most common sensors used at present are photaic and force. Other sensors are also used, e.g. capacitive sensors for proximity measurement, and these have all been grouped together under the one heading. The external senses shown in Figure 10.1, and their necessary hardware, are now examined.

PHOTAIC SENSING

In this type of sensing it is light that is being sensed. For the purposes of this text we will include in light that part of the electromagnetic spectrum containing both infrared and visible light, as this is the region normally used in robotics. The wavelengths considered range from the short, almost ultraviolet, to the longest, the infrared. The position of these wavelengths in the electromagnetic spectrum is shown in Figure 10.2.

Single photocell

In its simplest form photaic sensing can be effected by means of a single photocell. A photocell could be used to determine the presence or absence of a part. The sensor can be mounted on the robot gripper to determine whether a workpiece has been grasped or deposited. It could also be mounted external to the robot in the work area itself to indicate the availability of a workpiece for further work.

One common type of photocell used is the 'photoconductive' type. These sensors are made of semiconductive materials which lower their resistance when exposed to light. Thus photoconductive sensors are sometimes referred to as photoresistive sensors.

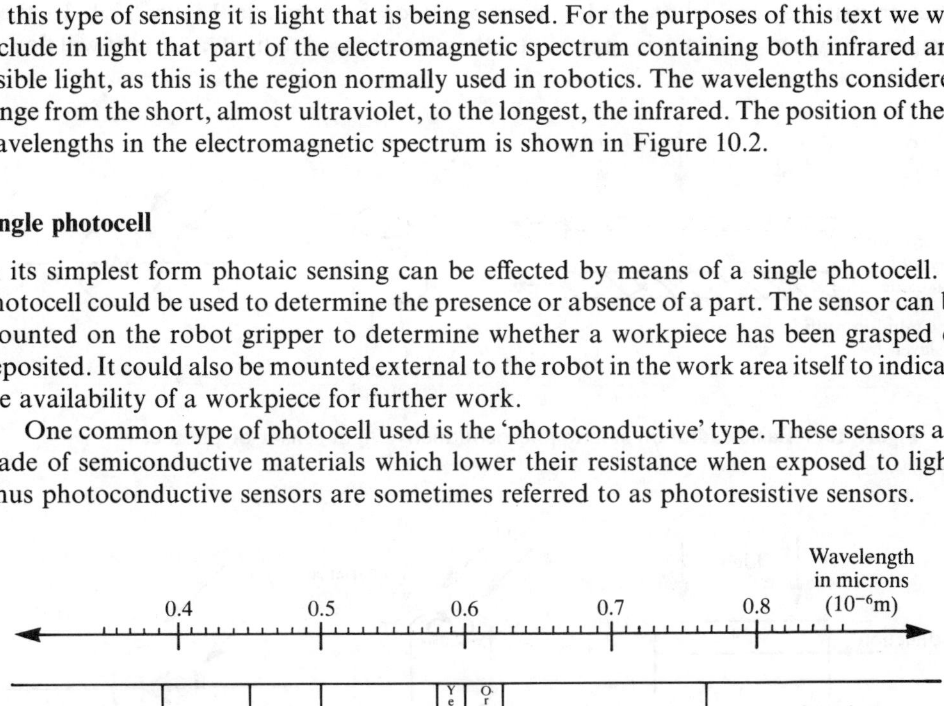

Figure 10.2 The part of the electromagnetic spectrum normally used for robotic photaic sensing

Figure 10.3 shows a photocell of the cadmium sulfide type, two construction configurations are shown. On the left, two metal electrodes sandwich a layer of cadmium sulfide. Alternative materials could be cadmium selenide for visible light, or lead sulfide or lead selenide for infrared sensing. To enable the light to reach the photoconductive layer an open mesh grid or a very thin transparent layer of tin oxide can be used as the top electrode. A slightly different configuration is shown in the center sketch.

When a voltage is applied across the cell in total darkness the metal salt is a poor conductor and a resistance of several megohms will be experienced. However, when light strikes the sensor then the resistance drops to a few ohms and when a voltage is applied a relatively large current will flow. With this type of cell the resistance varies inversely with the intensity of the light striking the sensor. Thus the intensity of the light can be measured by measuring the resistance. This type of cell is therefore useful not only as an on/off sensor, but can also be used for analog applications such as photographic exposure meters. In fact the principle of the resistance varying inversely in relation to the intensity of light falling on the material is applied in a more complex manner in the Vidicon tube discussed later. Also a number of individual photoresistive cells arranged as a matrix within an integrated circuit are used in the CCD camera also discussed later.

Photoconductive 'junction' photocells utilize the fact that most semiconductor diodes and transistors are photosensitive. Thus it is possible to have phototransistors, photo field effect transistors, and photodiodes. A photodiode is shown in Figure 10.4. In these components when a voltage is applied and they are exposed to light, a current will flow in

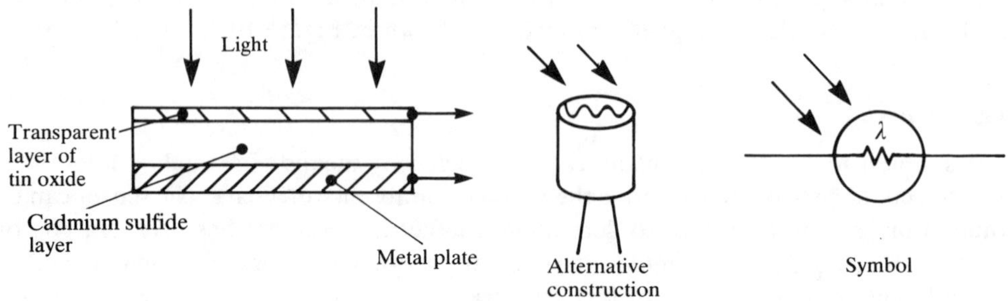

Figure 10.3 Basic cadmium sulfide photoconductive light sensor or photocell

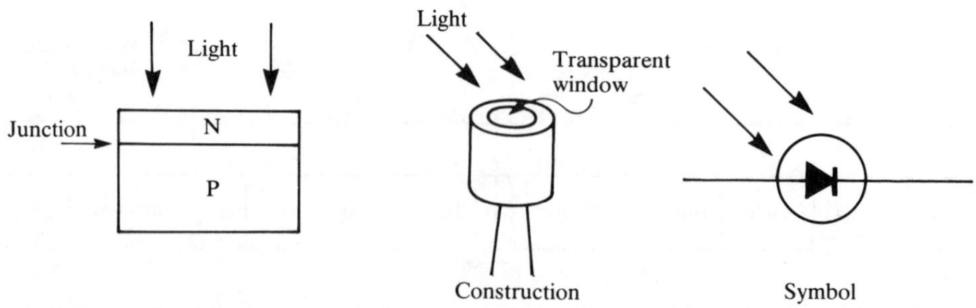

Figure 10.4 The photodiode – a photoconductive-junction light sensor

proportion to the light intensity falling on the semiconductor. The construction of these components is such that the junctions are exposed to as much light as possible. Once again, as well as being useful as a discrete component, photodiodes can be arranged in an array form on an integrated circuit to form the basis of a camera. These photodiode array cameras will also be considered later.

Both of the photoconductive sensors mentioned require to have a voltage applied to them from an external source. An alternative type of light sensor that does not require this external power supply is the photovoltaic light sensor. This sensor produces a voltage when exposed to light. Probably the most familiar manifestation of this is the solar cell that converts sunlight to electricity and is used to power items such as pocket calculators. Materials such as selenium, silicon and germanium are often used as the photovoltaic material.

Vision

The single unit photocells are most useful in robotics for detecting the presence or absence of a component, or a feature of a component such as a hole. To examine a component or indeed a scene with more sophistication, much more complex systems are required. These are commonly termed 'vision' systems and comprise a camera, electronic hardware, interfacing equipment, a computer, and software.

Vision systems for robots specifically, and automated manufacturing systems generally, do not have to exactly emulate human vision. It would in fact be prohibitively expensive to do so. Consider that the retina of the human eye is made up of about 100 000 000 discrete light sensitive cells spaced about 2 μm apart. In the central area of the eye the cells are in the form of cones that perform well at high light levels and are sensitive to color. The remaining cells are in the form of rods, these are sensitive at very low light levels, are not sensitive to color, but are highly sensitive to movement. The fact that the human eye is capable of resolving down to between 1 and 2 minutes of arc at high light levels means that at a distance of 1 meter it can resolve 0.3 mm or at 10 cm, 0.03 mm. The storage time for the eye is between 0.1 and 0.2 seconds, this is why we appear to see movement when a series of still pictures is flashed in front of our eyes at a rate greater than this, say twenty pictures per second. Cinema film is usually projected at 24 frames per second. In countries where power supply frequencies use 50 Hz, television pictures are effectively presented at 25 frames per second, and in the USA and Japan where 60 Hz is used the rate is 30 per second.

If we wanted to try to copy the performance of the human eye using a computer and camera for black and white vision, the following information provides some interesting results. Assume that each light cell in the eye constitutes one picture element, or 'pixel', and also that the intensity of light falling on that element is represented by a number, say between 0 and 255 where 0 is black and 255 is white, i.e. a 256 level gray scale. Then taking 8 bits of information as a minimum to describe each pixel we would need to process and analyze about 16 000 000 000 bits of information per second, this is obviously too high for conventional processing techniques. It is also unnecessary for practical robot vision. It is not necessary to have 100 000 000 pixels of information per picture, even in the human eye away from the center these pixels appear to be clumped together to form coarser definition elements. In practice the number of pixels and number of gray levels is limited to the minimum necessary for the anticipated task. The number of pictures required per second can also be drastically reduced, in the extreme a still picture will be sufficient, although moving components will require some picture refreshing.

The camera, in this case analogous to the human eye, of an artificial vision system may be of the vacuum tube or solid state type. In the vacuum tube type light impinges on a photosensitive screen which is scanned by an electron beam and the resulting signal is used for processing. In the solid state type discrete photoresistive elements on an integrated circuit are used and the signal from these is utilized for processing. In each case the amount of data for processing can be limited by electronic hardware, by software, or by the physical construction of the sensor itself. The computer is, of course, analogous to the human brain in that it accepts the information from the sensor, processes it, then makes a decision based on its analysis of the scene.

Areas of application of artificial vision

Qualitative and quantitative inspection. This large area of application of vision systems has many subdivisions. It is an attractive area for investment as it has been estimated that visual inspection accounts for around ten per cent of the total labor costs of manufactured products. Qualitative inspection involves using the vision system for checking a product on an accept or reject basis by examining the attributes. For example a flaw in the glass of a bottle, or a crack in a casting, would be attributes that would cause these components to be rejected. Other aspects of this would be the detection of foreign bodies such as mice or insects on a conveyor in a food processing plant, verification of a completed task in say an assembly sequence, or checking the degree of swarf build up on a drill bit. Quantitative inspection involves using the system to measure the dimensional features of a product or component. Examples of this are measuring the width of steel strip coming from a steel mill, gauging the diameter of a shaft during or after turning, or inspecting a component dimensionally on a conveyor belt.

Identification. Rather than check an attribute or measure a feature, the system is used here to identify and classify components or products. The vision process may involve identifying an individual item in a bin of jumbled components, reading characters from a label on a component traveling along a conveyor belt, or identifying a component by its shape and other characteristics among other components in a scene. Following on from the identification of a part the vision system may use this information as an input to a subsequent system, such as a robot, to allow the parts to be sorted and manipulated, or directed in a suitable manner, say for machine loading, assembly, or routing to a store.

Providing information for decision making, guidance and control. The visual data is used in this case to help the robot or automated system decide what to do next in the current situation. For example, laser vision is used in seam tracking in arc welding, here the visual data give immediate guidance to the robot controller to direct the welding torch thus enabling optimum quality welds. In more sophisticated applications vision can be used in conjunction with artificial intelligence techniques to determine the order of assembly of components, or to enable a mobile robot find its way through an unstructured environment.

Thus artificial vision should not be considered in the simply anthropomorphic manner as 'eyes for robots', but rather in a broader way as an integrated part of the overall manufacturing system, just as indeed an industrial robot is. Examined now are the essential components that are required for a vision system.

Vision system components

A complete robotic artificial vision system normally requires seven different elements:

1. camera
2. camera interface and initial image processing equipment
3. frame store
4. control processor or microcomputer
5. interfacing equipment to link the vision processor to the robot control system
6. scene illumination system
7. robot or other system capable of making a physical action in response to the visual data obtained.

The relationship between these elements is shown in Figure 10.5. Elements (1) to (4) will normally represent the contents of a commercial vision system, with the vendor also able to supply elements (5) and (6). Figure 10.6 shows a commercial vision system in operation.

The system elements are integrated in the following manner. Firstly the camera is the sensing element which in this case is an imaging sensor. The light from the illuminated scene is focused onto the sensor within the camera. This sensor then produces a signal which varies proportionately to the light intensity throughout the scene. The signal is transferred to the initial processing stage where it is transformed into a format suitable for further processing. As the data are being handled at a very high rate at this stage it is often

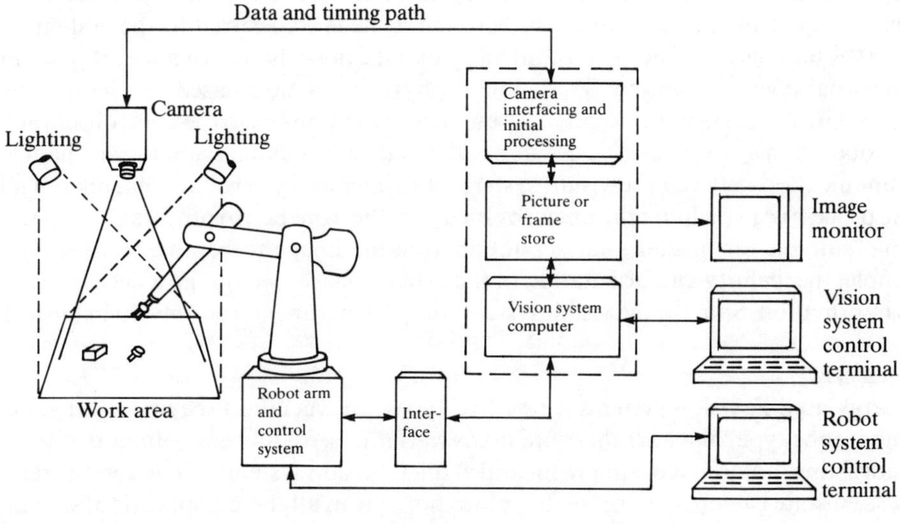

Figure 10.5 Elements of a robot vision system

Figure 10.6 A gray scale vision system for inspecting printed circuit boards. (Courtesy International Robomation/Intelligence Ltd)

necessary to use a storage device to hold the data before further analysis. This holding is carried out by the frame or picture store. At this stage a monitor can be used to view the scene observed by the camera. From here the scene is transferred to the system control processor for analysis. The vision control terminal can be used at this stage for setting the various parameters required. To enable a physical reaction based on the information obtained by this system the vision computer needs to be linked to the control computer for the robot. This requires to be done through an interface that can handle the communication between the vision system controller and the robot controller. Thus the basic process is: (a) acquire the image by means of the camera; (b) process this image into a format suitable for further analysis; (c) analyze the image to be able to determine, for example, the identity and orientation of the articles viewed; (d) make a decision based on this information. Some of the vision system components are now examined in more detail.

The camera

Cameras used in vision systems are of two types, i.e. vacuum tube and solid state. The vacuum tube type, of which there are many variations, is still very common in television broadcasting, home video equipment and artificial vision systems. However, the relatively new solid state cameras are becoming more popular in all these applications due to their advantages and steadily reducing cost. The photodiode array camera was probably the first generally used solid state type and is still popular. It is often used as a line scan camera. Among the more recent solid state types the CCD (charge coupled device) camera is most common; this type often being used in the form of an area array. Line scan and area array cameras are considered later, but before looking at the construction of these cameras it is worthwhile comparing some of the attributes of vacuum tube and solid state devices.

Scrutiny of vision system manufacturers' specifications will show that both vacuum tube and solid state cameras are widely used. However, although solid state cameras have many advantages over the vacuum tube types the latter are still very popular because of their:

1. relatively low cost
2. higher resolution
3. better gray level discrimination.

Some of the advantages of solid state cameras are:

1. Low weight. The solid state camera consisting of lens and LSI (large scale integrated circuit) sensor can be mounted remotely from the other associated hardware. Using LSI it is possible to include the video amplifier, the pulse generator used for synchronization, and scanning logic circuits, all within the one chip. This means that the camera only weighs a few grams and can be mounted easily on the end effector of a robot. The tube camera, however, can weigh a few hundred grams and therefore constitutes a considerable weight if arm mounting is to be considered.
2. For industrial usage the solid state camera is much more robust than the vacuum tube type.
3. Small size. A tube type camera will probably be at least 50 to 60 mm diameter by 250 mm long, whereas the solid state camera can be not much more than the outside diameter and focal length of the lens, say 20 mm diameter by 35 to 40 mm long.
4. The solid state camera has a longer life and is more reliable, having a longer mean time between failure rate than the tube type.
5. Solid state cameras use only a few milliwatts of power compared to the few watts required by tubes.
6. There is no risk of damage by high light intensities, whereas it is possible to damage tube type cameras by exposing them to intense light sources.
7. Solid state cameras have a broader temperature operating range at the lower temperatures.

A disadvantage of solid state cameras is that they can suffer from 'blooming'. This occurs when a very bright area in the scene saturates the relevant cells in the sensor and activates their surrounding cells so that the bright area tends to spread or 'bloom'.

The vidicon camera tube. A sectional schematic diagram of the 'vidicon' camera tube is shown in Figure 10.7 and an illustration of the camera in Figure 10.8. The principle of

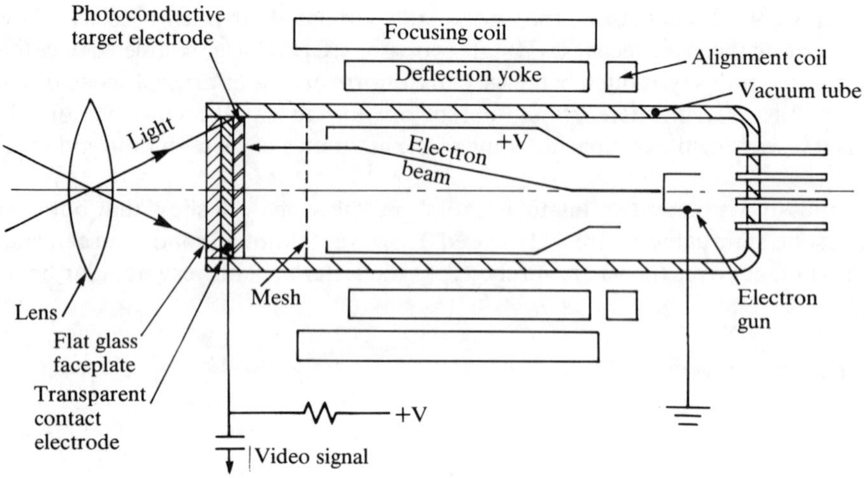

Figure 10.7 Schematic of the vidicon camera tube

Figure 10.8 Vidicon camera. (Courtesy Link Electronics)

operation of this camera is as follows. Light from the scene being viewed is focused by the camera lens, through a flat glass faceplate and transparent contact electrode, onto a photoconductive target electrode. The conductivity through the thickness of this target varies in relation to the intensity of the light at any particular point. Therefore if a voltage is created at a point across this electrode then the current produced will be indicative of the light intensity at that point. To produce the circuit an electron beam is used. The beam is fired from an electron gun at the back of the tube and controlled by magnetic fields produced by coils around the outside of the tube. The beam is focused and aimed onto the target electrode by a focusing coil and an alignment coil. A deflection yoke controls the scanning of the beam over the back face of the target electrode. A mesh electrode is used to slow down the beam before it reaches the target thus effectively making the potential of the rear face of the target the same as the cathode in the electron gun. Thus as the beam scans the target electrode the circuit is completed and a current is generated indicative of the light intensity on the target at each point. This current is picked up from the contact electrode and thus constitutes an analog video signal. The varying value of the current contains the information required to describe the scene being viewed. The electron beam scans the electrode a number of times to allow a full picture to be built up. Each scan produces a 'line' of picture and the more lines there are the greater will be the vertical resolution of the camera.

The solid state camera. In the solid state camera the light from the scene being viewed is focused by the lens onto an array of photosensitive elements. These elements are in the form of components of an integrated circuit on a single microchip. Illustrations of these chips are shown in Figure 10.9 and an enlargement showing a linear array of photosites is illustrated in Figure 10.10. These light sensitive picture elements can be arranged in a single line as in a line scan camera, or in an area array. The line scan camera simply looks at one line of a scene at a time as shown in Figure 10.11. This may be all that is necessary in some applications, say gauging the width of metal strip emerging from a rolling mill. If a picture of an object is required then this can be achieved by the line scan camera by mounting it over a conveyor belt and having the object pass under it. As the object passes through the field of view it is repeatedly scanned by the camera. Each scan is recorded in the computer memory until eventually a full picture of the object can be obtained for further analysis. In the area array camera, however, the whole scene can be static and the picture is focused onto a matrix of pixels as shown in Figure 10.12. Typically, line scan cameras have resolutions of between 256 and 2048 pixels, and area arrays between 32×32 and 512×512 pixels.

The photodiode array camera (based on information supplied by IPL Ltd) is the earliest form of solid state image sensor and it uses a matrix of photodiodes to sense the image. These photodiodes are part of an LSI which also contains other associated

Figure 10.9 A selection of photodiode arrays. (Courtesy IPL)

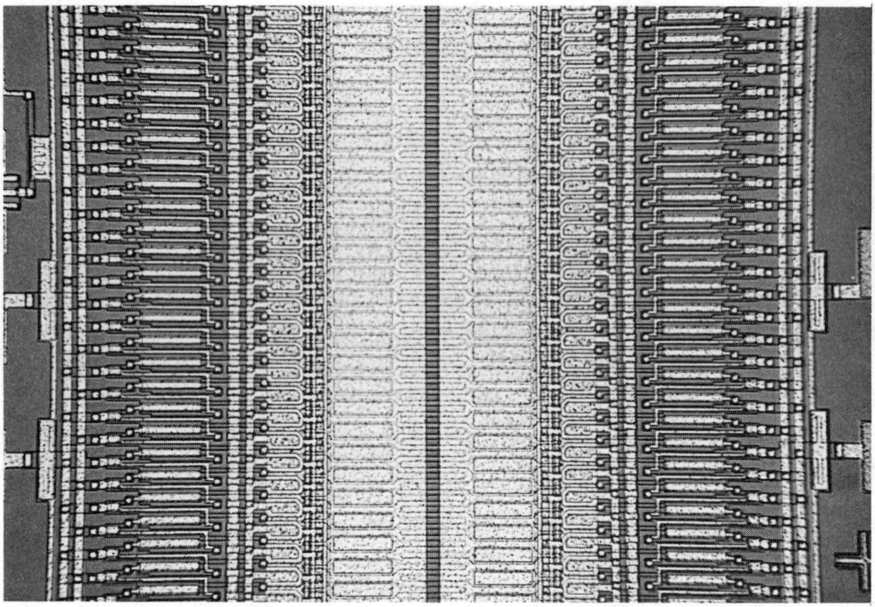

Figure 10.10 Linear array of photosites. (Courtesy IPL)

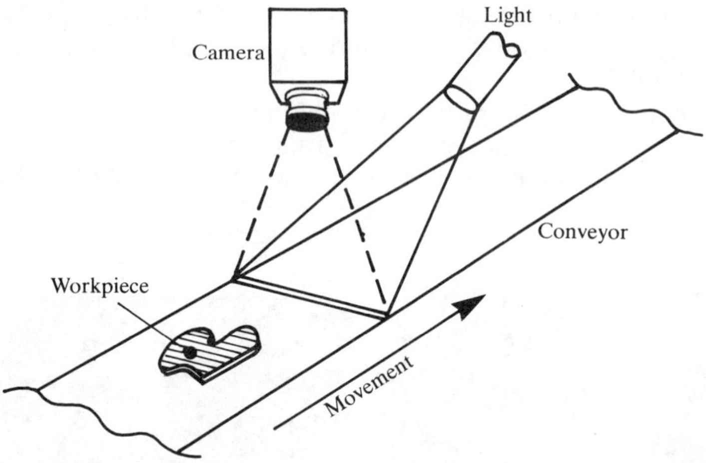

Figure 10.11 Line scan vision

elements such as the metal oxide semiconductor field effect transistors, or MOSFETs, which act as address switches. Although area arrays are produced, the most common form of the photodiode IC package is that of a linear array as used in the linescan camera. Typically the chips will be available in arrays of 50, 64, 100, 128, 256 and 1024 diodes.

A schematic section through a solid state camera and a representation of a section of a sensor is shown in Figures 10.13 and 10.14 respectively. Each photodiode has an inherent

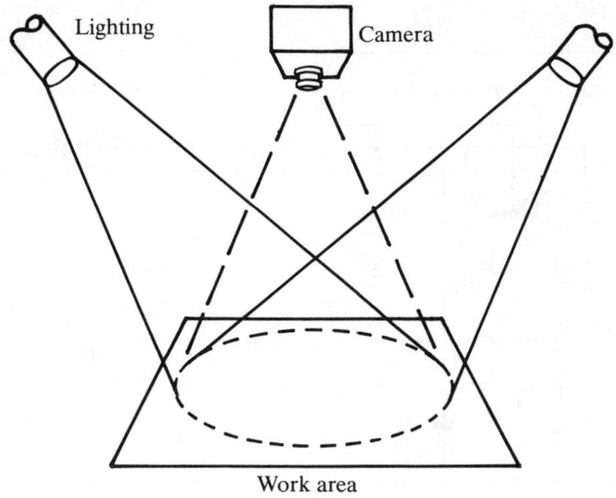

Figure 10.12 Area array vision

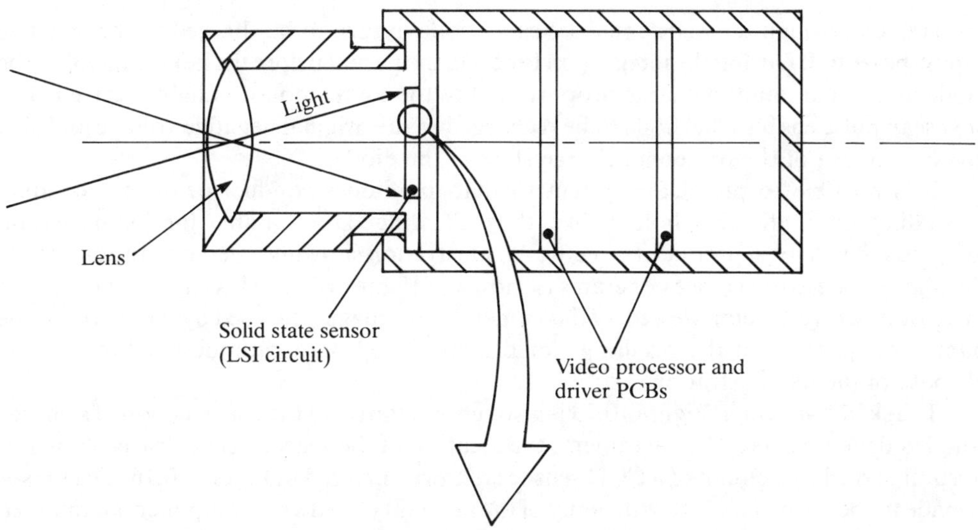

Figure 10.13 Schematic section through solid state camera

capacitance in parallel with it, which has a discharge rate dependent upon its leakage current. This leakage current is dependent upon the light intensity falling on the diode. The sensor operates in the following manner. The low power shift register is driven by two nonoverlapping clock signals C1 and C2, and a scan start pulse. The array scanning is initiated when this pulse appears at the scan start input terminal. This pulse is propagated from bit to bit, through the register, alternately by C1 and C2. Data are thus read out from the diodes at a rate equal to twice the clock frequency. The photodiodes in the array operate in what is termed a reverse bias, light integration mode. The first scan pulse to be

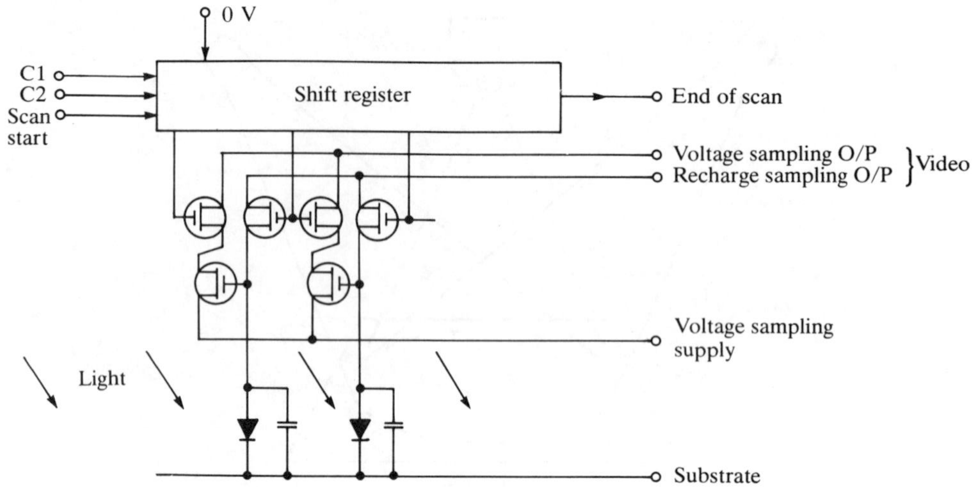

Figure 10.14 Block diagram of photodiode array

propagated through the register causes each diode in turn to be charged to the negative supply potential. During the integration time, i.e. the period before the next scan pulse, the diode loses an amount of charge proportional to the total amount of light received. The next scan pulse enables the diode to be recharged to the original negative voltage and also gives a readout of the preceding charge state of the diode.

It is possible to preset the system so that the diodes conduct or do not conduct depending upon the threshold value, thus allowing data suitable for binary vision (discussed later) to be produced directly. An illustration showing a demonstration of how the photodiode array camera operates is shown in Figure 10.15. This shows an object, in this case a steel rod, being viewed by the 'camera'. The image is focused by the lens onto the photodiode array and the resulting signal is displayed on a cathode ray tube for the purpose of the demonstration.

The sketch shown in Figure 10.13 is also representative of the construction of a charge coupled device camera (CCD) camera, the location of the sensor being clearly shown. A portion of the basic circuit of a CCD sensor is shown enlarged in Figure 10.16. The sensor is made in the form of an LSI with many of the necessary hardware components contained in the one chip. The electronic cells, one per pixel, can be arranged in either linear or area array form. Each cell has a photoconductive element 'R' as shown which is coupled through a field effect transistor switch FET 1 to a capacitor C1. By turning FET 1 on briefly, the capacitor will be allowed to charge up to a voltage which will be proportional to the light falling on the photoconductive element 'R'. Thus when all the FET 1 switches are switched on at the same moment then the pattern of voltages built up on the capacitors will be representative of the light pattern being projected onto the chip. Now when FET 2 is turned on briefly then the voltage on C1 will be transferred to C2, and similarly when FET 3 is switched on the voltage on C2 will be transferred to C1 of the next pixel component group. When FETs 2 and 3 are turned on alternately by applying voltage pulses the representative voltage pattern can be shifted along the row from pixel group to

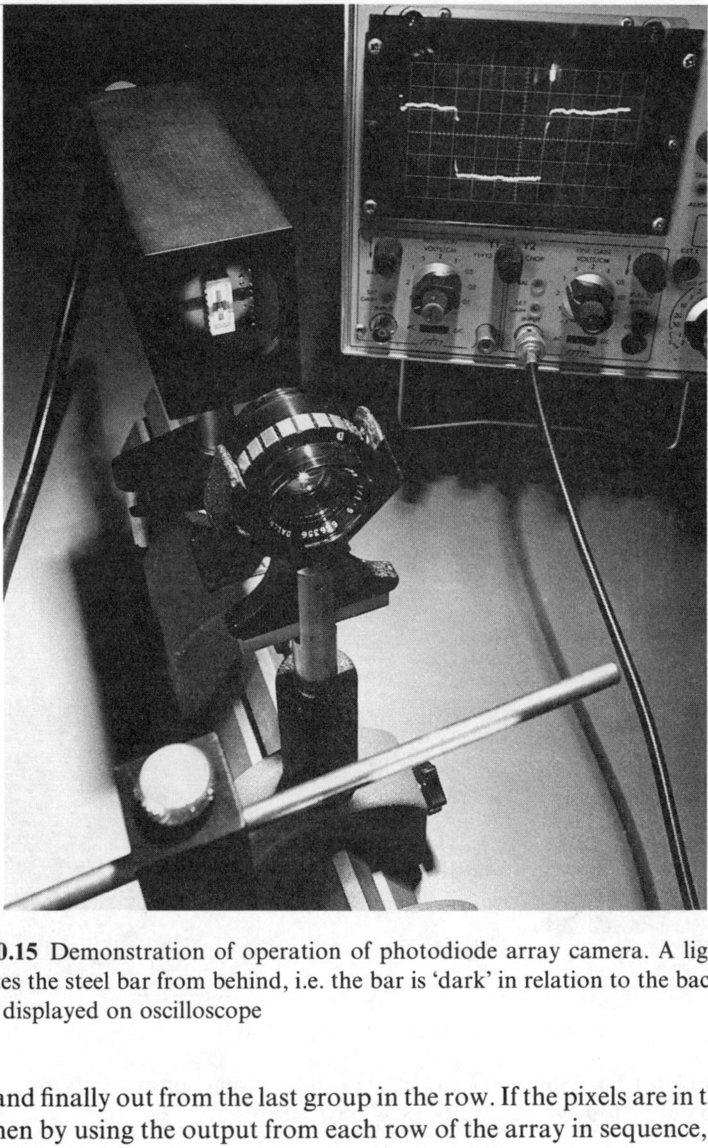

Figure 10.15 Demonstration of operation of photodiode array camera. A light source illuminates the steel bar from behind, i.e. the bar is 'dark' in relation to the background. Signal is displayed on oscilloscope

pixel group and finally out from the last group in the row. If the pixels are in the form of an area array then by using the output from each row of the array in sequence, a scan of the array is obtained. These data can then be used as a video input or taken, as in robotics, and used for further processing as in an artificial vision system.

Figure 10.17 shows the CCAT Systems' 'Sterling' CCD camera. This type of camera is small, light, and rugged, with low power consumption and high resolution and sensitivity. It is suitable not only for interfacing with robots but also for character recognition and surveillance. The image sensor has 604 by 576 picture elements. Useful pictures can be obtained down to about 1 lux, this is an illumination of one lumen per square meter. The minimum amount of light required for reading, writing, etc. is ten lumens per square meter, i.e. the light given off by ten 'international' candles at a distance of one meter from the area observed. It is 90 mm long by 50 mm diameter and weighs 250 g excluding lens.

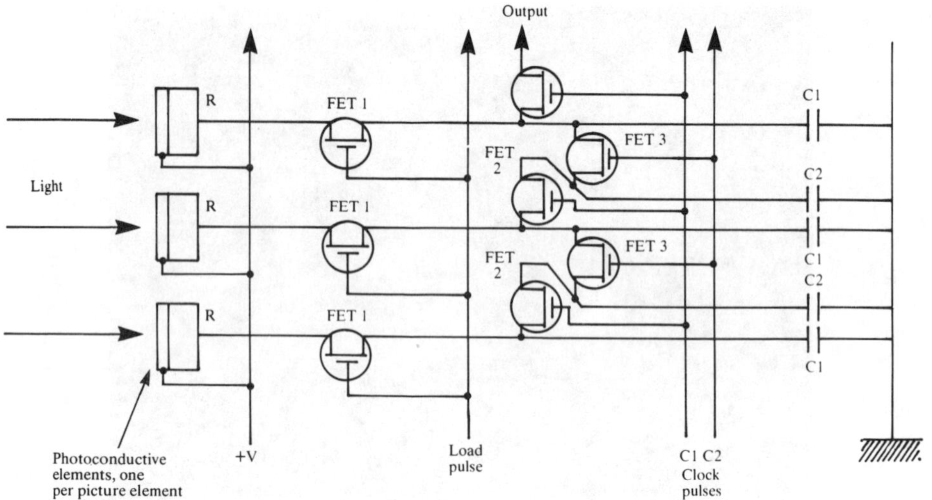

Figure 10.16 Portion of basic circuit of a CCD sensor

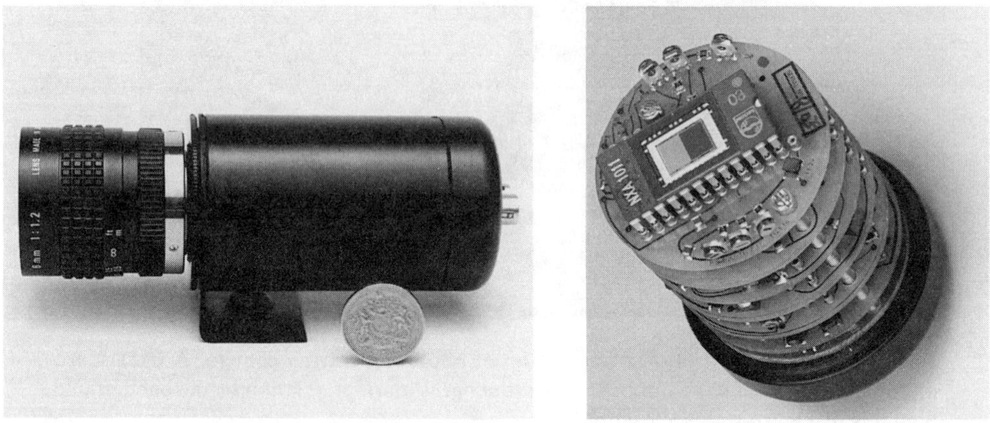

(a) (b)

Figure 10.17 (a) The 'Sterling' CCD camera; (b) internal construction of (a) showing CCD chip on the top left. (Courtesy CCAT Systems)

Camera interfacing and initial processing

The next stage in the vision system requires that the video signal from the camera be converted into a form suitable for manipulation by a computer. For this stage there has been dedicated hardware designed which enables the necessary fast processing times to be achieved. Speed is essential as in the factory environment production must be able to continue at optimum rate unimpeded by the processing time of the vision system. This means that high speed analog to digital conversion techniques are necessary.

This interfacing and processing stage requires a number of operations. Firstly, if the camera is of the vacuum tube a sampling procedure must be carried out. This requires that each line of video data, which is in a continuous analog form, be sampled, or 'chopped' to give a discrete number of pixels along its length. This may be at a resolution, in powers of two, from 128 to 512 pixels, and this is done for each line of the picture. For the video signal from a solid state camera this is much simplified as each pixel is derived from a unique, discrete, element on the integrated circuit sensor.

These points are now quantized. The process of quantization involves examining the light intensity at each pixel and allocating it to a specific gray level, in effect digitizing the analog signal proportional to the light intensity at that pixel. This is possible by creating a gray scale with, say, a value of 0 relating to black, 255 relating to white, and the shades of gray in between being allocated an appropriate value. Thus 8 bits of information will be required to define the brightness of each pixel in a 256 level gray scale. To reduce the processing time and complexity it is possible to reduce the number of levels, or shades, of gray down to 128 or 64. Taking this principle to the limit, the scene can be reduced to a binary image, i.e. black or white, by representing all gray values above a specified level as white, and all below this level as black. Only one bit of information is then needed to register the light intensity at each pixel, although much more attention needs to be given to lighting to enable a high enough contrast between background and object to be obtained.

In gray scale vision therefore, a very comprehensive picture of a scene can be built up. This means that a great deal of information needs to be subsequently processed to be able to analyze the scene adequately. For example, to be able to extract from the picture the edges of a component from a cluttered background, a comparison of various contrast changes over the whole scene has to be done. This is possible by comparing the gray levels of each pixel with those above, below and on either side.

The binary vision technique is much simpler, faster, and usually cheaper, although for it to work reliably the objects to be viewed must be in relatively sharp contrast to their background, and also should not be overlapping or tangled in each other. The binary image is obtained by using a technique called 'thresholding' which operates as follows. The objects viewed by the camera are reduced to their silhouettes by setting all the light intensities in the scene background to black, and all the object light intensities to white. This could, of course, be reversed depending on whether it was, for example, bright steel objects being viewed on a dull surface, or objects placed on top of a light-table thus showing up as black on a white background. The threshold value is now set by the vision system operator, see Figure 10.18. This value determines which intensities should be set to white and which to black during silhouette formation. The picture is now available in a digitized form, either in gray scale or binary, and at a certain resolution.

Figure 10.19 shows a picture at different levels of gray, and Figure 10.20 the same picture at different resolution levels.

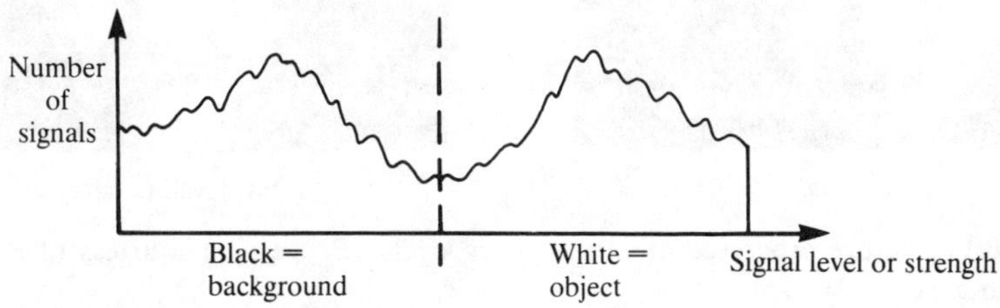

Figure 10.18 Setting of threshold value

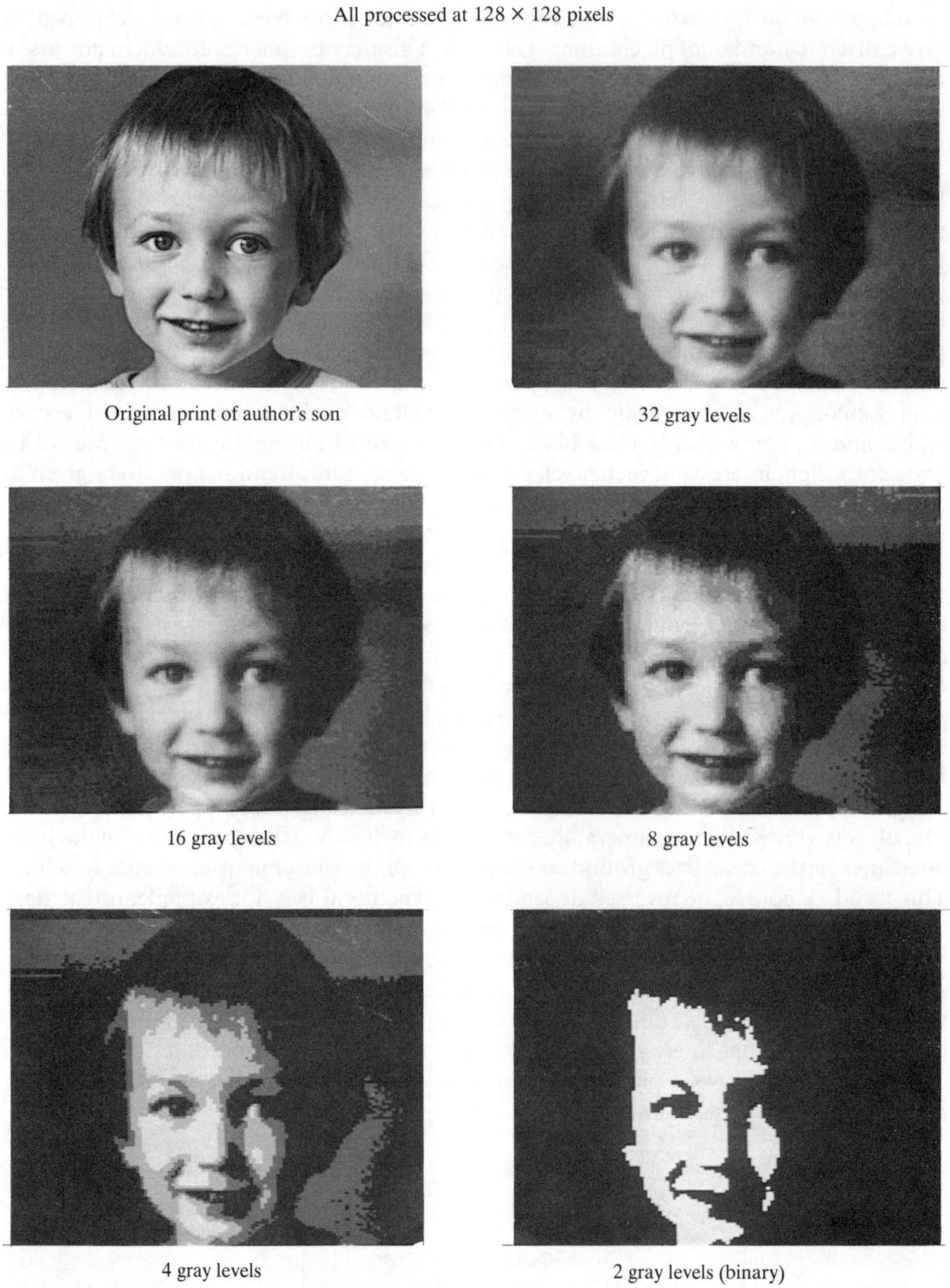

Figure 10.19 Picture processed at various gray levels. (Processing courtesy of Professor Bruce Batchelor, UWIST)

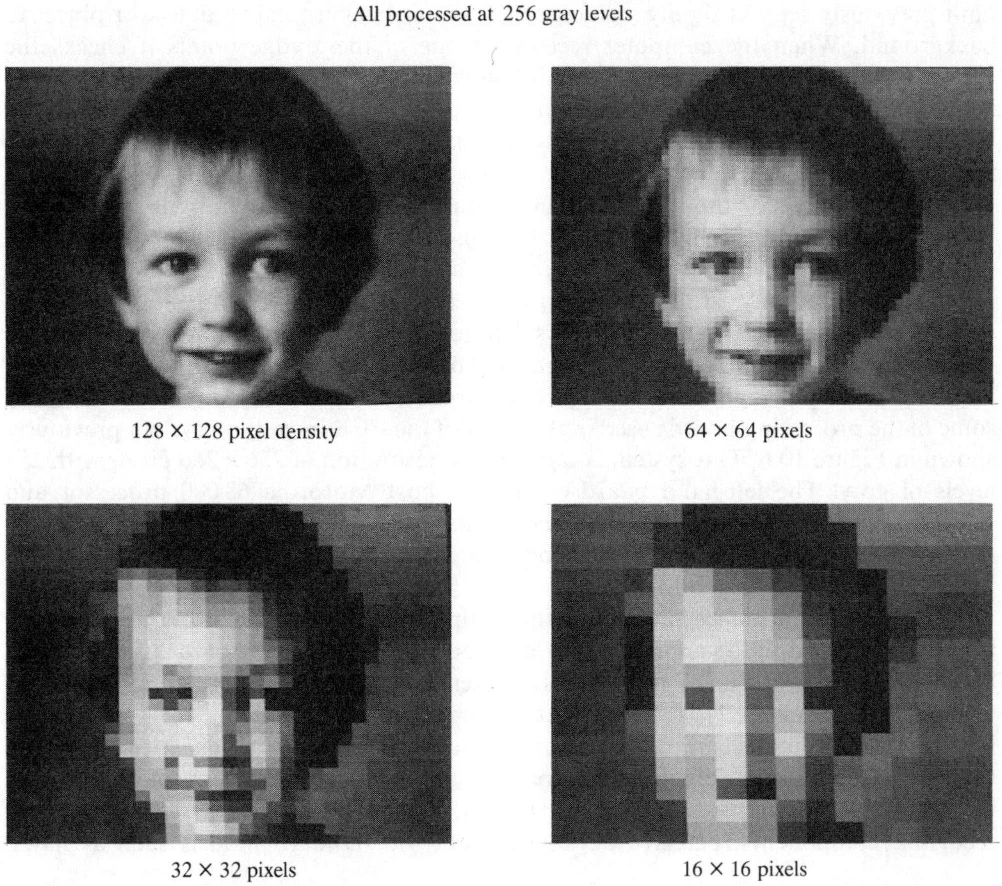

Figure 10.20 Picture processed at various pixel densities. (Processing courtesy of Professor Bruce Batchelor, UWIST)

The picture (or frame) store

The frame store is a facility to allow the picture obtained at the previous stage to be held awaiting further processing by the computer. This is not always necessary but is usually part of the system due to the extremely high information transmission rate from the video processor. The computer has to handle large amounts of these data and cannot always operate at a compatible speed with the video, the frame store therefore acts as a buffer between initial picture processing and analysis.

The vision system computer

The information now coming to the computer is in a digitized form suitable for further analysis. The first task is to extract, or 'trace', the profile or edge of the object from the scene. In some vision systems this can be implemented by dedicated hardware and may in fact be part of the preprocessing stage. In binary vision this edge extraction is carried out by systematically scanning the digitized image for the edges, probably starting at the top left and moving left to right down the frame. A change in signal governed by the threshold limit previously set will signify a change from either background to object or object to background. When the computer recognizes one of these edge points it checks the neighboring points on the line above. This is how it builds up a picture and determines if it is the start of a new component or a continuation of an already scanned one. When the frame scan is fully complete the computer has an 'image' of the edges of the object, and features such as holes, in its memory.

In gray scale vision the principle is similar but many more computations are required. This is because in gray level vision it is not an absolute value, determined by the threshold limit, that delineates an edge point. Here edges of objects are found by looking for a sharp rate of change of contrast. This means that each pixel point is examined in relation to perhaps four or eight of its neighbors before a value can be attached to it. This examination can be complex for the system may be used, for example, to find small flaws in a casting or for picking individual objects out from a jumbled heap. Figure 10.21 shows some of the processing boards used in the heart of the PCB inspection system previously shown in Figure 10.6. This system is capable of a resolution of 256×256 pixels with 256 levels of gray. The left-hand board contains a host Motorola 68 000 processor and EPROMs, etc. The camera interface board in the center digitizes the images to be displayed on a monitor. The right-hand board processes the image at forty million instructions per second (40 MIPS).

Once the image has been stored in the computer memory analysis of these data can begin. The most common approach is to examine the object and compare this image with 'templates' or 'models' held on file. These reference models will have to be taught to the vision system by showing samples to the camera and programming the computer to remember them by measuring and recording various attributes. Thus when the vision system examines a scene later it can compare the attributes of the component it views with those of the component held on file. It can then, depending upon the system being used, recognize components, perceive deviations from the standard models such as flaws,

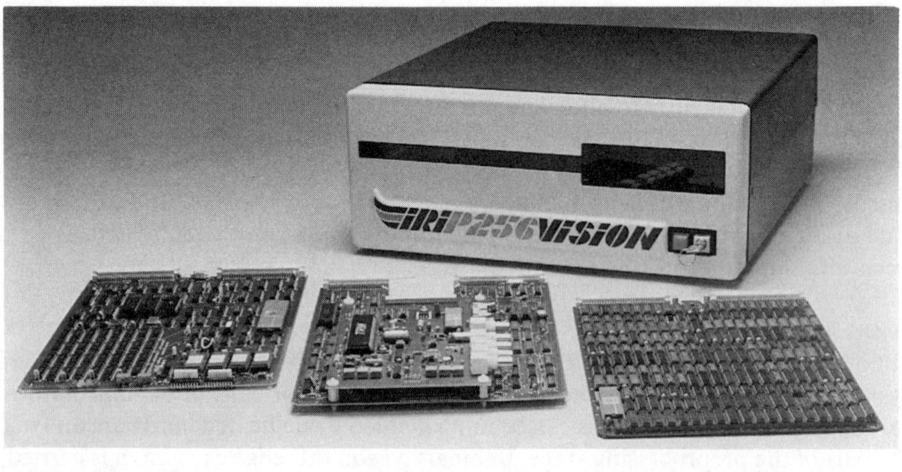

Figure 10.21 Vision system processing components. (Courtesy International Robomation/Intelligence Ltd)

identify the location and orientation of components within the scene, or carry out measurement of the object.

Consider, then, that the vision computer has now in its memory a record of the scene in the form of an area array of pixels. For this discussion it is assumed for simplicity that it is a binary system and therefore each pixel is represented by a '0' or a '1' in each memory location. The system can then assess the attributes of the scene or components in view. The following parameters are examples of some of these attributes and how the vision computer can recognize them.

Area. When it is considered that the binary image is made up of black and white pixels with, say, white pixels representing the object, then the sum of the number of white pixels in the image is a measure of the area of the object. This operation can be carried out by the processor as the image is being scanned thus giving the area measurement immediately the scan is complete. Should a specific measurement, rather than a relative value, be required then the actual size and spacing of the pixels together with lens magnification and object distance would have to be considered.

Perimeter. This operation can again be carried out as the picture scan proceeds. Where each scan line changes from black to white or white to black this is an indication of the component edge. By totaling the number of changes an indication of the perimeter length is achieved. The changes that occur due to holes within the object can be eliminated by software if necessary. If a measurement of the perimeter is required then once more pixel pitch, etc. must be taken into account.

Perimeter squared ÷ area. This is an indication of the compactness of the object. This feature does not change as the magnification or the scale of the image changes.

Centroid. The centroid is the position of the center of area of the object. It is the position given when the first moment of area about both the X and Y axes is calculated and then divided by the object area thus giving the x and y coordinates of the center of area. Figure 10.22 illustrates a rectangular component and some of the attributes mentioned here.

Number of holes. If the binary image is being considered as consisting of a black object on a white background, then every white area surrounded by black must be a hole. The number, position and size of these holes can therefore be calculated, stored, and used for identification or inspection purposes.

Maximum and minimum radii. These are the maximum and minimum distances from the centroid to the farthest and nearest points on the perimeter.

Minimum enclosing rectangle. This is obtained by observing the maximum and minimum points on the X and Y axes of the perimeter of the object.

The relationship of various of these parameters can also be used to determine the orientation of the component. For example, the angular relationship of the maximum

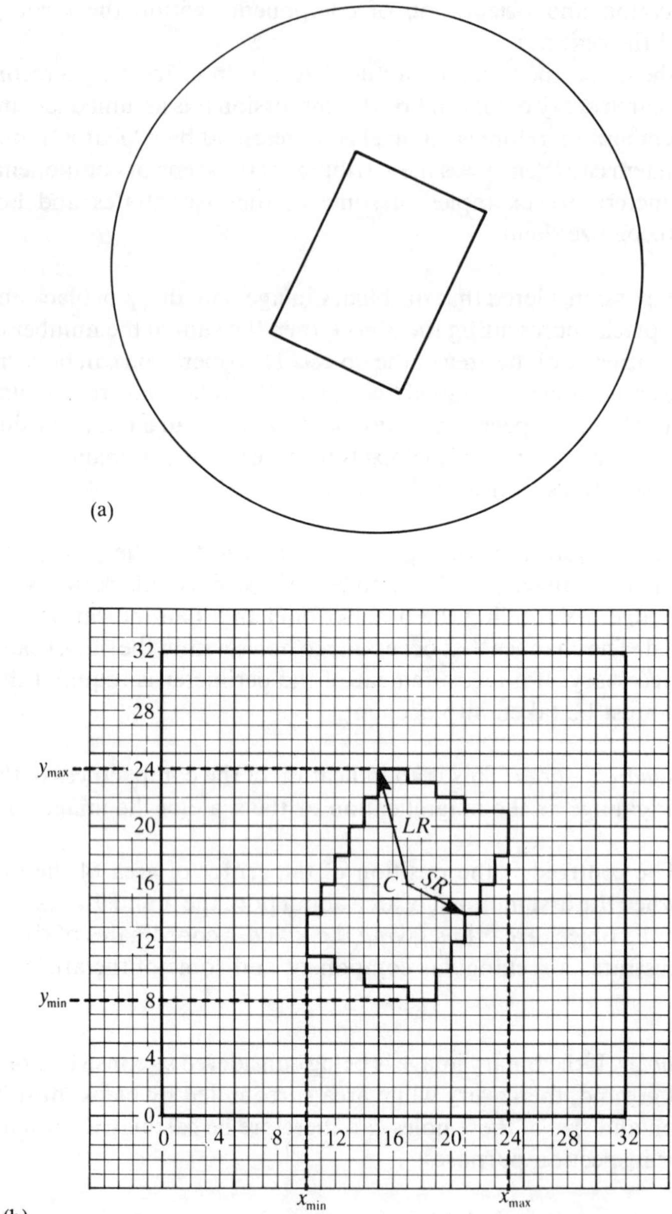

Area = 136 Perimeter = 40 Perimeter^2/area = 11.7

Longest radius = 8 Shortest radius = 4

Minimum enclosing rectangle = $16y \times 14x$

Figure 10.22 (a) Rectangular component in field of view of camera; (b) image observed on 32×32 pixel array radius to the *X* axis at a point about the centroid can be used, as also can the relationship of radius vectors from the centroid to the centers of any holes in the object. Thus the very asymmetry that necessitates the additional work of orienting the object is used to provide data which can be used to determine its existing orientation.

In most vision system applications at the moment discrete components are visible for viewing and analysis, i.e. there is no overlap with a neighbor. In a disorganized industrial scene this is usually an unnatural state, e.g. in a bin full of components there will be many overlaps and also the components will be lying in an infinite variety of orientations. This presents an extremely difficult problem for the vision system as only parts of objects will be visible and their edges will by lying at odd angles. The vision system will require to pick out discriminating features such as the shapes of the edges and the relative position of holes, etc. which will necessitate a gray scale system. There are now some commercially available systems that claim to do this.

Lighting

It is lighting that gives the scene the necessary contrast to enable the component, or component elements, to be differentiated from the background. It is possible that daylight or normal photographic lighting may be sufficient but care has to be taken in selecting the type and arrangement of the lighting system. In binary vision this requires that the background and object transmit to the vision sensor suitably different light intensity values, e.g. the background a white conveyor, or illuminated table on which the object shows up as a silhouette, or the background could be a matt black surface on which metallic components rest and reflect back to the sensor relatively high light levels. This would certainly be a very artificial situation in many factories where dust and grime are often present covering component and table or conveyor indiscriminately. This is where gray scale imaging can be advantageous as the varying contrast between surfaces can be used to derive the information required to determine the edges of the component. Under the proper lighting conditions it can also be used to identify flaws such as scratches and cracks in plastic goods or metal castings. Whatever lighting is used it should be strong enough to prevent interference by ambient light levels surrounding the work area.

One method that does not use normal general overhead lighting is that of 'structured' lighting. In this technique a thin plane of light is projected at an angle onto a moving conveyor as shown in Figure 10.11. The resulting narrow strip of light is observed by a line scan camera. As a three-dimensional object moves across the strip of light the apparent line of action of the light is displaced towards the light source. This means that the line scan camera no longer sees the reflected light. Thus by observing the change in length of the dark portion of the image, and scanning continually as the object passes under the camera, a picture of the component can be built up. Analysis of the light pattern can also be used to identify the three-dimensional shape of the component.

Generally, as in photography, manipulation and organization of light sources can be used to emphasize or reduce shades and highlights. This will facilitate identification of critical features such as component edges or flaws. The use of filtering may also be applied to create contrast between components and background. This may be done by, for example, using a red filter on the lens to enhance red components on a dark background.

Industrial vision system requirements

1. The cost should be as low as possible. This is related to the application, e.g. whether component identification or inspection, or the degree of disorder in the scene.
2. The system should be simple to program and easily understood. A sophisticated system should be able to be programmed by an operator with little training, e.g. direct from an instruction manual.
3. It should be easily reprogrammed. The system should be able to cope readily with product or component specification changes.
4. It must be fast enough in operation to be able to carry out all the necessary tasks without an increase in the work stage optimum cycle time. This may necessitate a very fast response as the robot operating on the information transmitted to it by the vision system will also have to operate within the same cycle.
5. The hardware should be rugged and reliable for use in the industrial environment. This is especially necessary if the device is to be mounted on the robot arm.
6. The system must have high integrity. Any errors the system makes may jeopardize seriously the following production operations. A low error rate is therefore necessary.
7. The system should be able to operate with simple scene illumination. The particular application may demand special lighting but generally speaking the simpler the lighting the less expensive will be the total installation.

Figure 10.23 shows a low cost binary vision system for teaching purpose. Figure 10.24 shows an industrial binary system, and Figure 10.6 has already shown a gray scale system. Color and 3D vision are potentially useful in some applications but commercially there is not a high enough demand at the moment to bring the limited work out of the research establishments into the industrial environment.

Laser sensing

Laser systems are used for both ranging and structured light 'vision' systems. The laser (an acronym for Light Amplification by Stimulated Emission of Radiation) produces a highly directional, monochromatic beam of light. This light is coherent, that is its light waves are all in phase. The active medium within which the light is amplified may be of a number of materials, e.g. the ruby laser is used to produce pulses of highly coherent light whereas helium and neon lasers produce a continuous beam. The laser system described below uses a cube of specially treated gallium arsenide which is capable of emitting infrared radiation when a current is passed through it.

Figure 10.25 shows the principle of operation of the 'EA Laser Ranger'. The infrared laser produces a dot of light on the object to be measured. The image of the dot is detected by the sensor. Because the sensor's field of view is fixed, the position of the dot between A and B is directly related to the object position. If the object moves, the dot changes position and so does its image on the sensor. This movement is registered and the new position calculated by the microprocessor. The measurement can be made 30 times per second.

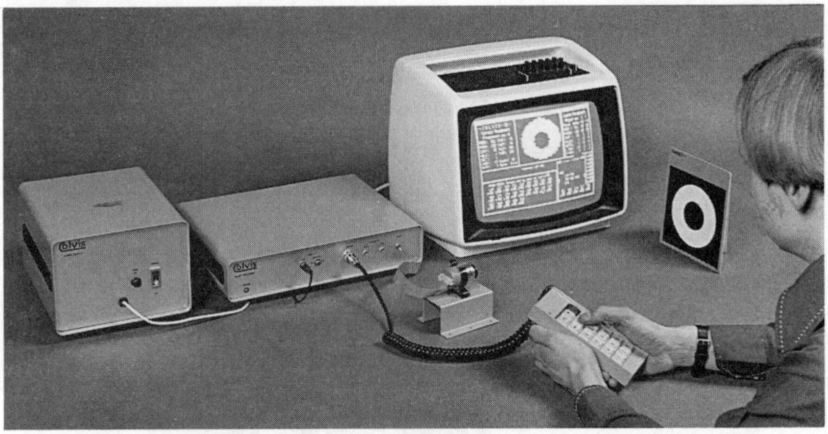

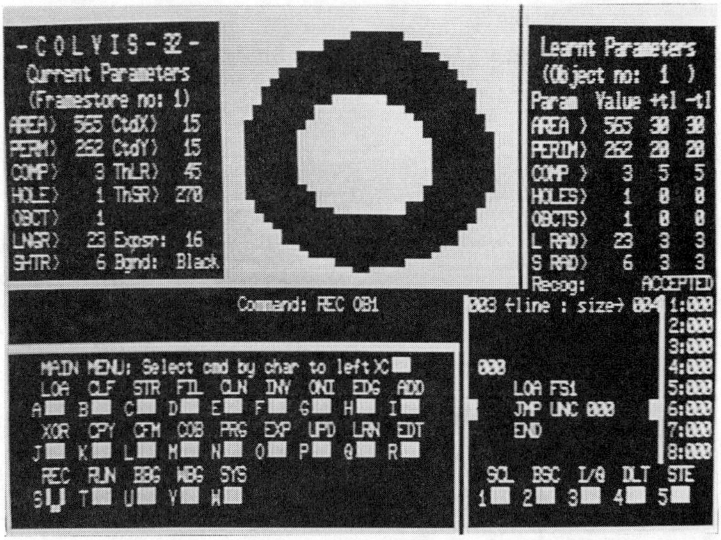

Figure 10.23 (a) Low cost binary vision system (originally supplied by Colne Robotics Ltd) comprising solid state camera, microcomputer, teach pendant, and power supply unit; (b) screen display from (a) – object is 'accepted' as learnt; parameters (top right) correspond with those currently in the frame store (top left)

Laser sensing is often used in robotic arc welding and it will be considered in this context in Chapter 13 on robot applications.

FORCE SENSING

Figure 10.1 showed force sensing subdivided into simple touch and complex tactile sensing. In simple touch sensing the robot will identify the presence or absence of a component by use of an on/off, i.e. binary, device such as a microswitch. In complex tactile

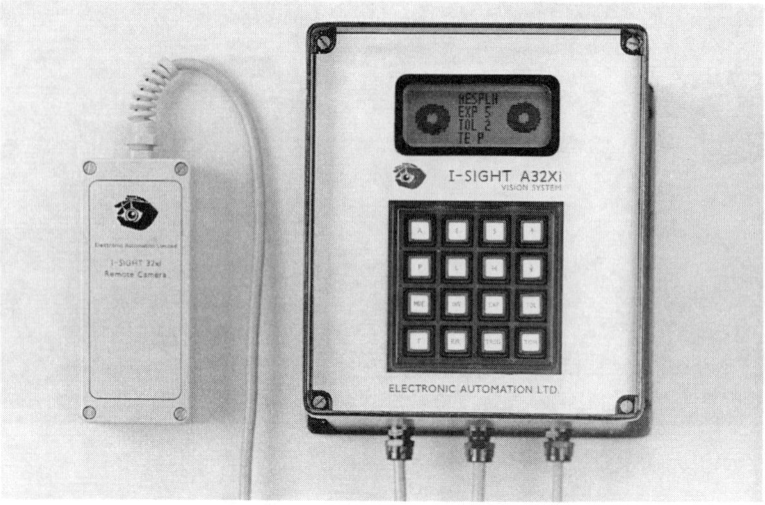

Figure 10.24 An industrial binary vision system – I-SIGHT A32Xi. (Courtesy Electronic Automation Ltd)

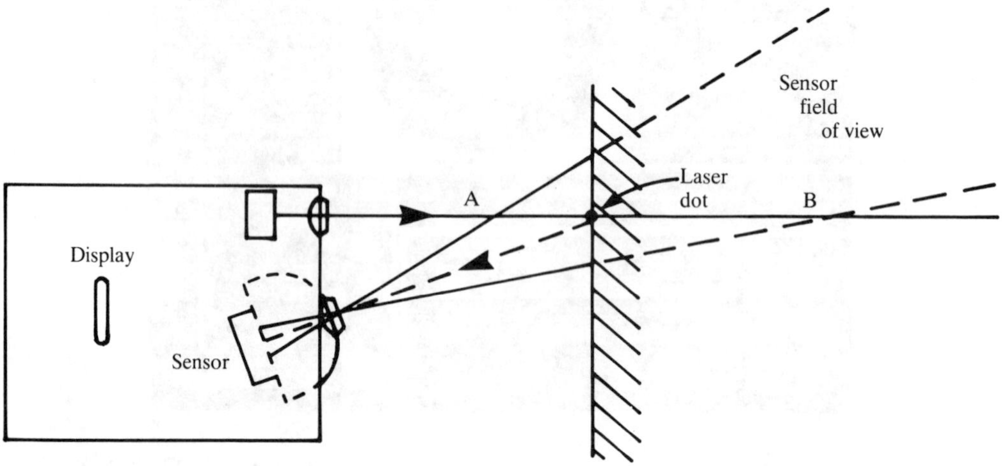

Figure 10.25 Principle of operation of laser ranging. (Courtesy Electronic Automation Ltd)

sensing the robot may be measuring the amount of deflection experienced by a finger, wrist or arm using strain gauges, to allow controlled tightening of bolts, lifting of light fragile components such as eggs, or heavy solid objects such as steel billets, or it may be identifying the shape of a component by using an area array of tactile sensors. Thus force sensing can range from the simplest binary to a complex analog array system.

Simple touch sensing

The most common sensor used in simple touch sensing is the limit switch. When the switch comes into contact with an object a microswitch contained within it is opened, or closed, by the force of the encounter. These switches and their use have been explained in the previous chapter.

Strain sensing

Strain sensing is usually carried out by mounting strain gauges on the robot member which is being monitored. Strain is the ratio of the extension of an element of the member to the original length of that element. For one element of a member subjected to a load the unit strain will be l/L where l is the elongation and L is the original length. It is usually expressed in terms of mm/m, in /in, or as a percentage, although in the case of the robot the signal from the transducer, or sensor, is fed directly to the robot controller. Two of the most common types of strain gauge (the bonded wire and the semiconductor strain gauges) are shown in Figures 10.26 and 10.27. These gauges can be used to measure compressive and tensile stresses and also torque.

In the bonded wire strain gauge a length of fine wire is arranged in the form of a grid and strongly bonded to a sheet of plastic or paper. This is then cemented onto the surface of the object being monitored so that the wire will be stretched or compressed along with the object. The gauge works on the principle that if factors such as composition and material are kept constant then the resistance of a wire varies proportionately to its length

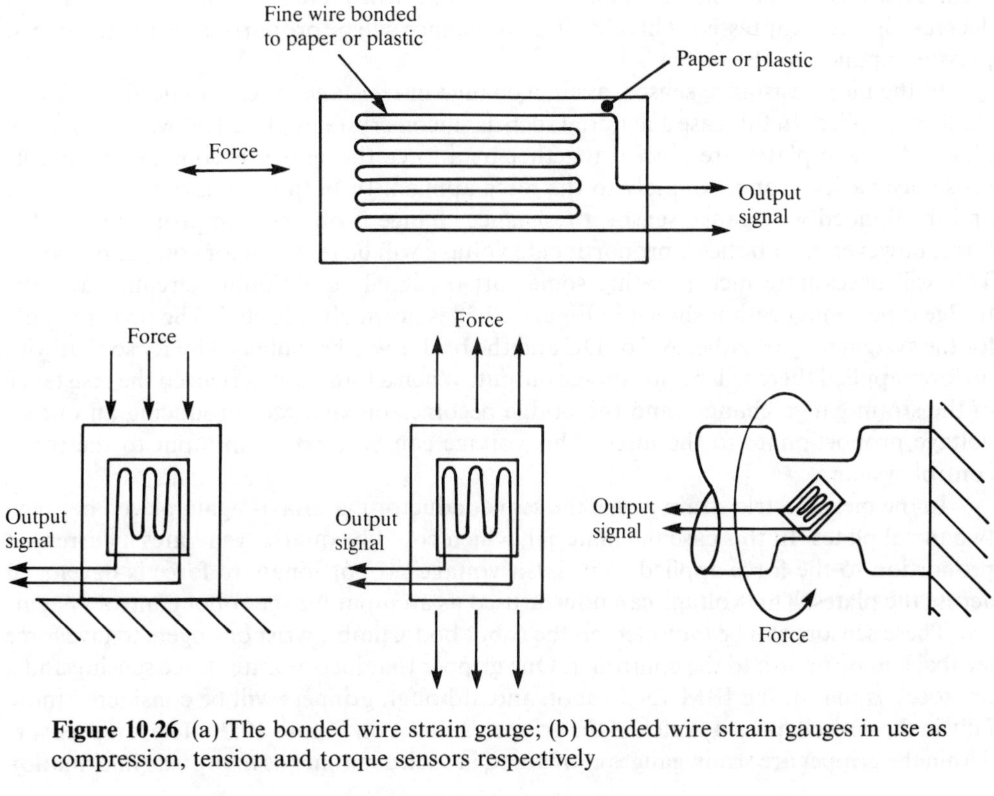

Figure 10.26 (a) The bonded wire strain gauge; (b) bonded wire strain gauges in use as compression, tension and torque sensors respectively

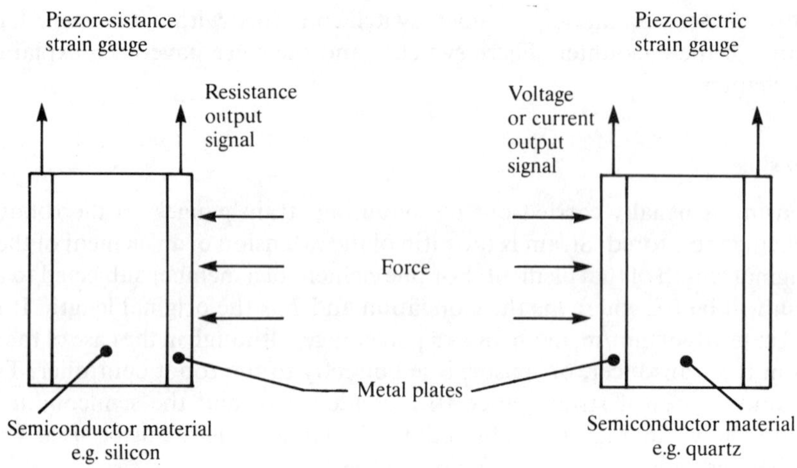

Figure 10.27 Semiconductor strain gauges

and inversely proportionately to its cross-sectional area. Thus when the object being monitored is subjected to a force, it will have its length either shortened due to compression, or increased due to tension. The bonded wire gauge will consequently have its wires increased or decreased in length and decreased or increased in cross-sectional area. This means that the resistance of the gauge will increase as it is stretched and decrease as it is compressed. This change in resistance will be proportionate to the force or pressure applied to the object.

In the piezoresistance sensor, again a change in resistance is experienced relative to the force applied. In this case a material such as silicon crystal is placed between two metal plates. As these plates are pushed together by a force the silicon is compressed and its resistance varies proportionately to the force applied. In both the piezoresistive sensor and the bonded wire gauge sensor a resistance change is observed in proportion to the force; however, in robotics a proportionate voltage will be required for control purposes. This will necessitate incorporating some sort of signal conditioning circuitry and the bridge type configuration shown in Figure 10.28 is normally adopted. The power supply for the system may be either AC or DC and the bridge will be balanced by R_1 so that with no force applied there will be no voltage output. When a force is experienced the resistance of the strain gauge changes and the bridge becomes unbalanced, producing an output voltage proportionate to the force. This voltage can be used as an input to the robot control system.

In the piezoelectric strain gauge the semiconductor material is again placed between two metal plates. In this case the material, which could be quartz, generates a current in proportion to the force applied to it, i.e. a voltage proportionate to force is developed across the plates. This voltage can now be used as an input for the robot control system.

These sensors can be mounted on the robot body, limbs, wrist or fingers to give force feedback information to the controller. One gripper that incorporates force sensing and a photocell is that of the IBM 7565 robot, and although grippers will be considered more fully in the next chapter, a sketch of the gripper is shown in Figure 10.29. The force sensors within the gripper are strain gauges configured in such a manner as to provide information

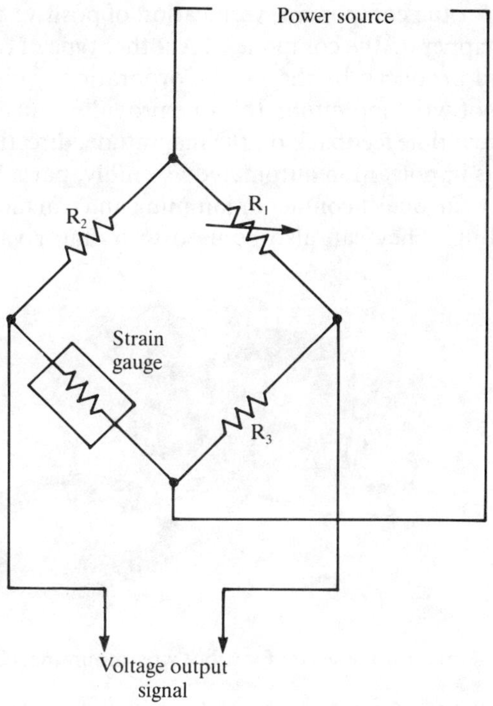

Figure 10.28 Basic bridge type signal conditioning circuit for strain gauges

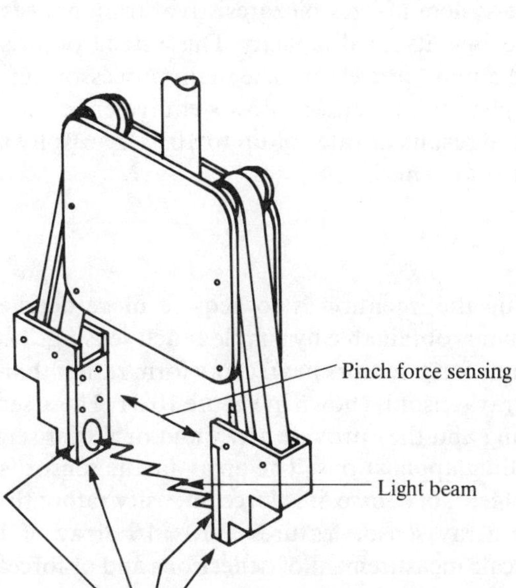

Figure 10.29 IBM 7565 AMS gripper

to allow search for the component, verification of positive gripping, and control of force exerted by the gripper on the component. Another type of robotic sensor is that shown in Figure 10.30 and produced by the Lord Corporation. This is one of a series of systems designed for robot wrist mounting, this one providing six axis force/torque sensing. The sensors deliver real time feedback on the magnitude, direction and location of forces and moments. This is important in automated assembly, parts handling, and certain process operations. They can detect contact or jamming and can monitor an end effector for force or torque build up. They can also be used to monitor variances in contour following

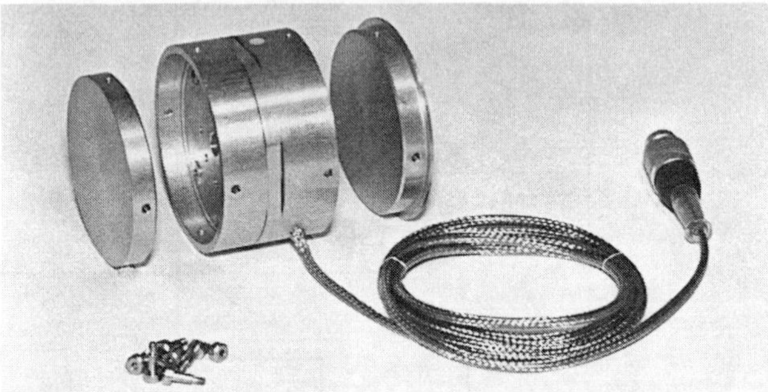

Figure 10.30 Force/torque sensor for robot wristmounting. (Courtesy Lord Corporation)

operations and to maintain constant tool pressure under dynamic as well as static conditions. The system utilizes piezoresistive strain gauges and each sensor is overload protected to five times its rated capacity. The system comprises the sensor, a preprocessing unit, cable, and a main processor. The main processor automatically resolves the forces and torques applied to the sensor into six equivalent cartesian force/torque components and transmits the results at rates of up to 100 Hz. A software programming language is included with the system.

Tactile sensing

In tactile sensing the intention is to acquire more detailed information on the robot environment than is obtainable by simple touch sensing. Tactile sensing can be said to be the use of multiple force sensors in an array form rather than simply at one or two points.

A tactile array sensor is shown in Figure 10.31. These sensors are on pads designed for gripper mounting and they provide array and or vector sensing for automated assembly and parts handling applications. The array tactile sensor is analogous to the solid state camera array except, of course, it is force intensity rather than light intensity that is being measured. The array sensor features a 10×16 array of 160 sensitive sites. Each site provides gray scale measurement of deflections and or forces normal to the touch surface. The touch element pitch is 1.8 mm. Transduction for the array is optoelectronic and mechanical deflection of the touch surface is read in sixteen 0.05 mm increments, i.e. 16 level gray scale, at each element. The transduction is carried out in two stages. Initially each element of the 10×16 array is deflected proportionately to the force acting upon it.

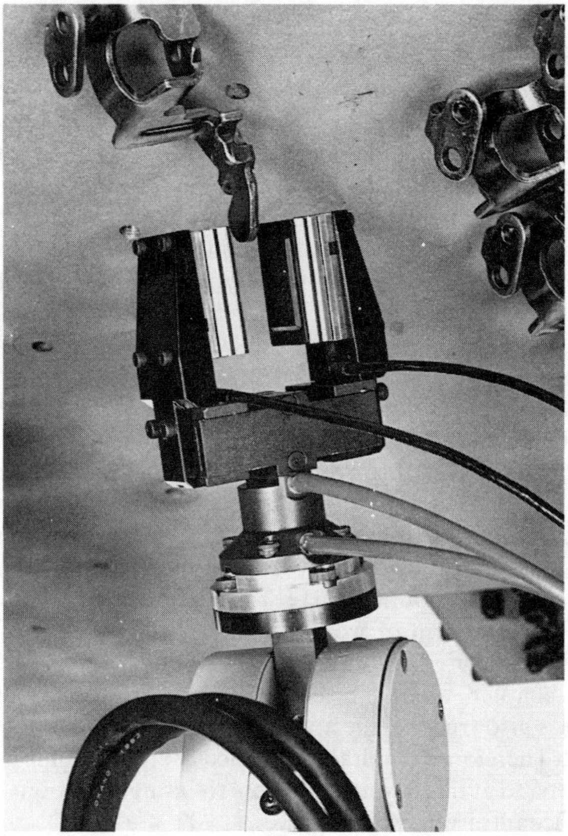

Figure 10.31 Tactile area array sensor. (Courtesy Lord Corporation)

Within the sensor this element movement is measured by a light emitter and detector for each element. As the element is deflected into the sensor it shadows a portion of the light. Thus the maximum amount of light is received by the detector when no deflection is present and the minimum amount when full deflection of the pin occludes most of the light. The range of light intensity is divided into the sixteen levels mentioned previously. The vector sensor which can be integral with the array sensor, measures forces along and about the three coordinate axes at the touch surface. A microprocessor-based interface supports the sensors.

OTHER SENSING

There is a wide range of other sensing techniques available to the robot designer, some of these being noted in Figure 10.1. The discussion here on these sensors is brief as they are either widely used but very simple in nature, or potentially sophisticated but not yet commercially available in a satisfactory form for use in industrial conditions.

Acoustic sensing

Acoustic sensing can be further subdivided into ultrasonics and voice recognition. Ultrasonics can again be split into simple ultrasonic ranging and more complex ultrasonic imaging. Ultrasonic ranging is a technique which has been in use for many years and there is a variety of commercially available ultrasonic sensors together with their associated electronic hardware. These systems are very useful for hobby and educational robotics but are not normally found in industrial applications. One problem of trying to apply ultrasonics in an industrial environment is the amount of noise surrounding the robot of a frequency likely to corrupt the signals. One typical source of this type of noise is pneumatic equipment which is widespread in factory applications. It is possible with techniques for coding and decoding the ultrasonic signal to filter out the unwanted noise, but this adds cost and complexity to the system.

The principle of ultrasonic ranging is simple and is shown in Figure 10.32. Very short pulses of sound in the inaudible 40 kHz range are emitted by a transmitter. These pulses spread out in the form of waves through the air surrounding the transmitter. When an object is encountered some of the wave energy is deflected back to the receiver. A microprocessor and associated circuitry counts the elapsed time between transmission and reception of a pulse and uses this to calculate the distance from the sensor to the object. Thus the sensor can be mounted on the robot gripper to give an indication of range from the gripper to a component or work surface.

A common transducer used in ultrasonic transmitters is the piezoelectric type. As noted earlier in the chapter, when subjected to compression the sensor will produce a voltage. The reverse is also true, i.e. the application of a voltage to the device causes it to compress or expand. Therefore if a voltage is applied at an ultrasonic frequency the device will compress and expand at this frequency. These resulting vibrations can be utilized by a diaphragm to produce ultrasonic sound waves.

Acoustic imaging is not yet commercially available due to various problems. As well as noise corruption there is the fact that resolution is limited to the wavelength of the signal. By increasing the frequency the resolution can be improved but this gives rise to other problems.

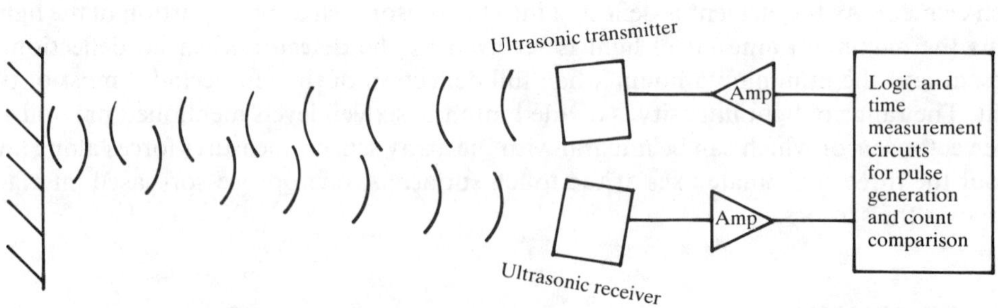

Distance between sensor and object $D = \frac{1}{2}S\Delta t$ where S = speed of sound and t = time between transmission and reception of pulse

Note: Single IC transceivers are available combining transmission and receiving circuitry in one unit.

Figure 10.32 Principle of operation of ultrasonic ranging sensor

Voice sensing is commercially available and, although not commonly used in industrial robot systems, is found on some numerically controlled machine tools. Voice sensing is used for programming and these systems require a voice profile of the programmer to be stored in the memory of the robot. This is necessary as every person has their own unique profile which makes it very difficult to enable these systems to cope with commands from a random instructor. This limitation can be beneficial, however, in that the robot will not respond to anyone whose voice profile is not stored in memory, thus ensuring only authorized personnel can exert control. The system will usually have a limited vocabulary, but by using permutations of the words a large number of commands can be recognized.

Magnetic sensing

The simplest example of this is the magnetic 'pick-up' and it is shown in Figure 10.33. The pick-up is in the form of a case containing a coil which is wound round a permanent magnet. The magnet produces a steady magnetic field around the pick-up and if the coil is

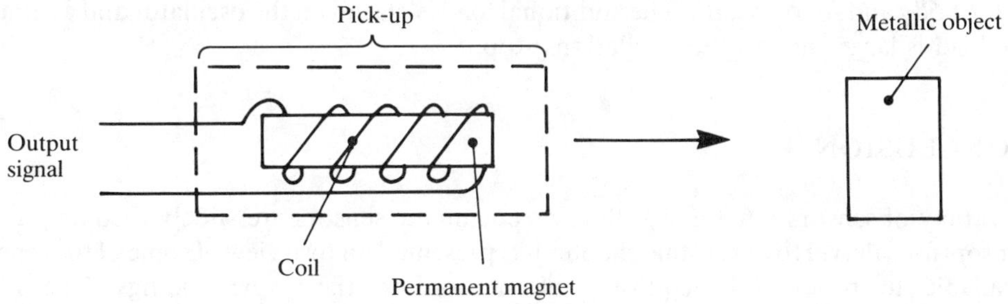

Figure 10.33 The magnetic pick-up

initially stationary no voltage will be induced in the coil. However, if the pick-up is mounted, for example, on the robot end effector, then as it approaches a metallic object the magnetic field is changed and a changing voltage will be induced in the coil. Over very short distances the value of this voltage will indicate the distance the end effector is from the metallic object. Usually, however, a threshold value of voltage is simply used to signify adequate proximity to the object.

Capacitive sensing

Capacitive sensing can be used for detecting the proximity of a nonmetallic object. Thus these sensors could be situated around the robot work area to detect the presence of a human within the safety boundaries. Usually an oscillator-type circuit is used. Oscillation of the circuit is maintained due to the feedback compensating for losses in the coil capacitor circuit. Adjustment of the circuit is such that if a further load is experienced by the circuit then oscillations stop. This in turn causes a change in current which can be used to operate a relay to cause some action, e.g. in the case of a person approaching the robot work area an alarm could be sounded or the robot immobilized. The load is imposed in

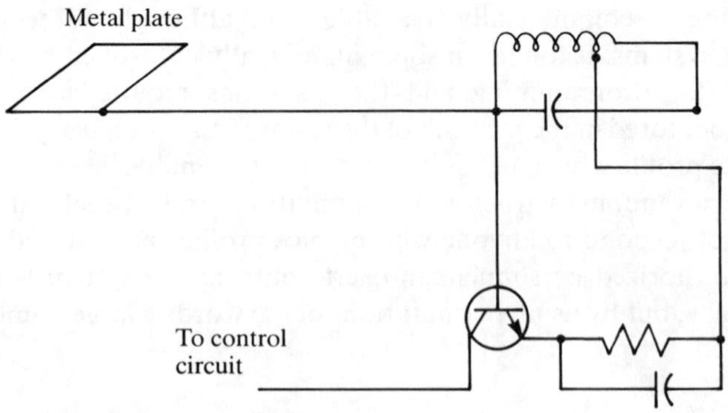

Figure 10.34 Capacitive relay type sensor

this case by the person approaching a metal plate connected to the circuit (see Figure 10.34). The person's body will now form one plate of a 'capacitor' consisting of the metal plate, the body, and the air between which acts as a dielectric. Thus the circuit is coupled to the body and so to ground. The additional load is placed on the oscillator and assuming the load is large enough the oscillations stop.

CONCLUSION

A variety of sensors remain, e.g. fluid or pneumatic sensors are widely used as process sensors for safety. However, this chapter has presented an overview of some of the sensors available to robots for acquiring information on their surroundings. The more sophisticated sensors such as vision and tactile systems are now commercially obtainable. They offer the robot great potential for tackling more complex tasks, working in more disorganized environments, reducing costs of the robot structure and ancillary equipment, and improved safety. Research and development of sensing is widespread and more efficient and useful systems will be developed. This will make robots increasingly attractive to employ in situations both within and outside the factory environment.

11

END EFFECTORS

INTRODUCTION

In the earlier chapters the construction of the basic robot has been considered. However, for the robot to be able to perform useful work, and indeed for it to conform to the definition of an industrial robot, it must have some means to effect manipulation of tools or materials. This it does by means of an appendage attached to the end of the arm, i.e. an end effector.

The end effector may adopt an infinite variety of configurations but basically it can be classified into two distinct types: (a) process end effectors such as paint spraying guns or grinding wheels, and (b) grippers which can range in sophistication from simple two fingered pneumatic grippers to complex 'intelligent' hands with their own sensory systems and microprocessors. The distinction between the process end effector and the gripper may appear to blur in some instances, e.g. when interchangeable tooling is used with end of arm adaptors, or special purpose vacuum grippers are employed for workpiece manipulation. However, the method of differentiation adopted here is that if the end effector is a tool directly used to change the state of the workpiece, then it is a process type. If it is used to grasp and directly manipulate or transport a workpiece, material, or tool, then it will be referred to as a gripper.

One of the essential characteristics of the industrial robot is its flexibility. Some robots are more general purpose than others, e.g. the revolute type can be applied to a variety of tasks whereas the SCARA type configuration is really specially designed for assembly. When robot end effectors are considered they are found to be usually much less general purpose and often custom designed. Process types will naturally be specific to the job in hand, for instance a welding torch or glue applicator, and if it is of the gripper type it will often be specially designed for the component being manipulated. General purpose grippers are widely available but will usually be purchased with a range of movement and grasp to suit a specific job. Thus when a robot is purchased the end effector will come under the classification of 'additional equipment' and will not be included in the basic robot cost. The end effector may be of relatively low cost, say $200 in the case of a simple pneumatic gripper, or it may cost up to twenty per cent of the price of the robot if it has to be specially designed, is complex, or is a particularly expensive process type.

The end effector will be bolted onto the robot wrist, which will be provided with a faceplate for this purpose. As there is still little standardization in this area, the user will probably have to make, or have made, an adaptor plate to fit between the wrist faceplate and the end effector. It can be possible in some cases to have compatibility and there are in fact designs that enable the robot to change its own end effector using couplings which allow for pneumatic, hydraulic, and electrical connections to be made between the wrist and the effector.

Generally the effector will be powered by any of the means previously mentioned for robot drive systems, i.e. fluid or electric, though pneumatic power specifically is the most popular because of its cleanliness, cheapness and speed. Where very large loads are involved hydraulic power is used although this will necessitate a more massive end effector. Electrically powered grippers are also available and although they have the disadvantage of requiring electric motors out at the arm extremities to drive them, they have the advantage of being easily controlled for sensory applications. Process and gripper type end effectors are shown in Figures 11.1 and 11.2 respectively.

END EFFECTOR DESIGN

One of the first points to consider when selecting or designing an end effector for a job is its weight in relation to the load handling capacity of the robot, and the weight, if any, to be lifted. In the case of process tooling such as spot welding guns the total weight is in the end effector itself, but with grippers there will be the additional material, component, or tool weight to be considered. Thus a robot with a 40 kg capacity may have only 30 kg capacity free after a 10 kg gripper has been added. Naturally the weight of the end effector must be kept to a minimum, not simply to ensure maximum payloads can be handled, but also to

Figure 11.1 Example of 'process type' end effector. An ASEA robot trims GRP car body components using a pneumatic tool. (Courtesy ASEA)

Figure 11.2 An ASEA IRB 60 handling wax patterns in the investment casting process using twin grippers. (Courtesy ASEA)

reduce the load on the tool and drive system. A long robot arm with a massive gripper or spot welding gun suspended from the wrist and moving at high speeds is not easy to control precisely.

As stated by Newton's Three Laws of Motion: (1) Every body continues in its state of rest or uniform motion in a straight line except if it is compelled by an external force, or forces, to change that state. Force is measured in newtons and equals mass multiplied by acceleration, where mass is in kilograms and acceleration in meters per second per second. (2) The rate of change of momentum is proportional to the applied force and takes place in the direction in which the force acts. Momentum equals mass multiplied by velocity. (3) To every action there is an equal and opposite reaction. Therefore a large robot with a load of say 30 kg attempting to move this load at a maximum horizontal speed of 2.0 m/s will give the payload a momentum of 60 kg m/s. To decelerate this load to rest in a time of say 0.5 s will require a force of $30 \times 2.0 \div 0.5 = 120$ newtons. There will also be the mass of the gripper and arm to consider, and although omitted for simplicity here, this could be appreciable in the large heavy duty robots, hence the need for good design to make the robot structure as light as possible while maintaining rigidity.

A number of points are apparent from the above example.

1. Once the robot arm is moving a heavy end-of-arm load, considerable force is required to stop it.
2. Acceleration and deceleration times are at least as important as maximum achievable speeds.

3. The load will continue moving in a straight line unless properly clamped to the gripper. This is not a problem in process tooling positively attached to the robot wrist, but in gripper design it must be ensured that there is a large enough coefficient of friction between the gripper fingers and the load to prevent the load being catapulted through the factory workspace when the arm stops.
4. The above example simply considered the horizontal movement of the load. In practice the load will be held a certain distance above floor level, therefore a vertical force must be exerted by the robot to maintain that distance against the force of gravity. When considering movement of loads in the vertical direction the influence of gravity must be taken into account. A body falling freely will be accelerated, due to gravity, at 9.81 meters per second per second (though for calculation purposes here this will be approximated to 10 m). Thus if the robot was to hold the 30 kg load stationary above the floor it must provide a force equal and opposite to the force of gravity. In this case that will be 30×10 kg m/s i.e. 300 newtons, this is the 'weight' of the load. Now if the robot was to require to move that load vertically from a state of rest it must obviously accelerate it. Should the vertical acceleration required be, say 5 m/s then the force required will be $30 \times (10 + 5)$ newtons, i.e. 450 newtons.

 In the extreme there are two contrasting ways in which a load can be held in a gripper when it is being lifted vertically, these are shown in Figures 11.3(a) and (b) (although of course in practice it is often a combination of these two approaches). In (a) the load is supported underneath by the gripper and therefore when lifting vertically the total force required is that due to the weight and the acceleration desired. However, if the load is grasped as in (b) then the force required to life will be the same, but a grasping force will also have to be applied to the gripper fingers to ensure the component does not slip. The grasping force necessary will be dependent on the coefficient of friction of the surface of the material being lifted, and the contact surface of the grippers. The static coefficient of friction is a ratio equal to the ratio of the frictional resistance experienced by a body when it is on the point of sliding along a specified surface, and the perpendicular force between the surfaces in contact. Thus if the load in (b) is 30 kg and it is to be held stationary and there is a coefficient of friction of 0.3 between the gripper surface and the load surface, then a force on each side equal to 300/0.3, i.e. 1000 newtons will be

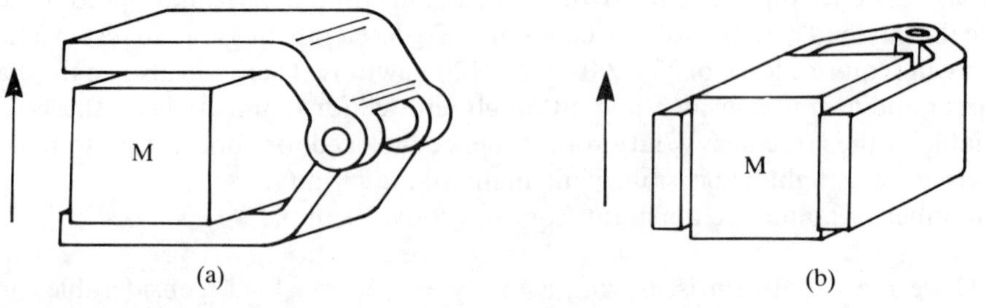

Figure 11.3 (a) Vertical force required with negligible clamping force; (b) vertical force required plus clamping force to resist slippage of load (this must also be enough to cope with forces caused by acceleration and deceleration)

necessary. Should the load be required to be accelerated upward once again, an additional horizontal force will be required to compensate for the additional forces due to acceleration, e.g. at 5 m/s the force necessary will be 1500 newtons per side.

There are other factors that should also be considered. For example, it is important to note that the center of gravity of the load should be kept as close as possible to the gripping areas. This is because the further out it is then the greater will be the clamping force required due to the turning force, or moment, created by the weight of the object and the distance from the center of gravity to the grippers. Also if the load is being held at the end of an arm that is moving with a circular motion there will be centrifugal forces to consider. Another factor is that once the minimum forces have been calculated they should be multiplied, possibly by two, to provide a factor of safety for the system.

VARIETIES OF GRIPPER CONSTRUCTION

The simplest type of gripper is that shown in Figures 11.4(a) and (b). It is pneumatically operated and is shown mounted on a pick and place unit. This system illustrates the use of a 'breakaway head' feature which allows the gripper to separate from the pick and place unit, or robot wrist, in the event of a malfunction. In a robotic installation this malfunction could correspond to a programming error or a collision with an unexpected object. Although popular, this simple gripper produces a nonparallel movement of the jaws which can give problems when picking up a variety of components of different widths. These problems can be lessened by using pivoting or vee shaped finger pads (see Figure 11.5). Parallel action grippers are designed to avoid these problems completely, although they may be more expensive due to their relatively complex construction. One schematic example is shown in Figure 11.6.

The jaw type grippers come in many shapes and sizes. Some may have extra-long fingers to reach into hot areas or lift hot components; some may be mounted in a double gripper configuration as shown in Figure 11.7 which can increase productivity considerably when loading and unloading machine tools. One side of the gripper unloads the finished component from say the chuck in a lathe, the gripper then indexes round and a raw blank, held in the other side, is inserted into the chuck. The gripper then moves out of the machine, deposits the component and picks up a fresh blank then continues the cycle. The Fanuc gripper shown in Figure 11.8 uses a simple finger construction which allows for a variation in part size, location and inclination to be tolerated in the component. The component is in this case displaced by the gripper to enable accurate deposition to take place. Some versions of the gripper have a rotation mechanism to enable the component to be rotated about the gripping center of the hand, thus allowing orientation of the component to a fixture, this facility in effect giving an addition degree of freedom to the robot. Further variations of the grasping type are cam operated devices for internal gripping, extra broad grippers which are virtually beams with individual grippers at each end, and inflatable grippers which are form-adaptable and used to grasp delicate or odd shaped parts.

Sometimes it is necessary to change end effector tooling or grippers frequently within the one work cycle. Here multiple indexable grippers or exchangeable tooling can be

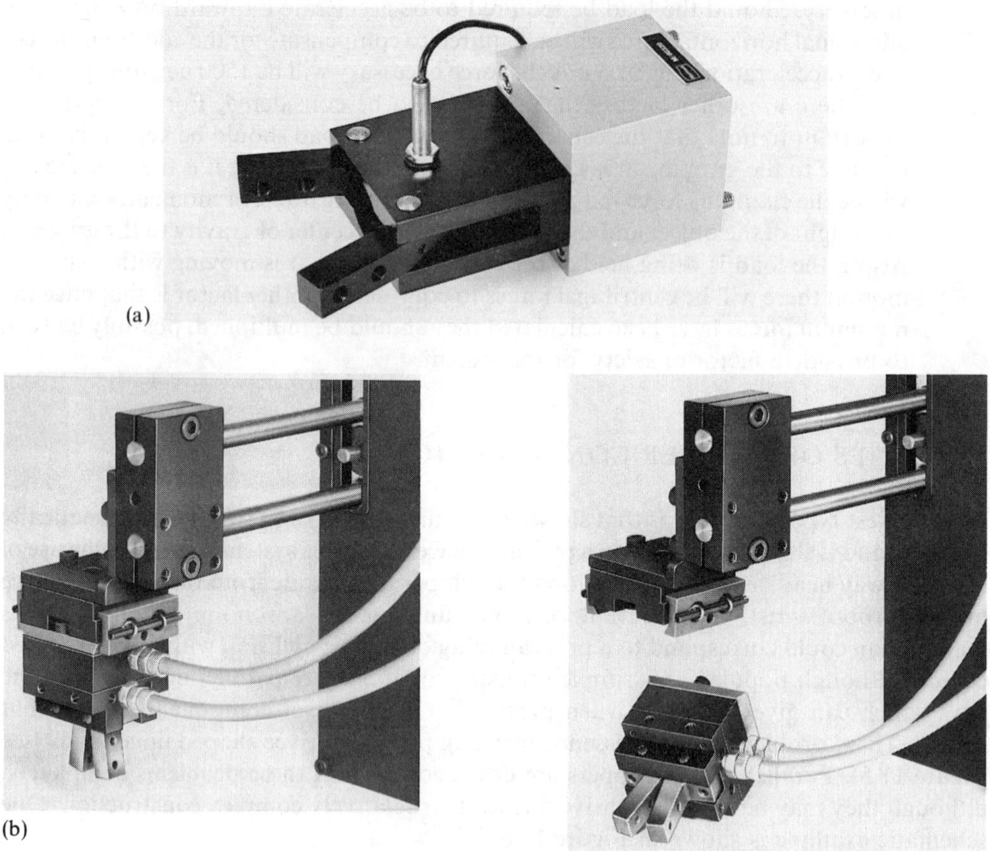

Figure 11.4 (a) Simple low-cost nonparallel action pneumatic gripper fitted with operation verification sensor (courtesy Norgren Martonair); (b) pneumatic gripper fitted with breakaway head safety attachment to minimize the possibility of damage to the robot (or, here, a pick and place unit) in the event of a malfunction (courtesy Arthur G. Russel Co. Inc.)

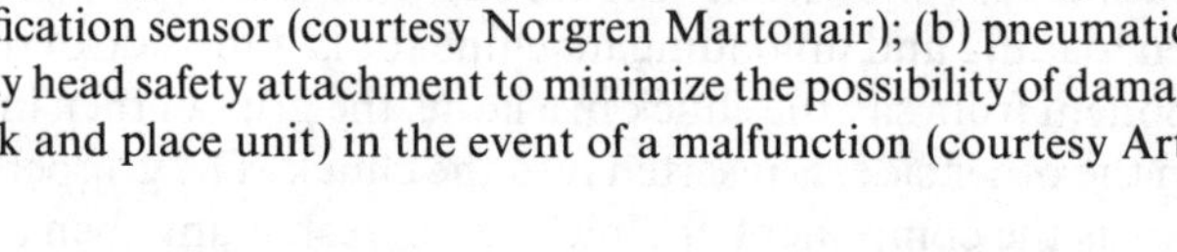

Figure 11.5 Finger pads: (a) vee shaped; (b) pivoting

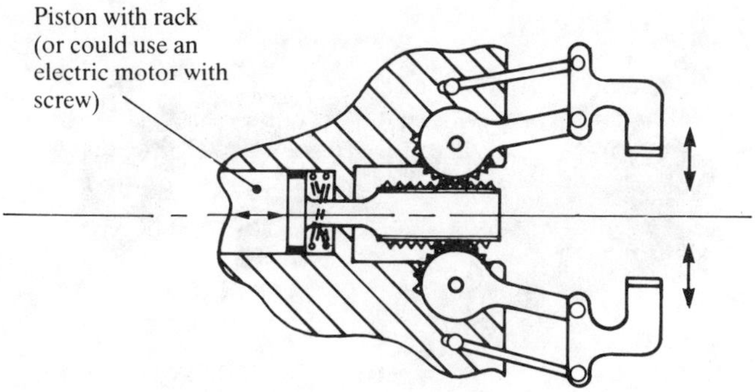

Figure 11.6 Example of gripper construction for parallel action

Figure 11.7 Double gripper configuration. (Courtesy Cincinnati Milacron)

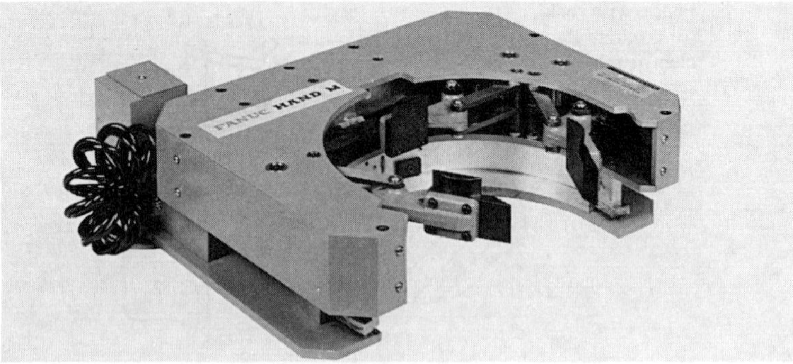

Figure 11.8 FANUC 'hand'. (Courtesy FANUC)

employed. Multiple indexable grippers are ideal where lightweight components and hence lightweight gripper constructions can be used. This decreases production cycle times as each gripper can be indexed into position as required without the need for the arm to move to a tool change point, an indexable gripper is shown in Figure 11.9. The tool change method is an alternative where the grippers are too heavy for multiple mounting or problems with precision are expected. Some sort of quick release mechanism, such as a bayonet coupling, is necessary here. Using this technique the robot wrist approaches a rack containing the grippers or tools, and couples with the required tooling. When using a bayonet coupling, a socket on the wrist can engage or disengage with a plug on the tooling by a push and twist action. This method is ideally suited for the situation when a variety of large, heavy, or awkward grippers or tools are necessary. However, overall productivity will be sacrificed as each changeover takes much longer than the simple indexing technique used on the multiple gripper.

Other methods used for lifting components and material include magnetic and vacuum grippers. Both types are usually specially designed to suit the particular task concerned. The magnetic grippers can be of the permanent magnet or electromagnet type and naturally can only be used if the components or material is ferrous. With the permanent magnet type there is no power source required hence it is ideally suited for explosive or hazardous atmospheres and material. Unless the workpiece is being clamped or otherwise restrained when it is deposited, some sort of stripper arrangement will be required to ensure the workpiece is released at the appropriate time. These magnets can be used for lifting individual sheets of material from a pile either by special design of the magnet, or by using additional magnets to levitate the sheets by magnetic induction. The electromagnets are activated by applying power to them; this implies the necessity of a DC power source but also indicates an opportunity for greater control. The electromagnet type grippers can therefore operate at higher speed and with greater flexibility than the permanent magnet type.

With the vacuum gripper, once again a large number of different configurations can be expected depending upon application. The vacuum will be produced either by a vacuum pump or by a venturi device. The vacuum pump has a high initial cost but low cost of operation, whereas the venturi is a simple device and is cheap to buy but is inefficient thus giving a high operating cost. The vacuum cups used on the gripper will probably be mounted such that independent vertical movement is possible so ensuring full

Figure 11.9 A 'multigrip' connected to the wrist of an IRB 1000 robot. (Courtesy ASEA)

contact over an uneven or sloping work surface. It is possible to build an array of vacuum cups and to use ball valves to ensure those cups not in full contact with the component or material are blocked off as necessary to enable the maximum vacuum to be achieved. A blow off facility is also often incorporated in vacuum grippers to assist quick release of components.

There are many other types of gripper used, their design being limited only by the imagination. Becoming more common are grippers with integral sensors such as microswitches or photosensors. More sophisticated types with tactile and force sensing are now commercially available and some of these sensors were discussed in the previous chapter. A few grippers are available as stand alone 'intelligent' units containing their own microprocessor system capable of interfacing with the robot computer. These grippers can monitor the conditions experienced by the fingers and take appropriate action as necessary.

A device designed to assist component insertion should be mentioned here though it is explained more fully in the chapter containing assembly applications. This is the Remote Center Compliance device which was originally developed at the Charles Stark Draper Laboratory in the USA. It is attached between the robot wrist and the robot gripper and was designed to cope with a slight misalignment of closely fitting components. It comprises two functional parts, one which enables rotation of the component about a remote center, and one which allows translational movement. Thus the position and orientation of the component is changed to enable smooth insertion.

GENERAL CONSIDERATIONS

This chapter has briefly reviewed the robot end effector and some specific design considerations, more general factors are now noted. Safety is of prime importance in gripper design and they should be designed to be failsafe. This applies particularly when any load is moved at considerable speed. A normal failsafe device that could be installed would be a spring operated clamp. The springs would be held back either pneumatically, hydraulically, or electromagnetically while fluid or electric power was applied, but if failure of the supply occurred then the springs would be free to operate the clamp. The clamp could also be triggered by either the first indication of a pressure drop, or an indication of movement of the component from finger sensors in the hand.

The decision on whether to use process tooling on the robot, i.e. presenting the tool to the work, or mounting grippers on the wrist and presenting the work to the various tools, depends very much on the weight and size of the work and tooling. It is also affected by the accessibility of the surfaces to be worked, how the work has to be presented to the tool or vice versa, and the complexity of the operation.

Normal good engineering practice should be applied to end effector design. The construction should be as simple as possible, maintenance should be simple and quick, and there should be no sharp edges or dangerous protrusions. The components used in the design should be standard wherever possible, they should be of proven reliability and require little servicing, e.g. sealed bearings. There should be a foolproof location and orientation method for the human operator to attach the end effector to the robot wrist. The desires for lightweight construction and for strength must be balanced. It may also be necessary to provide protection for the end effector if it is working in, for example, corrosive or high temperature environments. The use of shear pins should also be considered. These are pins that hold the end effector to the wrist and are designed so that if excessive pressure is applied to the end effector, say in a collision, then they will shear so preventing serious damage to the end effector and robot (see Figure 11.4).

In conclusion, there are many factors to consider in end effector selection and design. The mass, speed and acceleration of the end effector and load must be considered with all the attendant forces. Decisions have to be made on whether to mount process tooling on a gripper on the robot wrist, and whether to use multiple or interchangeable tooling. The working conditions must be taken into account, e.g. oily components will have a lower coefficient of friction. It may be possible to purchase an 'off the shelf' end effector, and these are available from a number of suppliers, or it may be necessary to design and build a special purpose gripper. This will obviously be dependent on the particular job being undertaken, e.g. in gripper selection the fragility and variation in size of the workpiece will be important. Speed of operation, range of effective movement, force capable of being applied, and sensory abilities should all be considered. Intelligent grippers utilizing many of the external sensing devices mentioned in the previous chapter are currently available. Research continues into creating universal hand-type grippers utilizing tactile sensors and operated by tendons similar to those of the human hand. These hands have a large number of degrees of freedom but at present they are not commercially available.

12

PROGRAMMING AND LANGUAGES

INTRODUCTION

Before a robot can tackle any piece of work, it must be given appropriate instructions on how to approach the task. This is done by programming or 'teaching' the robot. Robot programming is split into two basic types: on-line and off-line programming. Both have distinct advantages in various situations, the former being the first used and more common, the latter undergoing rapid development and showing much potential for the future.

In on-line programming the robot is programmed directly by the human operator. The robot can be observed as it responds to each instruction fed by the programmer. The position and velocity data required by the control system are read directly from the feedback devices on the robot arm. This type of programming has the advantages of being easy to learn and therefore quick to implement; offering the possibility to observe any likelihood of collision with other tools or equipment while programming is taking place; and not requiring as much computing power as off-line programming. It allows the original operator's skill and experience to be transferred efficiently to the robot, for example in spray painting or seam welding. The major disadvantage is that it can be very tedious and time consuming for long, complex jobs. On-line programming can also be said to restrict the full potential of computer controlled robots and this can best be understood by now considering off-line programming.

In off-line programming the robot is programmed indirectly and does not require the use of the robot in the preparation of instructions. Indeed, the programming may be carried out at a distance from the work environment using a computer terminal and a high level language. There are many advantages with this type of programming:

1. The robot can continue working productively while programming is taking place. This is much more efficient than on-line programming as no productive value added time is lost.
2. There is the potential for full integration with a computer aided design and manufacture (CAD/CAM) system. Simulation can be used to anticipate collisions,

etc., and the constructed program loaded downline to the robot. The robot could be integrated more easily into a fully automated environment where it would be working as one component amongst others such as NC machine tools and automatic guided vehicles.

3. It is possible to write programs able to cope with unusual or unexpected situations and to handle errors in the work situation, e.g. bad or misaligned components.
4. Variations in the work can be more easily accommodated, e.g. changes to the model of an electric motor assembly or different welding requirements due to a change of car model on a production line.
5. The advantages of adaptive and intelligent control can be more easily exploited making the fullest use of sensory abilities.

However, there are some disadvantages to off-line programming: much more computing power is required in the way of software and hardware; the real world contact is lost; the arm transducers themselves cannot be used to provide feedback information on the position of each joint or axis; and there is more programming skill required.

The methods of programming available can be envisaged as constituting a series of steps of increasing sophistication. This is shown in Figure 12.1. At the lowest level there are the methods used for programming the most basic of devices that would just attain the status of robot as defined earlier in the text. At the top of the diagram there are the programming methods that are being aimed at in some research establishments at the

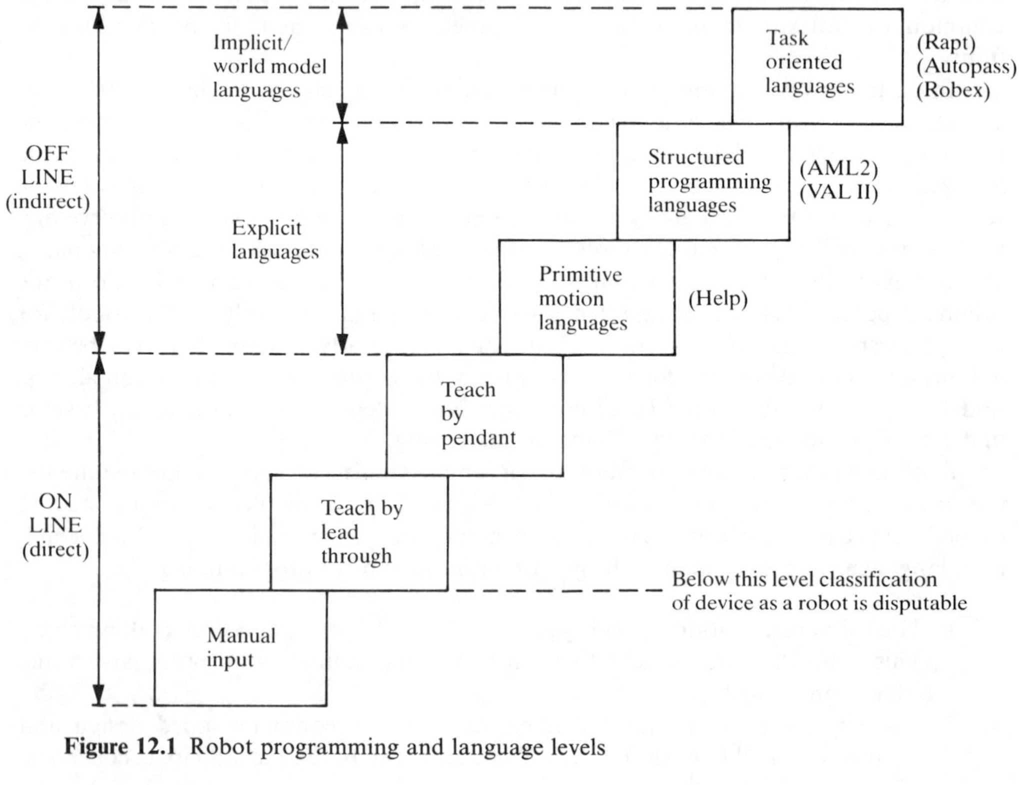

Figure 12.1 Robot programming and language levels

moment. These methods would allow the robot, using artificial intelligence techniques coupled with sensing abilities, to respond to a command such as 'assemble 500 off motor no. A123'. Each of the steps shown is now examined in more detail.

ON-LINE PROGRAMMING

Manual input

The simpler, lower cost, lower technology, non-servo-controlled robots use manual input programming methods. They may use cams, plugboards, or even fluidics or electronics for movement sequencing purposes, but the distance of travel and starting and stopping points are programmed by the setting of limit switches and mechanical trips and stops. These robots are usually pneumatic or hydraulic and due to their relatively nonsophisticated nature they are easily maintained and reliable. They are limited, however, to the pick and place type of operation, e.g. machine tool and press servicing, and the unloading of injection moulding and die-casting machines. They are normally of limited sequence and the smaller types are particularly fast in operation.

The most basic devices, just qualifying for the title of robot, are programmed by bolting metal blocks, or 'stops', to the axes of the robot thus arresting movement (see Figure 12.2). More complex robots are also available, for example when using cams to control the sequence of operations the cams or pegs may be mounted on a long cylindrical drum. As this programming drum rotates the cams make contact with microswitches. These switches then send signals which will operate solenoids that will open or close hydraulic or pneumatic control valves. The robot actuators respond to these fluid signals and move the arm axes in the required direction until further limit switches are encountered. These then send signals to close the appropriate valves before the arm members come into contact with mechanical stops. This ensures a smooth deceleration rather than a sudden halt which could be damaging to a larger robot. A programming drum may contain enough information for sixty different functions per cycle, changes in programming requiring an adjustment of cams or substitution of a new cylinder.

To allow more flexibility it is necessary to use programming utilizing electronic sequence control. This will enable the robot to change its sequence of operations depending on input conditions. For example, if the robot is placing bags of sugar on a

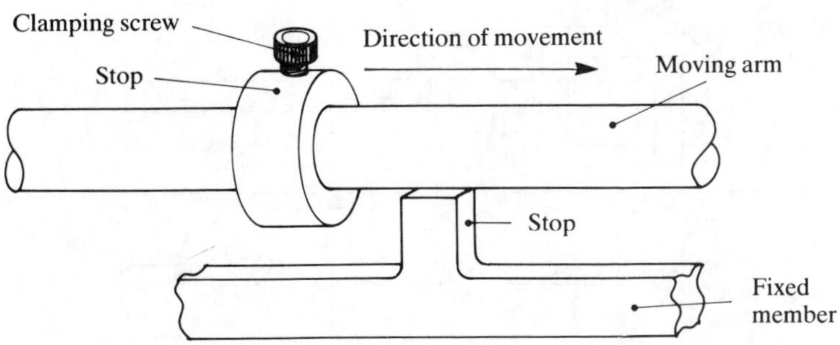

Figure 12.2 Methods of using mechanical stops – 1

scale, receiving a signal concerning the weight of the bags, then putting good bags on a conveyor and reject bags in a bin, then some sort of decision making ability is required. This is best achieved by using programming and control incorporating logic devices, the most popular now making use of a microprocessor. These robots can be quite sophisticated and could indeed be called 'hybrid' robots, as they may use mechanical stops moved by electric solenoids and be controlled by electronics. For example, in controlling rotation a completely manually programmed robot could only be expected to have a start and stop point determined by mechanical stops clamped in place by the programmer. However, by the use of rotary solenoids intermediate stops can be made or stopping points adjusted. The intermediate stops can be obtained by having fixed mechanical stops on the moving arm member, and an adjustable threaded stop operated by a rotary solenoid on the fixed arm member. When the solenoid is powered the adjustable stop moves out to arrest the movement of the fixed stop at the appropriate point in the work cycle. The precise position at which the robot movement is stopped is still set in the above method by manually moving the fixed stops and clamping them. However, by applying the same principle of using rotary solenoids or even stepping motors, to drive adjustable threaded stops, the exact stopping position can also be adjusted through the control system. See Figures 12.3 and 12.4.

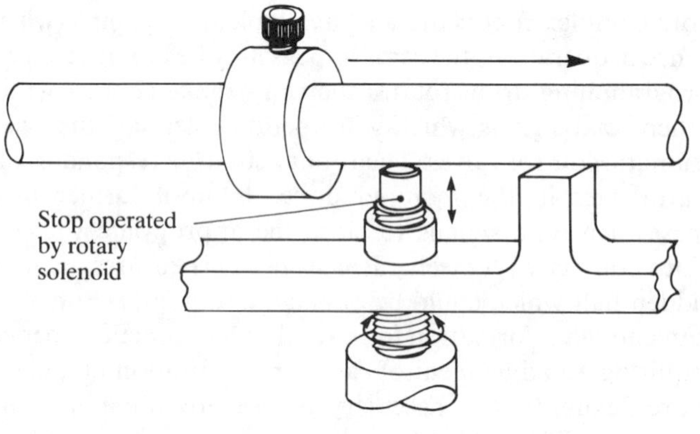

Figure 12.3 Methods of using mechanical stops – 2

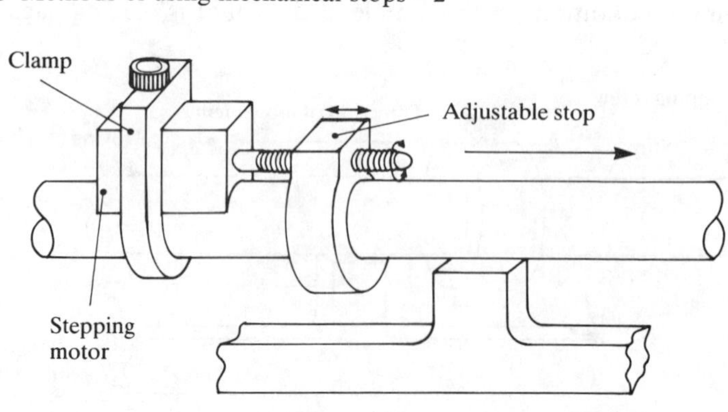

Figure 12.4 Methods of using mechanical stops – 3

Programming by the 'lead through' method

The next step upward in sophistication is to teach by the lead through method. This is ideally suited for continuous path programming, where a smooth flowing movement must be recorded, as in the application of a surface coating by spraying. Initially, paint spraying machines were developed using this method which was similar to the principles employed in the NC machines of the 1950s. The spraying machine was manually led through the movements required by the operator and readings from the transducers mounted on the arm were recorded on magnetic tapes. When the tape was played back these readings were converted into signals to actuators, which then caused the arm to repeat automatically the original movements. In these early machines no computer was required, although modern machines now have microprocessor-based systems to allow greater adaptability and control.

The original method of fitting a pistol-like attachment to the end of the end effector of the robot and pushing and pulling the arm through the required movements is still used. However, due to the safety requirements in some countries, for equipment used in flammable atmospheres, hydraulic robots are used instead of electric robots. These hydraulic robots are much heavier than their electric alternatives and are therefore more difficult to move, even if counterbalancing methods are employed. For this reason a technique of teach by lead through using a 'slave arm' was introduced. In this method a dimensionally equivalent arm is used which is constructed in a lightweight easily manipulated form. The programmer leads this arm through the required series of movements. As he does so the microprocessor samples, many times per second, the signals from the transducers and reads them into memory. The programmer may use a trigger or button on his pistol gripper to record cycle start and end points or points at which the spray gun has to begin or finish spraying. A large memory capacity is obviously required as there is a high number of position samples for each robot joint to be stored. On playback the robot can move at the speed used during programming or at some other more suitable speed, either faster or slower, depending upon the application. See Figures 12.5 and 12.6.

Teach by pendant

The hand-held pendant or console was originally used for point to point programming, but with microprocessor control systems for linear and circular interpolation continuous path movement can now be programmed. This is by far the most wisely used method today and the majority of industrial robots have a teach pendant associated with them. The pendant can be simple having just a few commands available, or very sophisticated like the pendant shown in Figure 12.7. This has twenty normal keys and three shift keys giving a total of eighty possible commands, couple this with the joystick and speed control attachment and we have a powerful programming device. By using this type of pendant the robot can be moved to a particular position by joystick control, then that position fed into memory by pressing one of the keyboard buttons. The speed at which the robot moves during programming is set in this pendant by using the speed control barrel just above the joystick. The speed at which the robot will move during its actual work cycle is programmed in through the keyboard. On a basic pendant the keys may simply indicate a

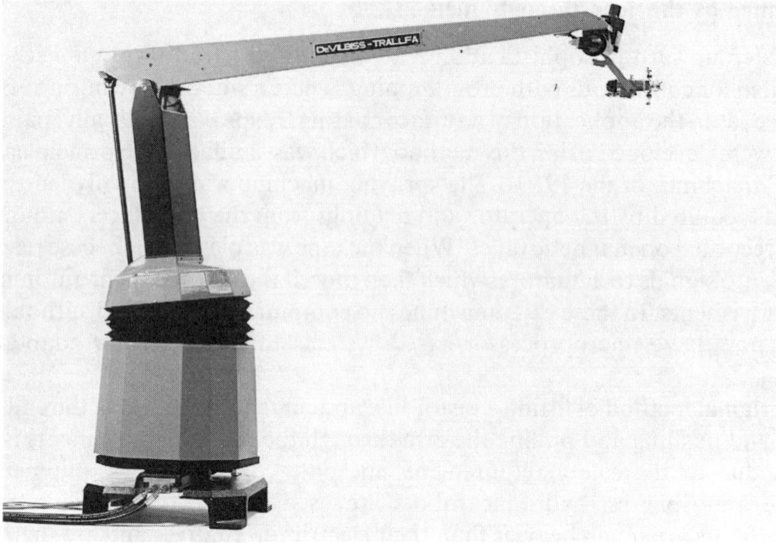

(a)

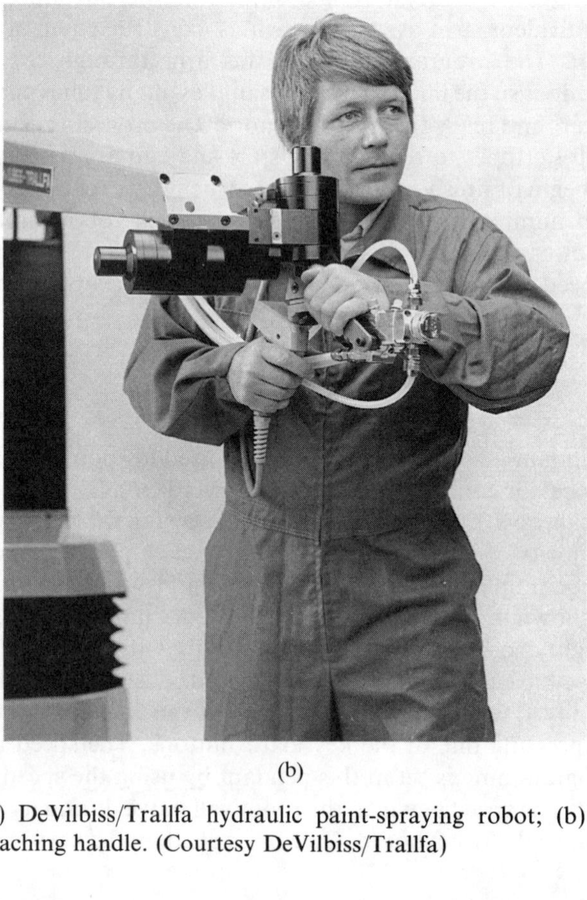

(b)

Figure 12.5 (a) DeVilbiss/Trallfa hydraulic paint-spraying robot; (b) programming Trallfa using teaching handle. (Courtesy DeVilbiss/Trallfa)

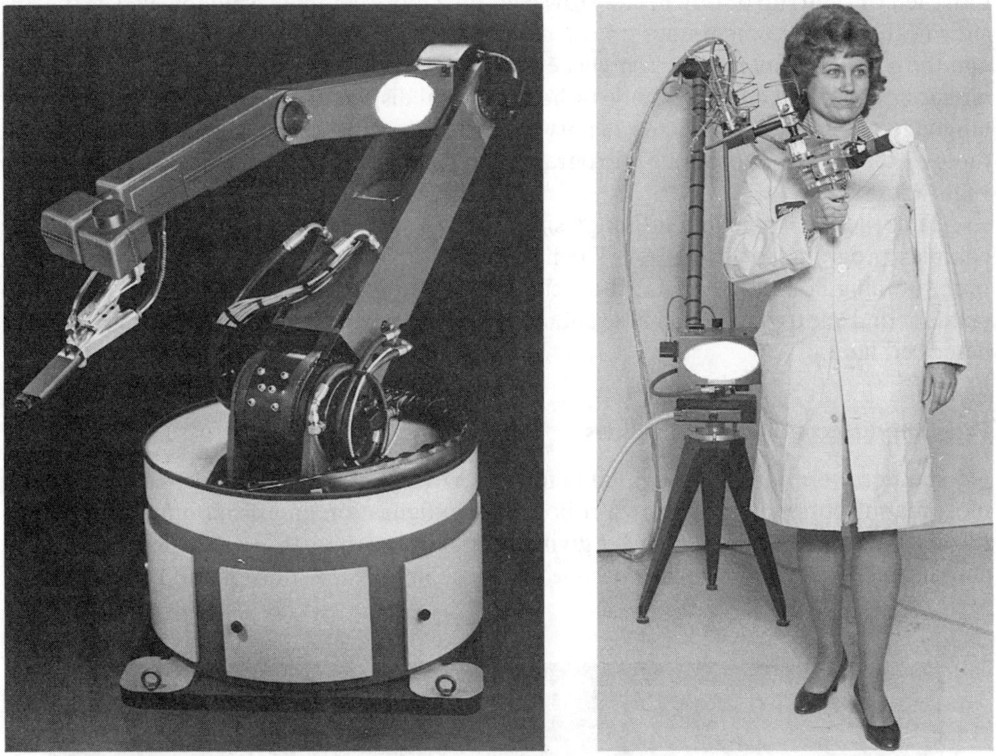

(a) (b)

Figure 12.6 Programming by lead through using slave arm; (a) hydraulic coating robot; (b) lightweight programming arm for robot in (a)

Figure 12.7 Reflex teach pendant

move left or right, up or down, etc. However, on a more complex pendant such as the one shown, three letter mnemonics on the buttons may be used for command identification and the command entered by a single keystroke. The pendant may have a one or two line, alphanumeric, light emitting diode or liquid crystal display. This enables a programming language to be used giving the facility for incorporating subroutines and conditional jumps etc. into the program. Programming using the teach pendant is shown in Figure 12.8.

Thus the teach pendant, although simple to use, offers the opportunity to devise quite complex programs for the robot. The next section therefore introduces some important programming concepts and facilities that can be utilized when using a modern teach pendant and control system. These concepts also apply to the off-line programming to be described later.

Programming concepts and facilities

No standard terminology is available for use when considering these concepts because most manufacturers use their own robot arm configuration and programming software. Rather than confuse the reader by giving all possible terms, those used here have been chosen to convey as simply as possible the more important concepts in programming.

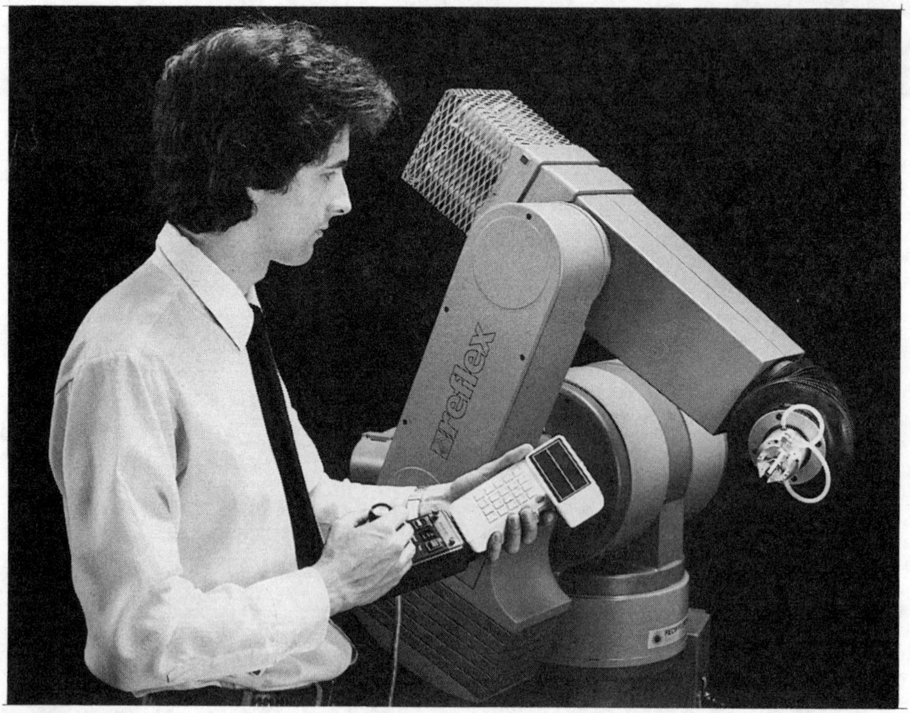

Figure 12.8 Programming by teach pendant. (Courtesy Reflex Automated Systems and Controls Ltd)

The robot operates in various modes. The three modes considered here are operating mode, programming mode and movement mode.

Operating mode
This mode describes the state of the control system and can be split into four types.

1. Immediate or manual mode. In this mode the robot reacts directly to the commands received from the operator through the pendant or controller. These movements and positions are not recorded in memory. This mode is used mainly to allow the programmer to move the robot to its calibration, or home, point and to a convenient position to start a work cycle. Taking the robot to its calibration point at every 'power up' is necessary on all robots that use incremental, rather than absolute, position feedback sensors. This is to ensure that the control system knows exactly where each axis is before a sequence of moves is started. The calibration is carried out by moving each joint individually, either by joystick or pushbutton, until markers on the robot arm are aligned by eye. This is precise enough to allow the control system to then take over and determine the home position accurately. This procedure will vary slightly depending upon the type of robot and control system in use.
2. Program or inactive mode. In this mode a programmer may enter a program into memory, or edit a program from memory, by using the pendant or keyboard. While this is happening the robot will remain unresponsive.
3. Learn or teach mode. In this mode the programmer instructs the robot, the robot executes the commands as they are given, and the control system stores the commands and points in memory.
4. Run or automatic mode. In this mode the robot will execute the program under the full direction of the computer control system. It will continue with the work cycle until complete or until interrupted.

Programming mode
This describes how the robot will move to a point, or through a work element, when programming under operating modes (1), (2) and (3) above.

1. Single axis mode. In this mode the programmer can move the robot one axis at a time until the end effector reaches the desired position.
2. Coordinated axis mode. In this mode the joystick commands will cause the joints to move in a coordinated fashion to enable the end effector to travel from point to point in a straight line or circle while maintaining the tool orientation constant.

Movement mode
This describes how the robot will move to a point, or through a work element, when running through a program under computer control, i.e. under operating mode (4).

1. Maximum velocity mode. In this mode the control system does not keep track of the position and orientation of the tool tip during axes movement. The axis that has farthest to travel is driven at its top speed, and the speeds of the remaining axes

are calculated and controlled to enable all joints to arrive at their destinations simultaneously. This has the advantage that the arm should always be able to reach top speed.

2. Coordinated motion mode. In this mode the joint movements are coordinated in such a way that the tool tip will travel in a straight line between two known points, or in a circle through three known points. Due to the restrictions imposed on the articulations during this type of movement, it is unlikely that the maximum possible speed will be achieved. This mode is ideal, however, for such operations as straight line insertion of components in assembly or welding a circular tube to a plate.

When programming the robot, the main concern of the programmer is how the end effector on the arm moves through space. This end effector may be the gripper, a cutting tool or some other device such as a spot welding unit. To be able to program the arm effectively it is necessary to understand some common terms and concepts. Those considered are:

position, orientation and location
tool axis and tool center point
reference frame, current frame, global frame and local frame
absolute location and relative location
pitch, yaw and roll movements.

- Position describes a single point in space using three coordinates, e.g. x, y and z, relative to the coordinate system being used. However, this is only a single point and a solid object might be lying in any orientation at this point.
- Orientation describes how the axes of a solid object are lying relative to the coordinate system being used by giving the angle of the major axes of the object relative to the axes of the current coordinate frame. These angles may be called alpha (α) for pitch, beta (β) for yaw, and gamma (γ) for roll, given in degrees. See Figure 12.9.
- Location describes the position and orientation of an object in space relative to the coordinate system.

- Tool axis is the major axis of the end effector or tool attached to the robot. It will often correspond with the center line of the gripper, or shaft of the tool, which in turn usually corresponds to the roll axis of the robot wrist. It is this axis which is used when defining tool orientation and it will always pass through the tool center point.
- Tool center point is the critical point on the robot when programming is taking place. It is this point that is programmed to move through space, e.g. if the robot is moving in coordinated motion mode it is this point that will follow a straight, or circular, path. The point lies along the roll axis at a certain distance (specified by the programmer) from the wrist faceplate. The point may be the actual tip of the tool or it may be an imaginary point somewhere beyond the end of the tip, depending upon the particular task being programmed. The robot control system

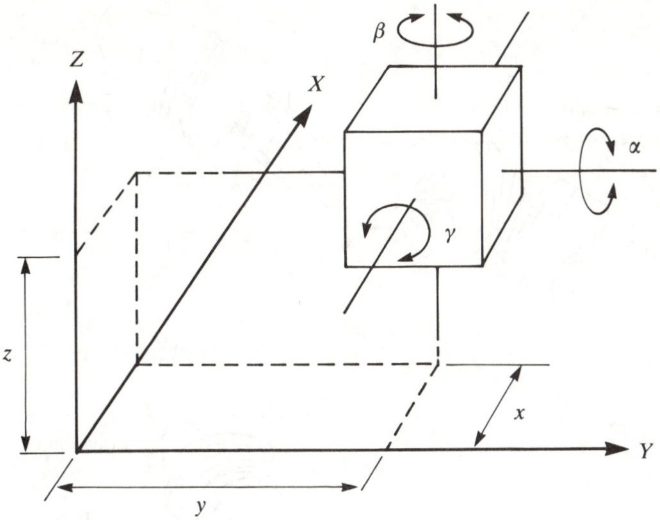

Figure 12.9 Location of an object in space

may be designed so that if the orientation of the tool axis is changed, by altering the pitch, yaw and roll coordinates, the tool center point will remain in the same position although the tool orientation will move. See Figure 12.10.

The tool center point location is therefore specified by stating six factors. The x, y and z linear coordinates for point position, and the α, β and γ angular coordinates for the tool orientation. These can be set by the programmer by moving the robot arm to the required location, using the teach pendant buttons or joystick, and then instructing the control system to remember that location. Alternatively the points may be fed into memory through a keyboard as in off-line programming.

- Reference frame. When the robot is being programmed, the locations of the tool movements must be recorded by the robot control system relative to some coordinate frame, otherwise the location coordinates would be meaningless. Thus the robot always has in its memory one or two coordinate reference frames from which to relate all its actions. These frames may be called the global frame and or the local frame. The particular frame in use at any moment in time may be termed the current frame.
- Global frame is the main reference frame of the robot. It may vary slightly from manufacturer to manufacturer but the following convention can usually be applied. The X axis points outwards from the center line of the robot. The Y axis runs to the left and right of the center line. Both X and Y axes are parallel to the floor and intersect on the Z axis. The Z axis is perpendicular to the floor passing along the center line of the robot waist axis. The point of intersection of the X and Y axes with the Z axis may be at floor level, or where the Y axis passes along the shoulder axis (see Figure 12.11).
- Local frame is a coordinate reference frame that travels with the robot arm as it

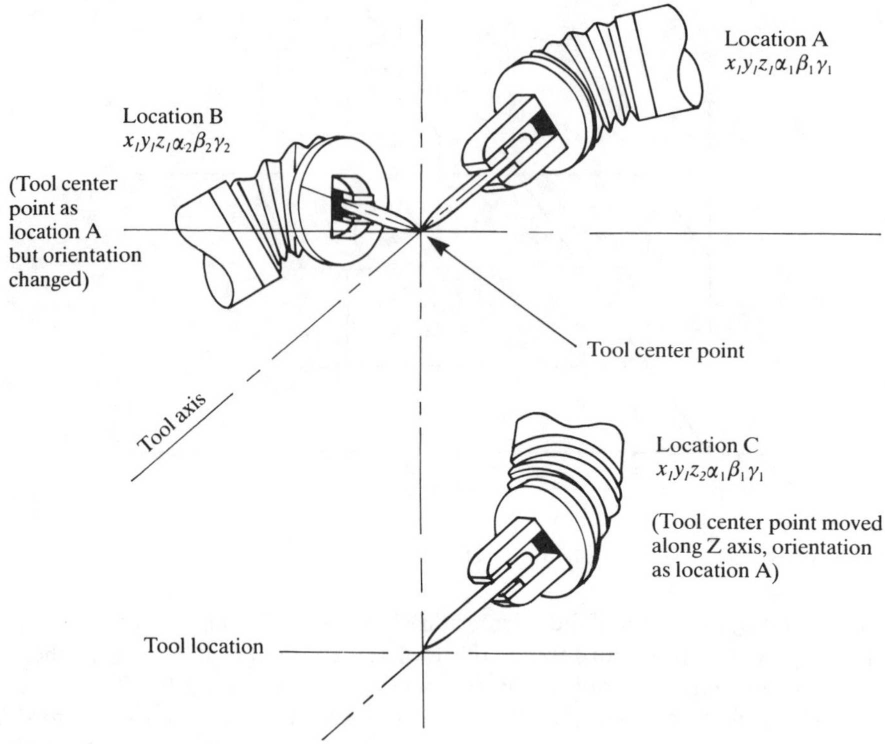

Figure 12.10 Tool location

moves from one location to another. It may be set, for example, relative to the faceplate on the end of the wrist. The tool center point may act as the local origin, and the tool axis act as the local *X* axis (see Figure 12.12). When the pitch, yaw and roll wrist axes are all zero, then the local *X*, *Y* and *Z* axes will lie parallel to the global *X*, *Y* and *Z* axes. However, when the orientation of the tool axis is changed, the local *X*, *Y* and *Z* axes will also change relative to the global axis. This facility is useful for carrying out identical operations in different locations within the work envelope of the robot. For example, the robot could be taught to insert a peg using a straight line movement into a hole in one particular location. It could then utilize the exact same program to carry out the task in another location, i.e. another point and orientation, by simply specifying a new local frame.

- Absolute location defines a point and tool orientation in space relative to the global reference frame. Once created this location is always the same. It is not affected by any movement or change in configuration by the robot arm between the definition of, and return to, the location.
- Relative location defines a point and tool orientation in space relative to the current frame in use when the point is created. Thus the location $x, y, z, \alpha, \beta, \gamma$, may define a location relative to the global frame, or to the local frame in use at that time. In the latter case the tool center point would, for example, move to a point

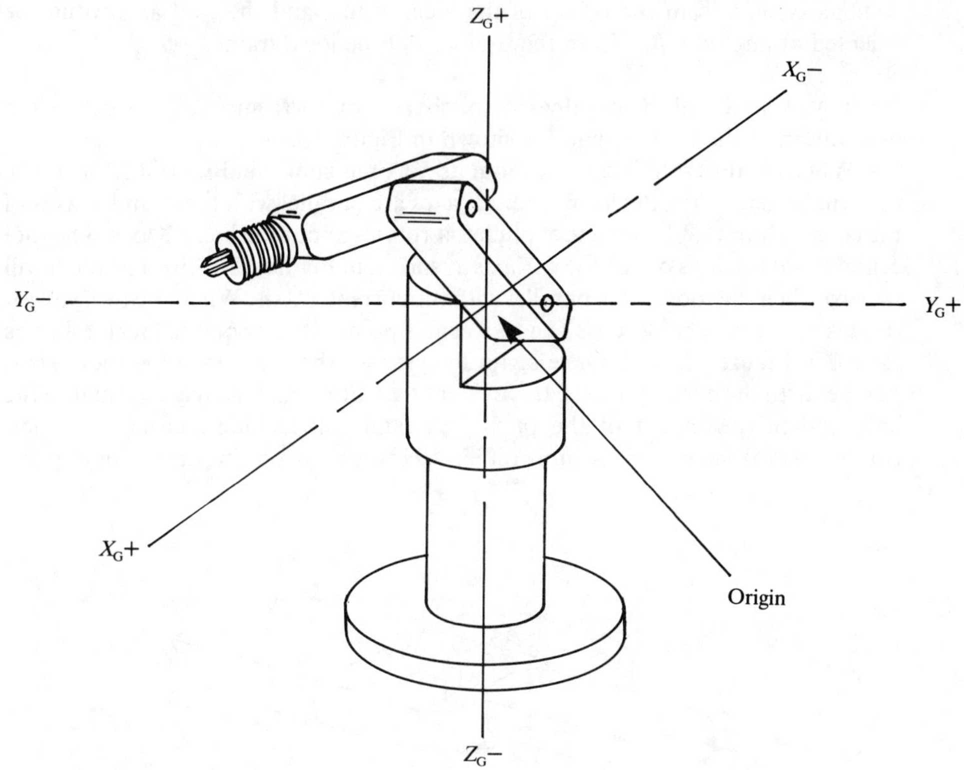

Figure 12.11 The global frame of a revolute type robot

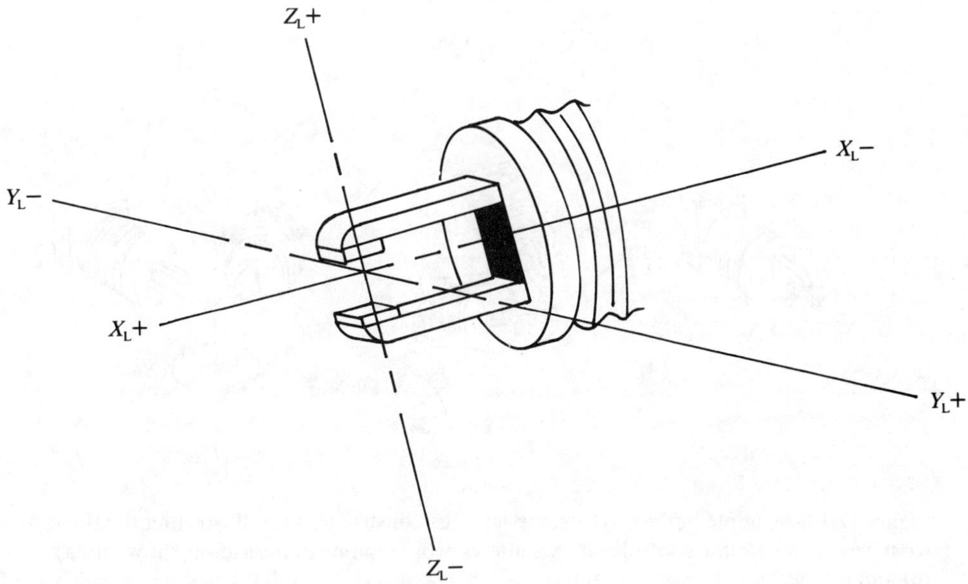

Figure 12.12 A local reference frame

distance x, y, z from the origin of the local frame, and the tool axis would be oriented at angles α, β, γ from the tool axis of the local frame.

- Pitch, yaw and roll. The values of pitch (α), yaw (β) and roll (γ) define the orientation of the tool in space as shown in Figure 12.13.

  When the tool rolls it rotates about its X axis as shown in Figure 12.13(a). The roll will be zero when the local Y and Z axes are parallel with the Y and Z axes of the current frame. When the tool pitches it rotates about the local Y axis. The tool center point remains on the local Y axis as shown in Figure 12.13(b). The pitch will be zero when the tool axis is parallel with the current X axis. When the tool yaws it rotates about its local Z axis. The tool center point remains on the local X axis as shown in Figure 12.13(c). Once again the yaw will be zero when the tool axis is parallel with the current X axis. It is evident that although it is easy to visualize the independent movement of the pitch, yaw and roll facilities, visualizing their combined movements will be more difficult. However, in on-line programming this

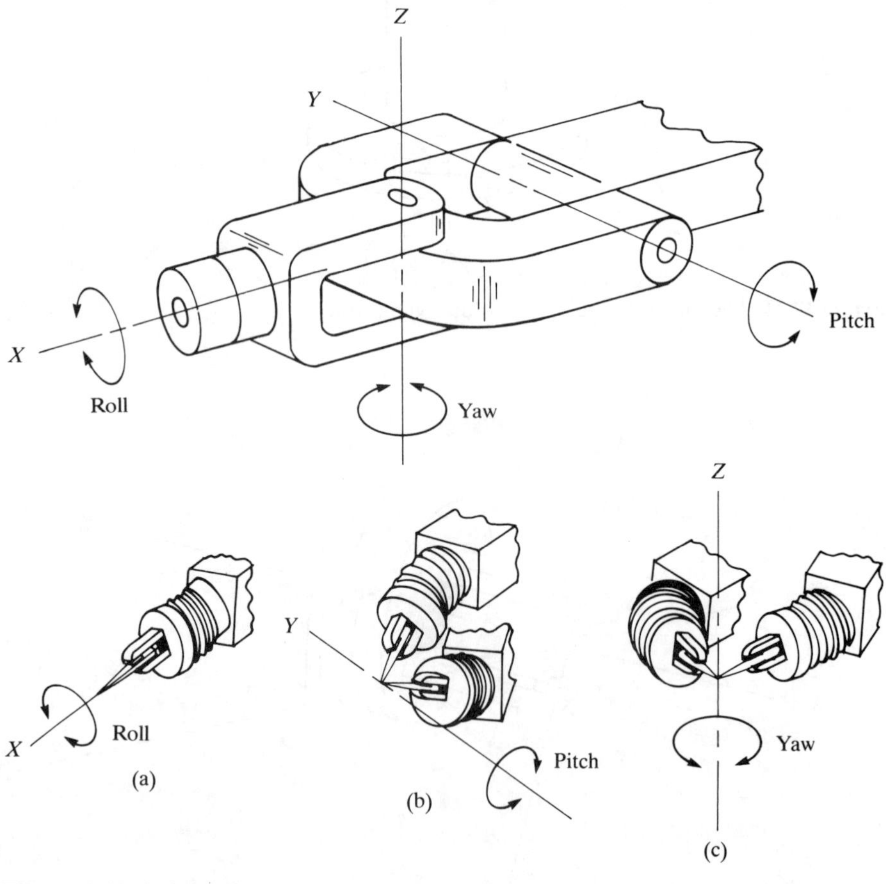

Figure 12.13 A simple but not very compact wrist construction for illustrating the three wrist axes. Appropriate control software allows programming of the axes as shown in (a), (b) and (c)

does not present a serious problem as the teach pendant, possibly incorporating a joystick control, will be used to move the tool to the exact location required before entering that location in the memory of the robot.

There may also be other functions available to the programmer and some of these are now noted.

- Subroutines are self-contained subdivisions of a program, and comprise a complete set of instructions for carrying out a specific task. These routines will be accessed by the robot control system as required. For example, in a palletizing operation a subroutine may instruct the robot to retrieve boxes from a conveyor and position them on a pallet in a four by four pattern. The robot may then be instructed to repeat this subroutine but this time on a higher plane on the Z axis thus allowing the next four boxes to be positioned on top of the existing ones.
- Loop. A set of instructions within a program that will allow the robot to repeat a task continuously until certain criteria have been satisfied. For example, the above-mentioned palletizing robot may continue to execute the subroutine until a stack of boxes five high is achieved. The robot will be programmed to then come out of the loop and possibly start work on a new pallet.
- Branching allows the robot to interrogate an input signal at a particular point in a program then, depending upon the result, to initiate an appropriate subroutine if necessary. Conversely the program may be interrupted by an externally generated signal, thus causing a particular subroutine to be initiated. In the palletizing example the robot may attempt to pick up a box from the conveyor but find that none has arrived. A branch routine will then be initialized to cause the robot either to wait, or to check if a box is available on another conveyor, or to notify an operator.

  The branch facility can be used in conjunction with the local frame concept mentioned earlier. This means that when a particular subroutine is initiated, the tool orientation and pattern of movement will be relative to the current frame defined at that time. See Figure 12.14.

  One further use of the branching facility is that of limited adaptive response. For example, with distance sensors mounted on the robot wrist, the arm can alter its position relative to the task as necessary. This would be appropriate when stacking variable thickness components.
- Stationary base line tracking is a facility that can be available for six axis computer controlled robots. It allows the robot to carry out a task on a workpiece moving through its work envelope irrespective of the speed and position of the workpiece. An example of this is the spot welding of car bodies on a moving assembly line. The position, orientation and speed of the tool remains constant relative to the workpiece throughout the work cycle. Fuller consideration is given to this facility in Chapter 13 within the section on spot welding.

Other facilities available include:

- Delay – this stops robot motion for a specified period of time

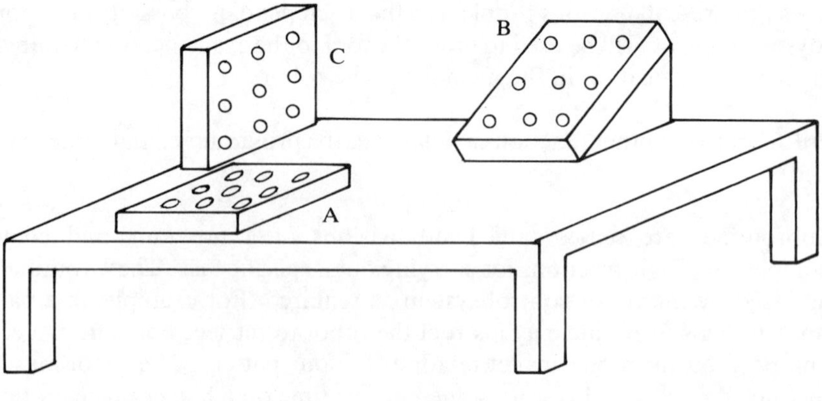

By using the one subroutine (developed for inserting bolts in object A) and utilizing the local frame facility, the robot can branch to objects C and B to carry out the insertion task. The location of one hole in objects B and C may be all that is necessary to define the whole operation.

Figure 12.14 Use of local frame facility

- Wait – this stops robot motion until an externally generated prompt signal is received
- Output – this stops robot motion and causes a signal to be output to a peripheral piece of equipment
- Edit – this allows the programmer to access an existing program and make alterations as required.

Many other facilities and aids to on-line programming are available but they are not described here as they vary from robot to robot and are constantly being improved and refined.

OFF-LINE PROGRAMMING AND LANGUAGES

An example of a workstation for off-line programming is shown in Figure 12.15 and Figure 12.16 illustrates some of the terms used in this section. A prerequisite for programming the robot off-line is the availability of a high level programming language. The instructions inside the robot control system and computer are in the form of machine code or machine language, which is in a binary configuration. This is obviously too cumbersome for a human operator to work with, a language that uses symbols to represent binary commands can be used instead. This is usually called assembly language and the translation from assembly code to machine code is performed by a program called an 'assembler'. However, this is also unwieldy for use in a factory environment and 'higher level' languages are required. These are the languages discussed here, and in them one statement will correspond to many machine code instructions.

Figure 12.15 Programming the IBM 7565 robot using a workstation capable of off-line programming

These high level languages allow shorter programs and are easier to understand by a third party once written. They take longer to execute than a machine code program and require more memory; however, this is not usually a serious problem in robotic applications. There are two methods whereby the high level language is translated into machine code within the robot computer. One is to take the entire program in high level language and then to translate it before executing it. This is called 'compilation' and is done by a program called a 'compiler'. It is efficient and quick but unsuitable if program changes are frequently required. The second method is to translate the program one statement at a time and then execute the statement immediately. This is called 'interpretation' and is done by a program called an 'interpreter'. It is particularly suitable for on-line work and when interaction with the operator is necessary.

Figure 12.1 shows some of the languages that can be used for off-line programming. The languages available can be split into two types: 'explicit' languages and 'implicit' or 'world model' languages. The explicit languages are machine oriented in that the programmer has to specify the tool or end effector positions and movements in detail. They can be further subdivided into two levels. These are the primitive motion languages that essentially resemble simple interpreted computer programming languages such as Basic, and the higher level, structured programming languages that resemble the easier to

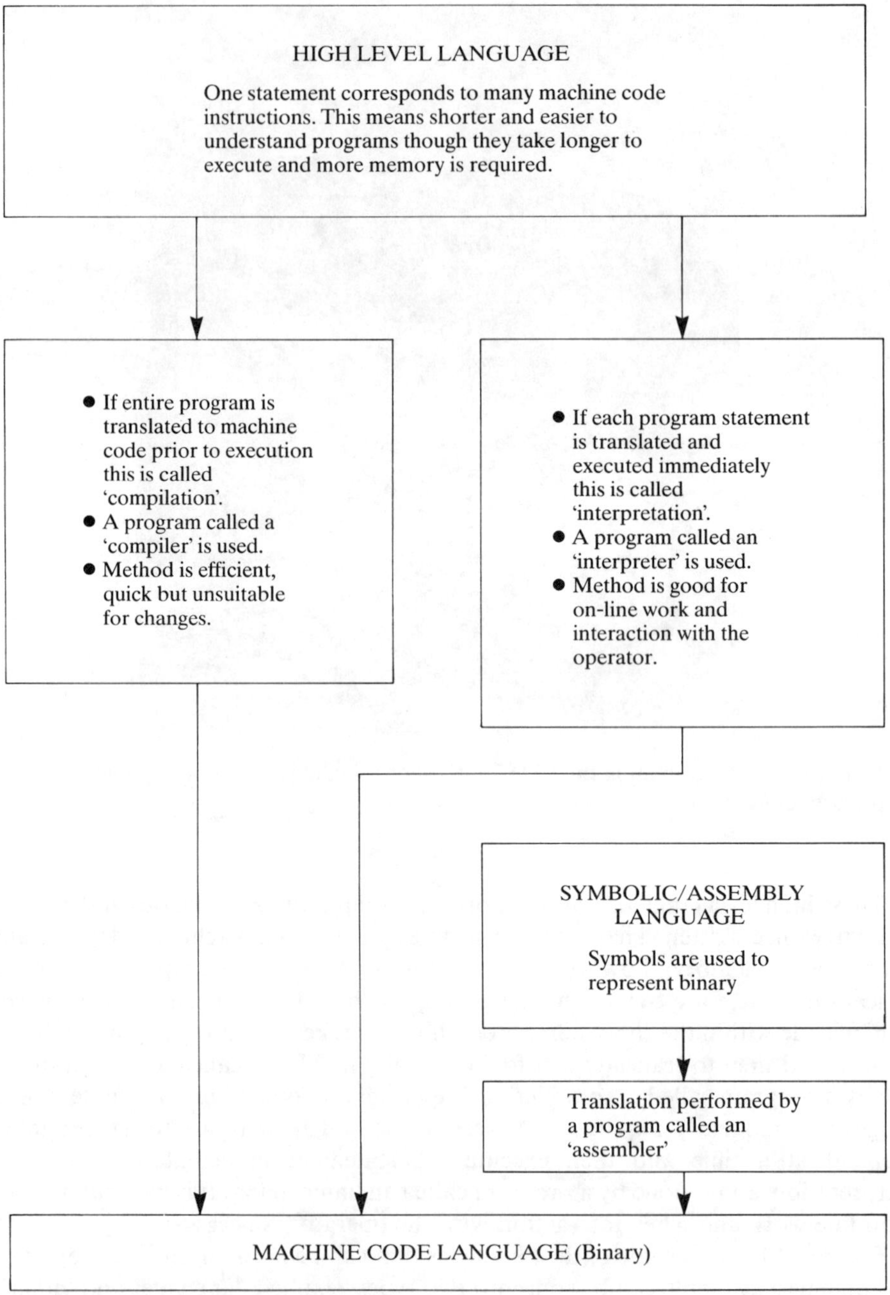

Figure 12.16 Computer language levels

understand and modify computing languages such as Pascal. The explicit languages are the type in commercial use today.

The implicit languages allow much simpler instructions to be used. These instructions will refer to the working environment of the robot, e.g. 'insert bolts at component locations a, b, c and d'. In this case the tool or end effector positions and movements necessary to accomplish the task are generated automatically by the robot computer. This requires much knowledge of the robot work environment and arm geometry to be held in memory, and the use of sensing such as vision will probably be necessary for full implementation. Research has now been under way for many years into these world model languages, and as computer hardware and software, artificial intelligence techniques and the integration of sensing develops, they should become commercially attractive in the future. Figure 12.17 indicates the difference in programming techniques between explicit and implicit languages.

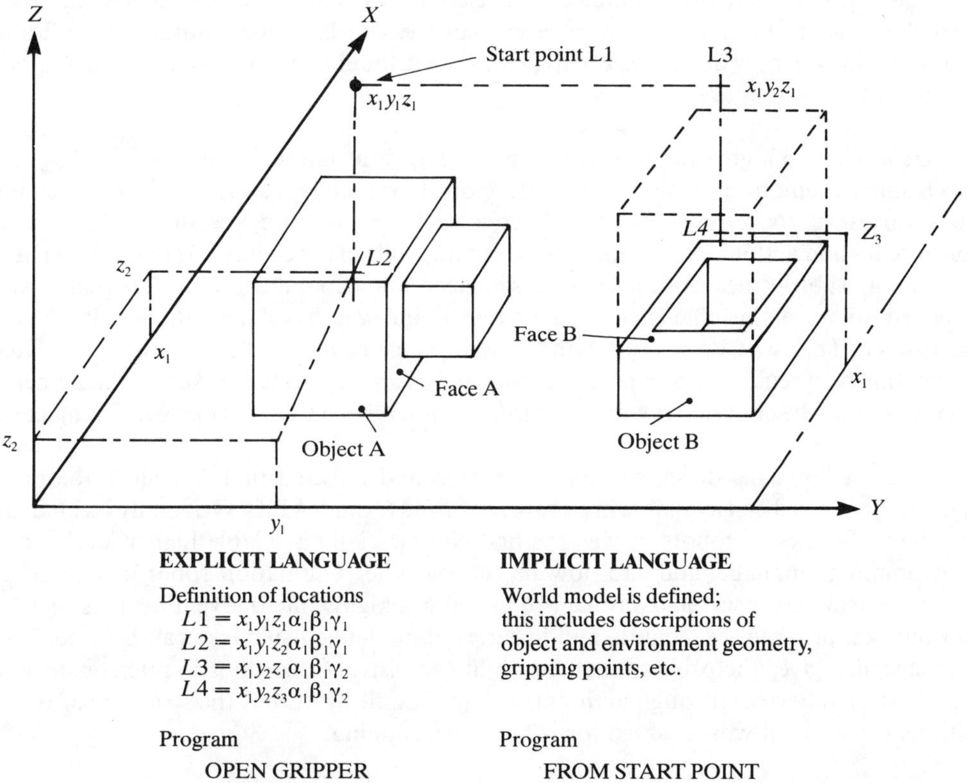

EXPLICIT LANGUAGE

Definition of locations

$L1 = x_1 y_1 z_1 \alpha_1 \beta_1 \gamma_1$
$L2 = x_1 y_1 z_2 \alpha_1 \beta_1 \gamma_1$
$L3 = x_1 y_2 z_1 \alpha_1 \beta_1 \gamma_2$
$L4 = x_1 y_2 z_3 \alpha_1 \beta_1 \gamma_2$

Program

OPEN GRIPPER
FROM L1
GO TO L2
CLOSE GRIPPER
GO TO L1
GO TO L3
GO TO L4
OPEN GRIPPER

IMPLICIT LANGUAGE

World model is defined;
this includes descriptions of
object and environment geometry,
gripping points, etc.

Program

FROM START POINT
MOVE OBJECT 1 ONTO OBJECT 2
WITH FACE A AGAINST FACE B

Figure 12.17 Difference between explicit and implicit language programs to lift object A, rotate it, and insert it in object B

Although it has always been possible to program industrial robots, it is only since around the beginning of the eighties that textual programming languages have become generally available. Work at the Massachusetts Institute of Technology resulted in the development of Wave, an experimental off-line language for programming robots for design purposes in the early 1970s. This work provided the basis for much of the following developments such as the commercially available languages VAL and AML. Also available at this time were programming languages developed for numerically controlled machine tools. Probably the most widely used and popular of these was APT (Automatic Programmed Tool). This language was therefore used as the basis of some robot languages and MCL, for example, is a development from this. Control systems used for NC machine tools have also formed the framework for robot programming systems. For example, the Cincinnati Milacron industrial robots are programmed by teach pendant using a version of the Acramatic NC control system. However, although hard copy of the computer instructions can be obtained via a teletype or visual display unit this is not a true textual language. Four of the more recent commercial languages suitable for off-line programming are now described briefly, although it should be noted that a large number of different languages are now available.

Help. This is a high level interactive language made available in 1982 by DEA (Digital Electronic Automation). It was initially developed for their measuring machines then for their cartesian coordinate PRAGMA robots. Locations are specified in cartesian coordinates and actions can be initiated by calling built in procedures. It has looping and branching, subroutines and parallel task execution capabilities. The language was designed to have the efficiency of a compiler and yet have the interactivity of an interpreter. This was done by having, within each program element, the translater alternating between acting as a pure compiler and as an interpreter to execute the element. The system can be configured with up to four arms having a total of twelve different axes.

VAL. A language developed by Unimation and released in 1979. Since the mid-eighties it has also been available in a more advanced form, VAL II. Originally created for the Puma™ series of robots, it was the first commercially available higher level robot programming language, and can now be run on other Unimation robot models. It is interpretive in nature, is relatively easy to use, and is highly interactive. Facilities such as subroutines, program branching, and libraries of predefined subtasks can be used. The language also gives the programmer the ability to instruct the robot to interrogate and signal external devices through an input/output module. VAL has the structure of Basic with new command words added for robot programming.

MCL. An extension of APT, MCL (Manufacturing Control Language) was developed by the McDonnell Douglas Corporation and became available in 1981. It was written for the programming of work cells in which there may be a number of pieces of automated equipment, including robots, under the control of one or more computers. The language can be used to control any type of robot, but a postprocessor is required in each case to change the MCL code into a form compatible with the robot. It can be used to control the activities of every device in the work cell using the postprocessor technique. Real time decision making, vision processing and program verification can all be handled.

The MCL control system is thus using a hierarchical structure in which programs are written at one level and then moved downward in the structure to be handled by postprocessors which change the programs to a form understandable by the appropriate device.

AR Basic. This was developed jointly by the American Robot Corporation in the USA, and by Rediffusion in the UK, for the Merlin (USA)/Reflex (UK) robot. This language is simple to learn and can be easily understood by anyone with a knowledge of Basic. Some additional commands are, of course, necessary to enable a robot to be controlled effectively, e.g. for opening or closing grippers or interfacing requirements. As shown in Figure 12.17 an explicit programming language requires two parts, one to define locations, and one to give instructions. An example of a simple program in AR Basic is shown in Figure 12.18. The program is to enable the robot to carry out a dip coating operation by picking up a component, dipping it into a vat of liquid, removing it from the liquid and allowing it to drip, then drying it over a heater.

AML. AML was issued by IBM in 1982, the latest version at time of writing being AML2. It was developed for use on the IBM 7565 assembly robots but a subset of the language can be used on their SCARA type robots and run on the IBM personal computer. The language software can also be purchased separately. It is a well structured interactive language. It was designed to be powerful and yet to also have simple subsets that are structured to be used by programmers with a wide range of experience. The interactive and interpretive structure facilitates debugging and fine tuning when the program is eventually run on-line, after the off-line programming has been completed.

Karel. A language developed by FANUC for use with the Karel Control System. This is a general purpose, sophisticated language. It uses symbols, has many mathematical functions, and is designed for integration with sensors and other computerized systems. Self-diagnostic functions are included and it is available with a 'Manufacturing Automation Protocol' interface. MAP will be discussed in Chapter 17.

The powerful and general purpose nature of both AML and Karel, coupled with the resources of IBM in the former, and FANUC and GMF (General Motors Fanuc) in the latter, make them both possible influences towards an industry standard.

There are many other languages in existence e.g. Rail (Automatix Inc.), Jars (NASA), and RPL (SRI International). There are also implicit languages still under development.

SIMULATION

Visual simulation involves creating a computer model in software of the robot, the workplace and the material. The robot is programmed at an off-line programming workstation with a sophisticated graphics terminal. The actions of the robot can then be viewed on the visual display unit. This will show, three dimensionally, in either wire frame or solid modeling, the actions of the robot or robots in real time. Thus this system can be used by the programmer to observe how the robot will operate in the real world.

ENGLISH INSTRUCTION

Dip the base of each workpiece into that vat of liquid. Hold it there for 30 seconds. Take it out and let the excess liquid drip off. Then hold it under the dryer for another 30 seconds.

LOCATION DEFINITION FILE

Point/tool	Type	X	Y	Z	Roll	Pitch	Yaw
WORKPC←INIT	abs	20.12	−25.94	−9.12	0.00	81.00	0.00
VAT	abs	1.34	−30.30	−9.36	0.00	81.00	−85.00
IN←VAT	rel	−3.6	0.0	−1.20	0.00	0.00	0.00
DOWN	rel	−1.90	00.00	−0.60	0.00	0.00	0.00
DRYER	rel	12.89	−29.08	15.99	0.00	−7.45	−40.00
RESTING←PLACE	rel	2.08	30.97	13.44	0.00	81.00	80.00
GRIP	tool	05.23	0.0	0.0	0.0	0.0	0.0

Instruction file

```
 10 SET←TOOL TO GRIP                      ! Use grippers with this program
 20 SET←SPEED TO 20                       ! Set the arm to move at 20 inches/sec
 30 START:                                ! A statement label
 40 PRINT "HOW MANY WORKPIECES?"
 50 INPUT TOTAL                           ! Operator to specify how many
 60 !                                       workpieces are on the table.
 70 !                                       The tables are not always full.
 80 DEFINE GRIPPER AS D-OUT (21)          ! Assign the name 'GRIPPER" to
 90 LET GR←CLOSE=1                        ! digital output port number 21
100 LET GR←OPEN=0                           and define the variables
110 !                                       GR←OPEN and GR←CLOSE.
120 DEFINE DRYER AS D←OUT (22)            ! Assign the name "DRYER" to digital
130 LET DRYER←CN=1                        ! input number 22 and define the
140 LET DRYER←OFF=0                       ! variables DRYER←ON and DRYER←OFF.
150 !                                     !
160 LET WORKPC=WORKPC←INIT                ! Initialize the point "WORKPC"
170 LET GRIPPER=GR←OPEN                   ! Issue I/O signal to open gripper
180 !
190 !                         BEGIN LOOP
200 !
210 FOR I=1 TO TOTAL
220 MOVE TO WORKPC                        ! Move gripper over workpiece
230 MOVE TO DOWN, IN FRAME WORKPC         ! Get ready to grasp the workpiece
240 LET GRIPPER=GR←CLOSE                  ! Grasp workpiece
250 MOVE TO VAT
260 MOVE TO IN←VAT, IN FRAME VAT          ! Place workpiece in vat
270 DELAY 30                              ! Let workpiece soak
280 MOVE TO VAT                           ! Raise out of liquid
290 DELAY 10                              ! Let liquid drip off
300 MOVE TO DRYER
310 LET DRYER=DRYER←ON                    ! Issue I/O signal to turn dryer on
320 DELAY 30                              ! Dry the workpiece
330 LET DRYER=DRYER←OFF                   ! Turn the dryer off
340 !                                     !
350 MOVE TO WORKPC                        ! Place workpiece back on
360 MOVE TO DOWN, IN FRAME WORKPC         ! table.
370 LET GRIPPER=GR←OPEN                   ! Let go of workpiece
380 !
390 LET WORKPC.X=WORKPC.X+2.5             ! Redefine the point WORKPC to define
400 !                                       the next workpiece position.
410 NEXT I                                ! Return to beginning of loop
420 !                                       to pick up next workpiece
430 !
440 !                         END LOOP
450 !
460 MOVE TO RESTING←PLACE                 ! Move the arm away to a waiting
470 !                                       position
480 !
490 PAUSE "Task is complete. Start again?"  ! Operator to decide.
500 GO TO START
```

Figure 12.18 A simple program in AR Basic. (Courtesy of Reflex Automated Systems and Controls Ltd)

Robot simulation has also a number of other uses:

1. It can be used as an aid in robot design by reducing the need for physical prototypes and allowing simulated trials to be carried out. The robot design development time and cost can be reduced.
2. The prospective robot purchaser may utilize it to help choose a suitable robot. Robot selection would be facilitated by being able to visualize the robot geometry and work envelope, estimate cycle times, and consider various other parameters such as joint movement limitations. Assuming the simulation package contains a library of a number of robot models the buyer can therefore make comparisons between them before making a final decision.
3. Robot workcells can be simulated and modified to achieve the most efficient workcell layout for each application. Potential collisions can be detected and robot arm movements optimized.
4. Simulation is an efficient method for education and training in robotics. This allows the student to program the robot and experiment with layouts without the dangers inherent in using the physical machine.

Although much research is still being carried out in this area a number of commercial packages are available, some of these are now briefly noted.

Robographics. This is a facility produced by Computervision to run on their CAD systems. As explained above this is a typical system that allows various permutations of robots, end effectors and workcell arrangements to be considered by simulating them graphically. There are three separate modules: kinematic, path generation and output generation. The kinematic module is a mathematical model that enables the geometric configurations and movements of a particular robot to be analyzed. A separate model is required for each different robot and library of various models can be built up. The path generation module programs the robot and displays and validates the programmed path. The output generation module takes the above information and generates output which can be directly down loaded to a robot controller.

By using this simulation package the robot installation designer and programmer need have little knowledge of the language associated with the particular robot. This is because once the robot motion is defined, the program from the simulation system can be directly output to the robot through the postprocessor for that robot model. This means that the same programmer can, with ease, try out various robots that use different languages, for highest efficiency in different work situations. Figure 12.19 illustrates a wire frame robot simulation which shows the tool path of the end effector. Figure 12.20 illustrates how color shading can be added to achieve a solid representation of a robot installation. Both displays would be animated.

Robot-Sim. This is from the Calma subsidiary of General Electric and is integrated with Calma's Design, Drafting, and Manufacturing (DDM) software. Three-dimensional models of robots and peripherals can be created, modified and manipulated. It has a dynamic capability that can predict the force and torque in joints and arm deflections due

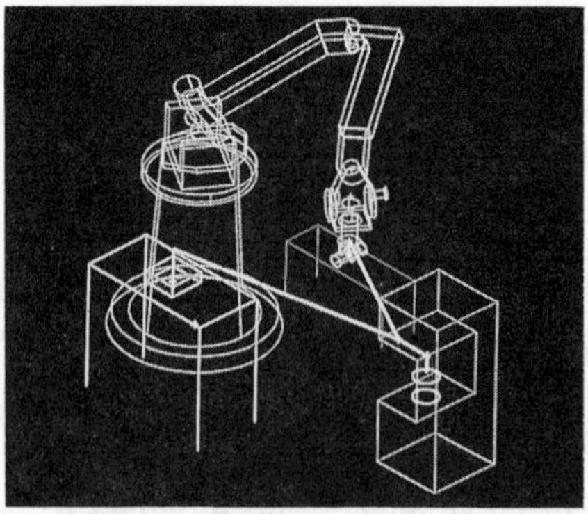

Figure 12.19 'Wire frame' robot simulation showing path of end effector. (Courtesy Computervision)

Figure 12.20 Color shaded animated simulation of robot work cell. (Courtesy Computervision)

to payload. As the robot motion program cannot be passed directly to the robot, the text output can be used by the programmer as input to another package Robot-Pro. This in turn produces output that can be postprocessed into a robot controller language.

Robocam. A package from Silma Inc. offering several different software modules. It has a library of standard robot models that can be inserted into simulated workcells.

MEMS (Mechanism Modelling System). Offered by Intergraph in collaboration with GMF Robotics Corporation. It allows robot program simulation, editing and verification in colored, three-dimensional animation. Collisions can be identified, cycle times estimated, and postprocessing of the output into a robot control language is possible.

Other popular systems are Catia developed by Dassault Systems, Grasp from BYG Systems Ltd and, one of the first commercial simulation packages, Place from McAuto a division of McDonnell Douglas Information Systems.

In conclusion, although not directly related to robot programming, it is worth noting that as well as simulating the robot and its immediate environment, it is also useful to be able to simulate the entire factory. Various packages are available for this purpose that assist in plant layout, identification of production bottlenecks, and generally facilitate achievement of the optimum organization for a particular manufacturing system. Examples of these are See Why by Istel, and Mast by Citroen Industries.

III

APPLYING ROBOTS

13

APPLICATIONS

INTRODUCTION

From the analysis of industrial robot applications included in Appendix 1 it can be seen that robots are already employed in a wide variety of tasks. As their capabilities increase and their relative costs decrease so their potential range of applications broadens.

In a text of this length it is not possible to cover every application; nor is it necessary, as topical case study information can be gleaned from magazines and journals such as those mentioned in the bibliography in Appendix 2. However, some applications will be considered briefly in this chapter.

It may be noted that assembly has been given extra attention. The development of faster robots specially designed for assembly, and the suitability of the assembly process for automation, has led to a steady increase in robotic assembly applications. For example, in the UK the proportion or robots in assembly grew from 3 per cent in 1982 to 10 percent in 1986. One particular area that is developing is that of electronic product assembly, and this is therefore considered in some detail.

ROBOTS IN INJECTION MOULDING

The injection moulding process is probably the most widely used and economic process for producing plastic components. There are several variations of this technique but the most common one is now described; refer to Figures 13.1(a) – (e). Granulated plastic or plastic powder falls from a hopper into a heated cylinder in which moves a reciprocating screw plasticizer. This screw acts as a combined plasticizing and injection unit. As the plastic is fed along the cylinder it passes along the zones shown. After passing through the feed zone it passes through the compression zone where the reduction in the screw flight depth forces the plastic to compress. The work done here is converted to heat as the plastic shears so making it a semifluid mass. Finally, in the metering zone additional heat is applied by conduction from the heating bands on the cylinder outer surface. There is a chamber just in front of the screw, as this becomes filled it forces the screw back thus tripping a limit switch. This activates an hydraulic cylinder that forces the screw forward and injects the fluid plastic into the closed mould. A one-way valve prevents the plastic

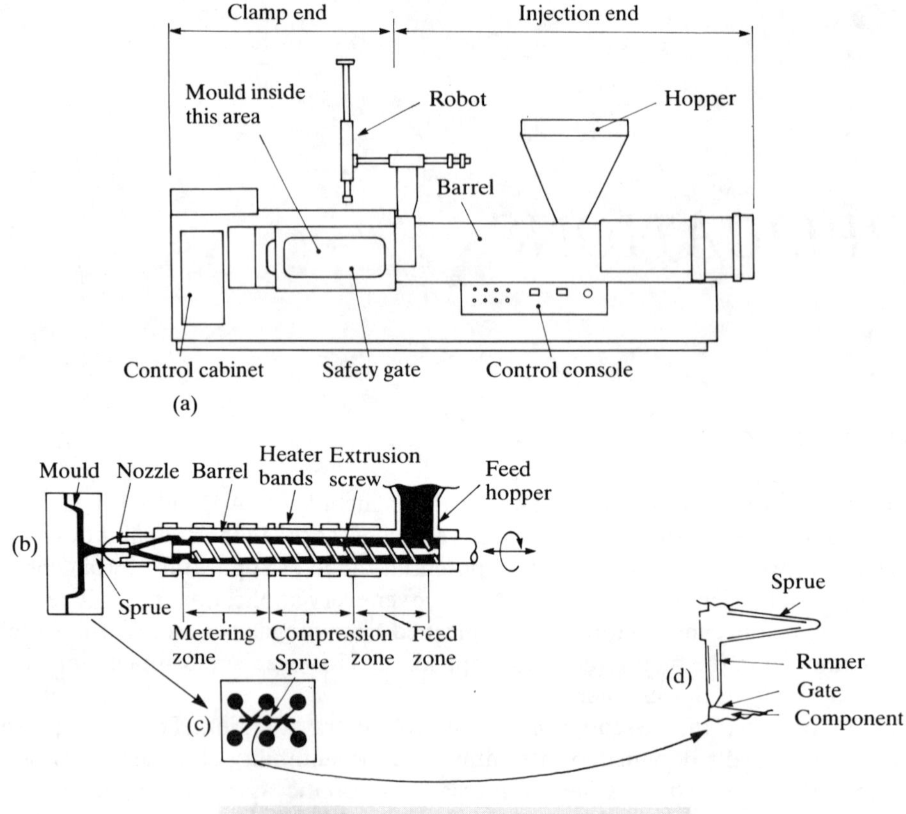

Figure 13.1 Injection moulding: (a) injection moulding machine; (b) principle of operation; (c) and (d) typical mould cavity layout; (e) typical injection moulding 'robot' (courtesy Haynes & Fordham Ltd)

flowing back under pressure into the screw chamber. The moulding tool itself is composed of at least two parts and its function is to input the desired shape to the material and cool the formed component. It opens along the mould parting line to allow the component to be ejected. Within the mould the molten plastic flows into the die through a sprue. It then travels along a runner system and through a gate into the actual component cavity. This sprue, runner and gate system is usually cooled along with the component and thus has to be removed from the finished articles either by hand or mechanically.

From the statistics of robot application, injection moulding is apparently an important area, but in fact very few of these robots are servo controlled or resemble the type of robot with which this book is predominantly concerned. For example, of the 266 robots installed in the UK in 1984 and 1985 only 5 were servo controlled devices. These non-servo-controlled robots are basically pick and place units under microprocessor control for sequencing. They are usually mounted above the moulding machine and clamped rigidly onto the machine die plate. They are of cartesian coordinate construction using straight line vertical and horizontal movements.

A typical sequence of movement would be as follows:

1. The injection moulding machine opens, ejecting the finished components complete with runner system.
2. The robotic unloader arm moves vertically downwards, grasps the component, then retracts to allow the die to close and allow the moulding of new components to commence.
3. The arm moves horizontally to either the front or the rear of the machine to deposit the finished components onto a conveyor, chute, or pallet.

It is possible to provide the arm with a 'nipper' to remove the sprue and runner system while traveling, the sprue and runner being released for disposal before depositing the finished component. The control systems of these robots usually allow interfacing with other pieces of equipment for safety and increased efficiency. Various types of end effector are used, from simple grippers to specially designed vacuum grippers for transparent lenses.

As mentioned above, the injection moulding process is also amenable to the application of servo controlled robots although this is still uncommon. These robots are much more expensive and therefore more economically justified when utilized to service two machines. They can also carry out more complex tasks such as die servicing.

The advantages of applying robots to injection moulding are: decreased moulding cycle times giving higher production; more consistent cycle times giving better quality components; and removal of human operators from a hazardous and often unpleasant environment. Hazards can arise from the moving tool platens with clamping pressures up to 5000 tonnes. Also the plastic is injected at temperatures of up to 300°C and unexpected escape of this at high pressure is obviously dangerous. In addition the environment can be unpleasant due to the heat generated and the sometimes noxious fumes produced. It may be noted in conclusion that die casting machine servicing, although not considered here, is a very similar application. In the high pressure die casting process molten metal is injected into a die in a comparable manner to that of injection moulding, although the machine and machine operation are themselves quite different.

ROBOTS IN SPOT WELDING

This is the most common application of servo controlled industrial robots mainly because of their widespread use in the automotive industry. Motor car manufacturers have applied robots in spot welding to improve productivity and release human operators from what is a tiring, awkward and unpleasant task. Spot welding is the simplest form of resistance welding and it is worthwhile at this point to consider briefly the principles of welding.

Welding is a process in which two materials, in this case metals, are permanently joined together through localized coalescence. This results from a combination of heat, pressure and metallurgical conditions. It is possible under the correct conditions to achieve welds with high temperature and no pressure, or normal temperature and high pressure, or varying degrees of each. The two types considered in this chapter are spot welding and arc or fusion welding. In spot welding heat and pressure are used to produce coalescence. In arc welding heat only is used to melt the material to be joined and the additional material used to fill the joint.

A typical arrangement of electrodes and work for spot welding is shown in Figure 13.2(a). Electric current is passed across the electrodes and heat is produced as a consequence of the electrical resistances of the workpieces and the interfaces between them. A specific amount of pressure is applied at the beginning of the weld cycle to hold the workpieces in contact. This controls the electrical resistance at the interface until the proper temperature is achieved when it is then increased to aid the coalescence. In robotic spot welding the electrodes are contained within a spot welding gun mounted on the robot wrist as shown in Figure 13.2(b). The electrodes are water cooled and pressure is applied via a pneumatic or hydraulic cylinder, they also have their areas reduced at the tips to produce welds typically between 1.5 mm and 13 mm in diameter. The electrodes are first closed on the work, pressure is applied and current flows in a controlled manner to produce a weld. This weld consists of a nugget in the location shown in the previous figure. The electrodes are then opened and the welding gun is moved by the robot to a new location. The guns are massive and unwieldy for human workers and the cables carrying the current, of up to 1500 A, are thick and heavy, but these factors present no problems for robots.

The manipulation of the spot welding gun to achieve the welds in the appropriate position, and hold the gun in the correct orientation, demands a robot with full six degrees of freedom. It also requires sophisticated control to produce these welds on a fabrication line that may be moving continuously.

Most major automotive industry factories throughout the world have now installed robotized assembly lines for spot welding. The reasons are numerous. This industry, because of its high volume and high labor costs, is always eager to automate wherever possible. Due to increasingly discerning purchasers, cars are required in many permutations of style, number of doors, colors, trims, etc. This means that although the total volume is of mass production proportions, say over 1000 completed cars per day from one plant, a great variety of derivatives are required within these numbers. Reprogrammable automation is therefore very attractive. Robots can achieve consistently more precise welding of higher quality than a human operator trying to cope with the unwieldy nature of the welding gun. This should result in fewer welds coupled with a

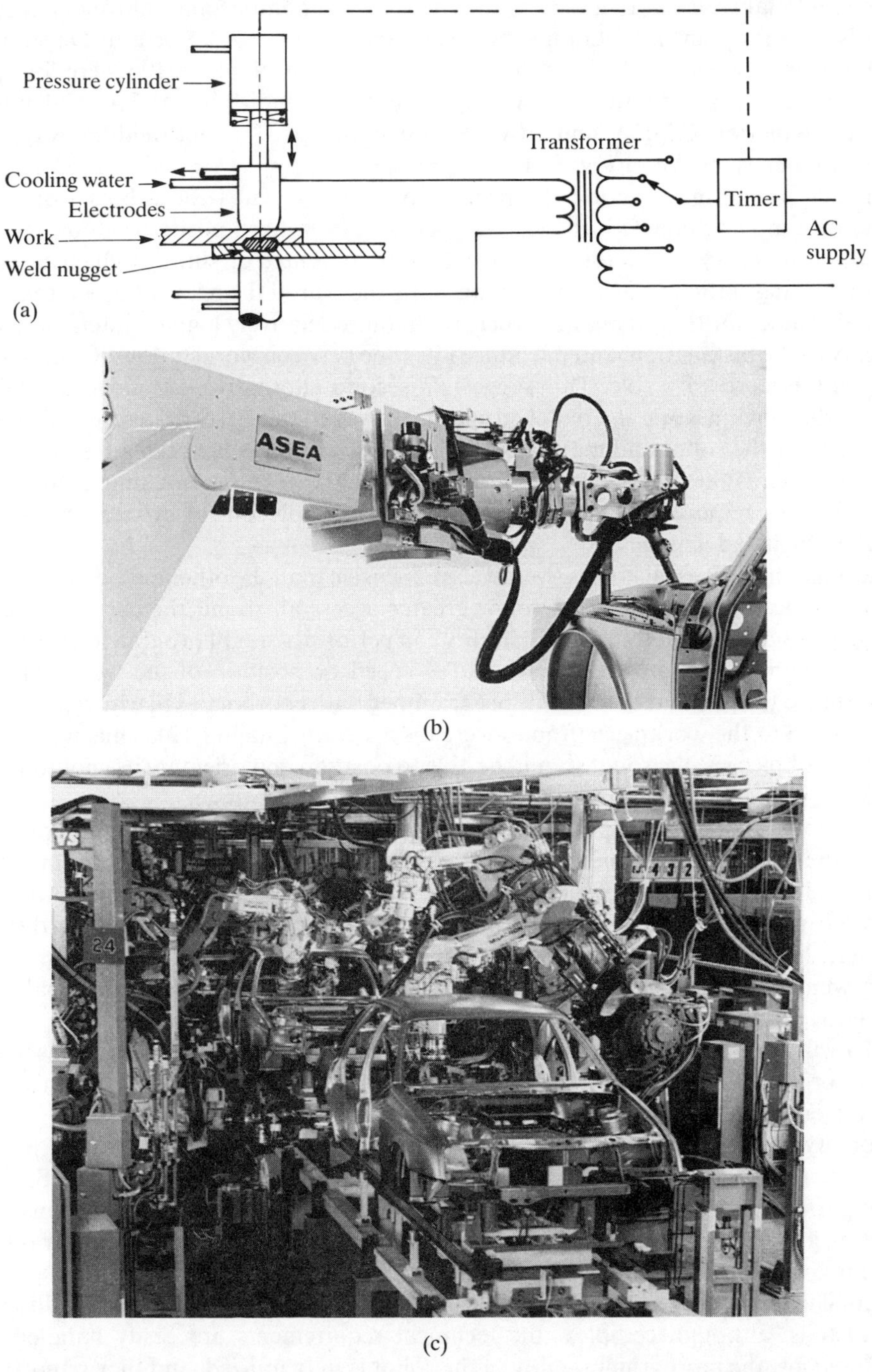

Figure 13.2 (a) Basic configuration of spot welding system; (b) spot welding gun mounted on ASEA IRB 90S/2 robot (courtesy ASEA); (c) typical spot-welding line using Cincinnati Milacron robots (courtesy Cincinnati Milacron)

higher strength fabrication. Evidence of the suitability and importance of robots in spot welding is shown by an installation some years ago for the Ford Sierra in Dagenham where 120 robots were installed for the body construction of which 106 were for spot welding. Generally at the time of writing, robots produced by ASEA, Cincinnati Milacron, Unimation, KUKA, and NIMAK, are all popular for spot welding. A typical spot welding line is shown in Figure 13.2(c).

For automobile manufacture the production line on which the robots work will probably be a moving one with the robots operating from fixed positions alongside the line. One alternative system to this is to have the work mounted on shuttles which move to each spot welding station in turn and remain static there until the relevant operations are completed. These shuttle systems are expensive, require relatively high maintenance, add complexity to the installation, and introduce idle time between workstations. Continuous working is therefore advisable. This suggests a second alternative, i.e. move the robot parallel to the moving work at a matched speed. However, this necessitates an additional drive system for the robot and may cause interference problems between adjacent robot workstations. Thus the first technique, termed stationary base line tracking, is the most popular. It does require extra sophistication of the robot control system but this is outweighed by its advantages.

Stationary base line tracking is overall less expensive than the other methods, it incurs less maintenance, and it allows the robot greater access all round the car body being welded. The robot with full tracking capability can perform a taught program on the work moving through its station irrespective of the speed or position of the work. This is achieved due to the fact that the robot is programmed with coordinates in which the frame used is relative to the workpiece (frames were discussed in Chapter 12). Thus no matter where the workpiece is, the robot should be able to position and orientate its spot welding gun about a particular point in exactly the same manner each time.

The robot knows the workpiece position and velocity by analyzing data from process sensors in the work area. One piece of information comes when the particular car model is identified as it passes a sensor, e.g. a pattern of microswitches or a laser reader. Another microswitch would be triggered and a signal sent to the robot controller when the car body enters the tracking window of the robot, this signal may also be used as a datum. The tracking window is that work envelope within which the robot is physically capable of maneuvering and positioning the welding gun in the required manner. The robot controller will then know where the spot welds have to be put at any point in time as it will also monitor feedback from a further sensor attached to the conveying system. This may be a shaft encoder, tachogenerator, etc. providing information on the speed of the conveyor. By comparing the conveyor speed, the cycle initiation signal and the points stored in memory, the robot will know exactly where to go to carry out its welds. Special branching routines can be built into the program to cope with anomalous situations, e.g. the welding gun becoming stuck to the material or a car body moving prematurely out of the tracking window.

In conclusion, spot welding is a job ideally suited to the servo controlled industrial robot. That is, although complex, the technical requirements are easily handled by existing systems, the reprogrammability of the robot is fully utilized, and they remove an unpleasant and difficult task from human workers.

ROBOTS IN ARC WELDING

Arc welding is a job that, when done by humans, requires considerable training, skill and concentration. It has been said that it is more an art than a science as there are often so many variables to be considered in order to achieve a good quality weld. It can also be a very unpleasant job due to the heat, radiation and smoke produced, which may necessitate the worker being isolated from the rest of the workforce. It is a job that requires well trained, skilled workers; however, it is becoming increasingly difficult to find people both capable of that skill and willing to work under those conditions. Also, demands for improved productivity mean that 'arc on' times, i.e. the amount of time welding is actually taking place as opposed to the idle time for refreshments and relaxation, require to be increased. Thus there is a combination of factors indicating the use of industrial robots for this application.

There are many types of arc welding but the technique most suited to robotization is gas shielded metal arc welding (GSMAW) and a brief review of this method is now given.

GSMAW is commonly referred to by a number of names depending on the type used, e.g. MIG, CO_2, MAG, etc. In the process an electric arc is struck between the electrode, which is the filler wire, and the workpiece. This arc is hot enough to melt the two components to be joined at the seam and the filler wire. Thus a weld pool is created of molten metal. The welding torch follows the seam moving the weld pool along with it and leaving a cooling solidified joint behind. The arc and molten metal are protected from the atmosphere by an externally supplied gas flow. The shielding gases may be inert such as argon and helium, hence the term 'MIG' for metal inert gas welding, or active such as oxygen or carbon dioxide giving the term 'MAG' for metal active gas welding. In the 1950s it was found that CO_2 made a very effective and inexpensive gas shield on its own. Thus today welding of ferrous metals is carried out using, for example, 100 percent CO_2 for steels, and mixtures of CO_2 with argon or helium for such as stainless steels.

Therefore in GSMAW the parent metal and bare wire are melted by the arc to form the weld pool which is protected from atmosphere by gas fed coaxially with the wire through the welding gun nozzle. The wire is fed at a constant speed selected to give the required current. The arc length is varied by adjusting the voltage of the power source. This technique is therefore suited to automatic control and integration with an industrial robot system. Figure 13.3(a) shows the basic principles of the MIG welding process.

The use of robots in arc welding first occurred around 1974 with commercial availability about 1978. However, it is only recently that the gathered experience, and improved control systems, coupled in some cases with sensing, have made the process economically attractive. A robot arc welding system will usually comprise the robot arm, the work handling unit, the welding torch, the welding unit power source and wire feed unit, the robot power source and control cabinet, and the system control panel. For welding cells, as opposed to moving lines, the robot and the work handling unit are often mounted on the same base to provide constant stability and accuracy. The work handling unit may have a number of axes actuated by servo motors, these motors will have their movements coordinated with those of the robot providing a multi axis system, a workholder of this type is shown in Figure 13.3(b). Typical installations are shown in Figures 13.3(c) and (d).

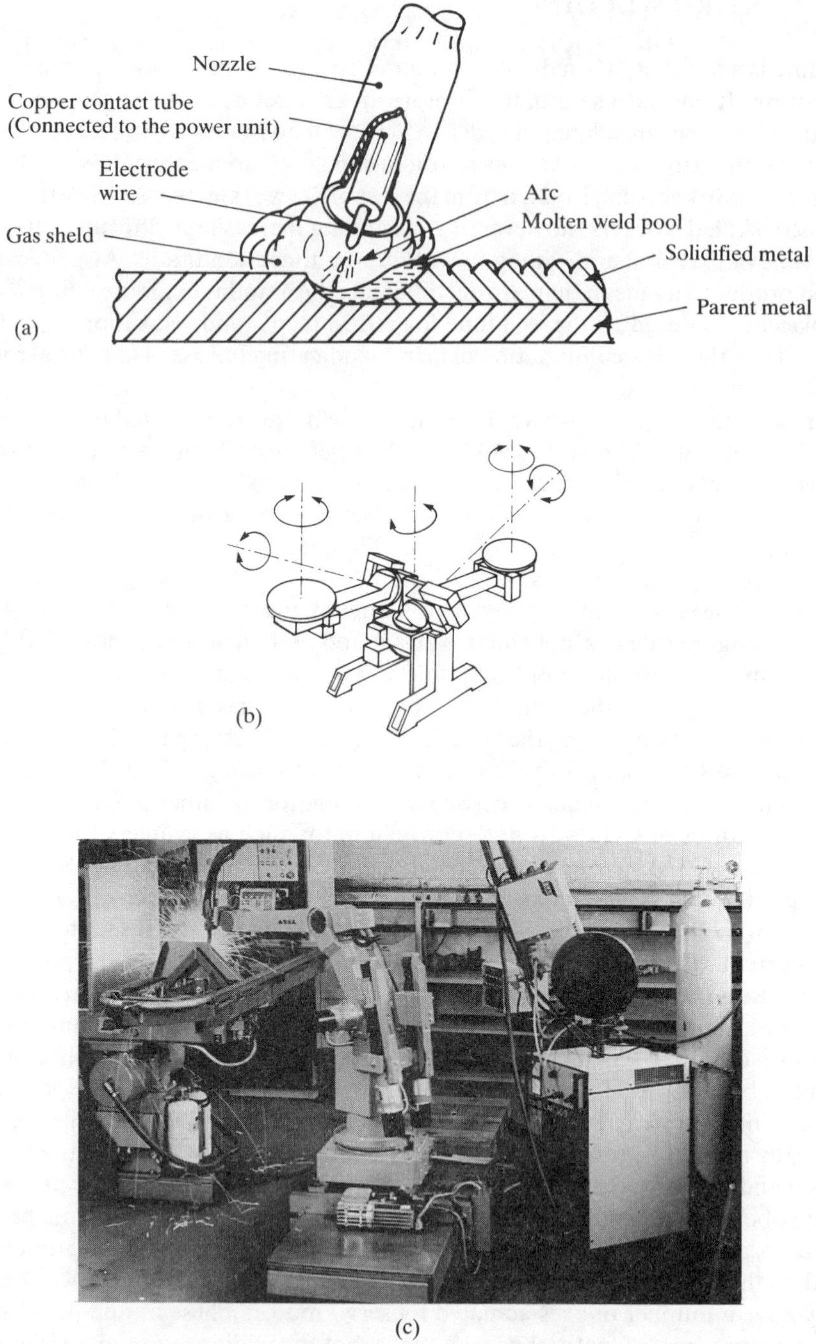

Figure 13.3 (a) Principle of MIG welding process; (b) example of servo-controlled workholder (courtesy ASEA); (c) and (d) typical arc welding installations (courtesy ASEA and Cincinnati Milacron)

(d)

Figure 13.3 continued

Although in many cases robotic arc welding may be identified as a desirable investment, it is not a simple application to implement. The MIG process normally demands high precision both in the welding robot and in the work presentation. For example, if a 1 mm diameter filler wire is being used then the net accuracy of the seam placement relative to the wire should be + or − 0.5 mm thus necessitating the use of a robot with considerable precision. Often, even if the robot is in itself adequate, the work presented to the robot does not conform to that anticipated. For example, the parts being joined may be out of tolerance, or the work may be poorly fitted together, or due to heating of the work as welding progresses the job may move, in each of these cases the seam will be displaced from the anticipated position. To overcome this some sort of seam finding and tracking facility is required. This facility is now available either as part of the robot system or as 'add-on' equipment.

Seam tracking for robotic arc welding can be implemented in a number of ways. For example, one system uses the robot controller to oscillate the welding torch over the seam while welding. As it does this the changes in the arc current are monitored by the controller. The robot's computer uses this information to continually correct the robot's position in response to these changes and modifies in advance the ensuing points in the program. Other systems use lasers to provide a greater flexibility. They allow seam finding as well as seam tracking to be carried out automatically. They are suitable for use when welding thin sheet and they can measure errors and changes in seam gap, height, and lateral position.

The ASEA Seam Finder system is shown in Figures 13.4(a) and (b). The following information has been supplied by ASEA. As can be seen the system is mounted onto the MIG welding gun. It consists of an optical (laser) sensor and a microprocessor which evaluates the sensor signals and transmits the results to the adaptive functions in the robot control system. When searching for the seam, the joint is defined in three dimensions and the welding gun is positioned simultaneously. A complete search in three dimensions and location of the gun takes approximately 1.5 seconds. A search in two dimensions, however, is often sufficient. The search is performed without actuation of the arc. The robot is programmed with information relating to the material thickness and the search type, edge or overlap joint, or fillet joint, etc. The sensor is equiped with a laser diode which emits infrared radiation. Figures 13.4(c) and (d) illustrate the principle of operation. For range finding the subject is illuminated with the laser beam, the point illuminated being registered on a sensor via a system of lenses. The displacement registered on this sensor depends upon the distance between the subject and the sensor. When searching for a lap joint or an edge the sensor is drawn over the joint. A search stop command is generated when the step in the distance measured at the joint is registered by the sensor's microprocessor system. This principle is adapted for other types of joint.

Another system is the Meta Torch the following information being supplied by Meta Machines. This is a compact device mounted on the robot arm and it consists of two lasers, a solid state camera, and the MIG welding torch, see Figure 13.5. The lasers emit stripes of structured light which are projected onto the seam approximately 16 mm in front of the arc. The image formed by the stripes on the workpiece is picked up by the CCD chip of the camera via an optical filter system which removes much of the glare of the welding arc. A microprocessor-based compiler analyzes the positional information and signals the robot to modify its programmed path accordingly. The system also interfaces with the other equipment and continually adjusts the weld parameters such as wire feed rate, speed and weld voltage to be controlled to achieve optimum quality.

ROBOTS IN ASSEMBLY

To create a finished product a number of manufacturing processes must be carried out. Firstly, the raw material used undergoes a primary process such as casting, rolling, forging, etc. Next the component is subjected to secondary processes such as machining or further working, e.g. piercing and blanking components from sheet metal. It is also possible to make a component at this level from the raw material by using such processes as injection moulding, die casting, or powder metallurgy. Finally, individual components are brought together in a specific manner to form a finished product. This last stage is the assembly process. Components such as transistors, ICs, resistors and unpopulated PCBs can be brought together to form finished PCBs which are in effect subassemblies. These subassemblies occur at many different levels; for example, in a motor car a spark plug is a subassembly. This plug becomes part of the larger engine assembly, which in turn becomes a subassembly for the completed car.

Components can be joined together by a vast range of techniques, the most common of which are as follows, the particular type chosen depends upon the application and assembly process used.

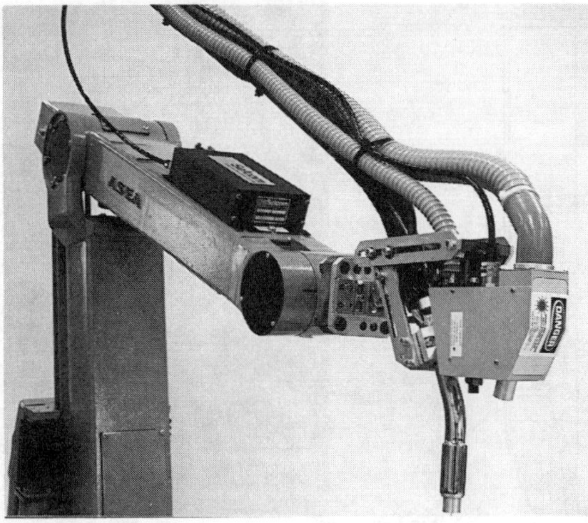

(a)

(b)

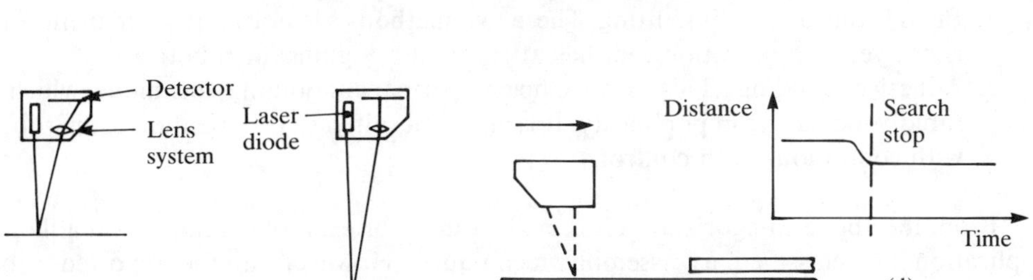

(c) (d)

Figure 13.4 (a) ASEA Seam Finder; (b) ASEA seam finder applied; (c) range finding; (d) lap joint/edge searching principle. (Photographs courtesy ASEA)

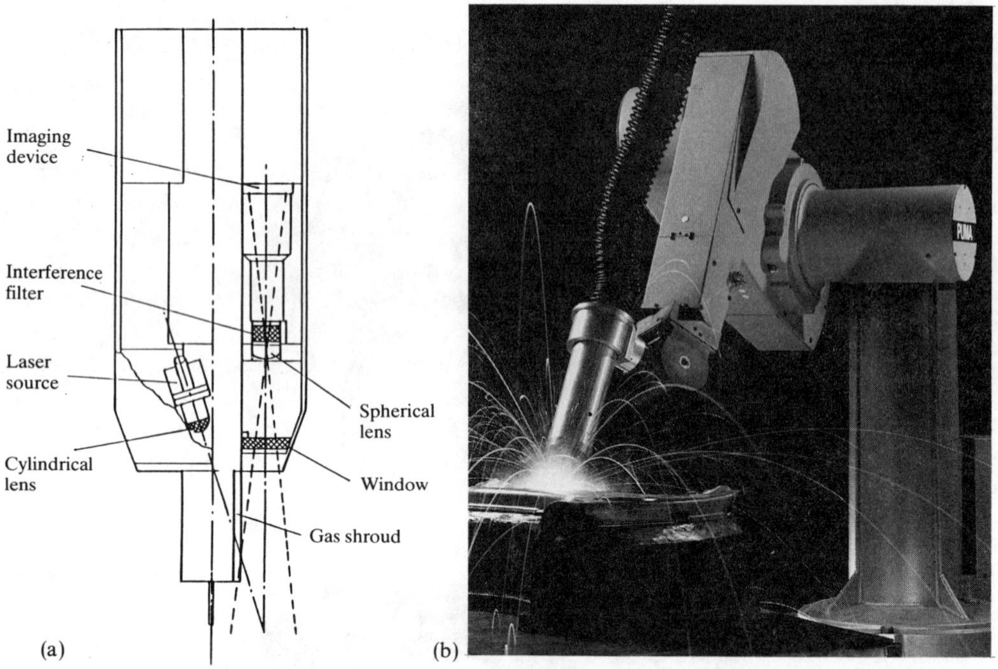

Figure 13.5 (a) MetaTorch; (b) MetaTorch in operation. (Courtesy Meta Machines Ltd)

1. Mechanical fastening. This includes permanent methods such as riveting, staking, crimping and bending. It also includes semipermanent methods utilizing studs, bolts, nuts, pins, split pins and screws of various types. Industrial robots have been applied successfully in this area.
2. Soldering and brazing. These are essentially semipermanent methods as the joined parts can be separated by reheating. The application of robots in this area has normally been confined to the application of solder in electronic assembly, although even for this other techniques such as flow or wave soldering are usually more economic.
3. Welding. The main variations of this permanent joining process applicable to robotics, i.e. spot and arc welding, have already been covered.
4. Press fitting and shrink fitting. These two methods are normally permanent. This is not yet an application that has attracted any significant robotization.
5. Adhesive bonding. This is an efficient permanent joining technique which is rapidly increasing in popularity. It is also well suited to robotization using robots with continuous path control.

From the above categorization it is evident that, for example, welding and adhesive application are classified as assembly techniques. However, in the reported robot application distribution statistics these are usually considered as separate areas. Thus, in this text, assembly in relation to industrial robotics will be limited in its scope. Here assembly will be taken to mean the bringing together of discrete components or subassemblies and combining them in a specified manner to form a finished product. It will not include categories (2)–(5) above even though in the fullest definition they are 'assembly' processes. An exception to this would be if, say, a plastic product was being assembled and there were particular points in the process where two components had to be joined by ultrasonic welding or a spot of adhesive, etc.

The market for assembly robots has long been recognized but until recently the high speeds and accuracies required were unobtainable at reasonable cost. However, in the past few years the use of industrial robots in assembly operations has accelerated. This is because software and hardware technology has now advanced to such a level as to make the construction of robots specifically for assembly operations commercially attractive.

Generally speaking, assembly costs can account for around 50 per cent of total production costs, and between 20 and 25 per cent of total manufacturing labor costs. Thus if carrying out a Pareto analysis of costs it is evident that assembly is an area to be tackled. When further analyzing the costs it is apparent that many arise from manual labor carrying out repetitive tasks – ideal work for a robot. The tasks are often complex, however, and this can give rise to problems. Other costs arise from scrap and rework due to human error or ingress of dirt, etc. This can also give rise to poor product quality which, if not overcome by expensive additional inspection, can incur higher costs to the firm in dealing with customer complaints, returned goods, warranty claims, etc.

In the past the method of overcoming the problems associated with manual labor was to design and build special-purpose assembly machines dedicated to a specific task and product. These machines were mechanically controlled by cams or utilized pneumatic actuators controlled by pneumatic or hardwired logic. The problem with these machines was that they were very expensive and they became useless as soon as the product or assembly they were producing became obsolete. The shorter life cycles of many products now make it very difficult to justify the investment in one of these machines. Some piece of equipment that could be easily modified to suit each new product was needed, and it is the industrial robot that has satisfied that need.

Thus assembly is one application that presents a variety of opportunities for robot justification, i.e. a reduction in labor costs, improvements in product quality, removal of monotonous jobs, and the ability to spread the investment cost over more than one product. Given then that a typical assembly robot now has a precision of 0.05 mm, a total cycle time of less than 5 s, and can now be purchased at a reasonable cost, assembly by robot in industry will increase rapidly.

Assembly robots

The potentially large market for assembly robots and the special requirements of the process have led to a number of specially designed machines. Assembly work is usually carried out in a quite small work area at high speed. Good design of the product should enable it to be assembled on a 'stacking' basis, although the robot precision required will probably remain high. The robots discussed below are typical of those presently used in industry.

The Unimation Puma™

This robot is worth a special mention as it was one of the first servo controlled robots specifically designed for assembly work. It is an electrically driven revolute type robot and an example is shown in Figure 13.6. The Puma robot arose from the General Motors' Programmable Universal Machine for Assembly project in the late 1970s in which Unimation Inc. was a GM subcontractor. It was originally designed to occupy about the same workspace as a human and to be able to lift loads of up to around 2 kg, GM having estimated that 90 per cent of their motor car parts weighed less than 5 lb. It has also been a very popular robot for research purposes due to its versatility. The robot is now produced in a variety of sizes for use in many different applications.

Figure 13.6 Unimate Puma™ 200 assembly robot. (Courtesy Unimation)

The Selective Compliance Assembly Robot Arm (SCARA)
This is an extremely popular and efficient robot configuration for assembly. It is of simple four axis construction as shown in Figure 13.7. The robot design was originally the result of work carried out at Yamanashi University in Japan led by Professor Makino. The work was supported by a research consortium comprising a number of major Japanese companies. The design is such that a peg being inserted vertically downward into a hole is able to move horizontally by the amount of lateral compliance in the arm. Thus if a peg is lowered toward a chamfered hole and is slightly out of the exact required position, then the compliance will allow a slight lateral movement to allow insertion to take place without wedging. The arm is thus designed to be very rigid in the vertical direction, i.e. its tilting compliance is small, and have a higher degree of compliance in the horizontal plane. The robot is therefore ideally suited to the stacking type of assembly which should result from good design and can constitute 80 per cent of all assembly work. The partially cylindrical work envelope is also well suited to the assembly task.

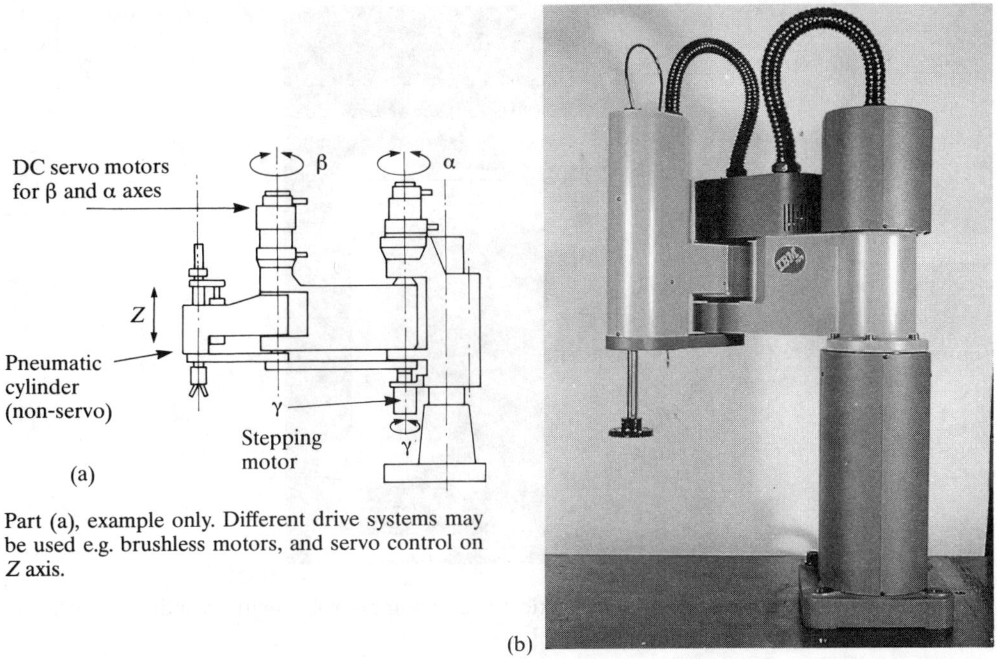

Figure 13.7 SCARA robot: (a) principle of operation; (b) IBM 7575 SCARA robot (with servo control on Z axis (courtesy IBM)

In the basic design shown in Figure 13.7(a) the alpha and beta axes are driven by DC servo motors and position the component in the horizontal plane. The gamma axis, i.e. wrist rotation, is driven by a stepping motor via a toothed belt. It is this axis that orientates the component in the horizontal plane. The vertical Z axis provides the insertion action. In this case it is a pneumatic non-servoed action controlled by manually set mechanical stops. A number of companies now sell SCARA type robots, e.g. the IBM 7575 robot shown in Figure 13.7(b) where all four axes are servo driven. Many variations and improvements on the basic design are constantly appearing.

AdeptOne assembly robot. This SCARA type robot shown in Figure 13.8 uses a high level language and control system, i.e. VAL II at time of writing, and utilizes direct drive motors as described in Chapter 8. The robot thus combines a very simple construction with sophisticated control. It is very fast and has an almost completely cylindrical work envelope. All axes are servo controlled. In common with most newer robots the software associated with the machine also allows for easy interfacing with other equipment and in particular vision systems. It allows networking (discussed later in Chapter 17) and real time trajectory computation in response to signals from external sensors and other devices. Another feature is that each robot is provided with its own individual mathematical model, or signature. This means that the exact dimensions between the axes of each machine are measured and recorded to allow precise control of the robot for off-line programming, thus improving the robot accuracy.

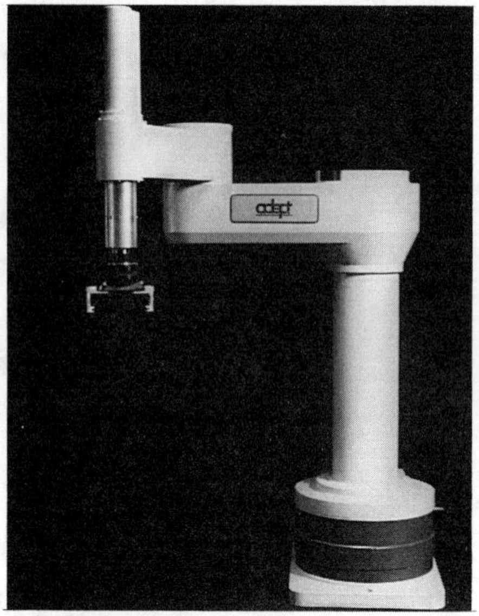

Figure 13.8 AdeptOne direct drive vertical axes assembly robot with cylindrical work envelope. (Courtesy Adept Technology Ltd)

ASEA IRB 1000

This unique 'pendulum' design, shown in Figure 13.9, was conceived specially for medium volume small parts assembly. The robot consists of a hanging arm with a vertical axis and two horizontal rotational axes intersecting at right angles. This gives the robot the pendulum type action. The wrist has roll, pitch and yaw functions. The robot construction reduces inertia allowing higher acceleration and higher amplification in the servo control system. It also provides high repeatability and versatility within a compact work envelope. A particular attribute of this robot, also now available on other robots, is the multiple gripper system that can be provided. Up to six individual double-acting grippers can be controlled simultaneously, each with its own tool center point or 'frame'. Standardized ancillary equipment is also provided such as a vision system, a screwdriving system, a component magazine, and other equipment such as conveyor systems, bowl feeders etc.

Microbo Castor

This is shown in Figure 13.10 as an example of a precision assembly robot. The robot is claimed to have an accuracy of 0.01 mm, a repeatability of 0.02 mm and a 1 kg maximum payload. An interesting aspect is its modular construction which allows it to be extended or modified to suit the particular application. It is driven by hydraulic and electric DC servo motors.

DEA PRAGMA arm

One of the early robots successfully designed for assembly work it is shown in Figure 13.11. It was originally designed for multi-arm operation and its own program-

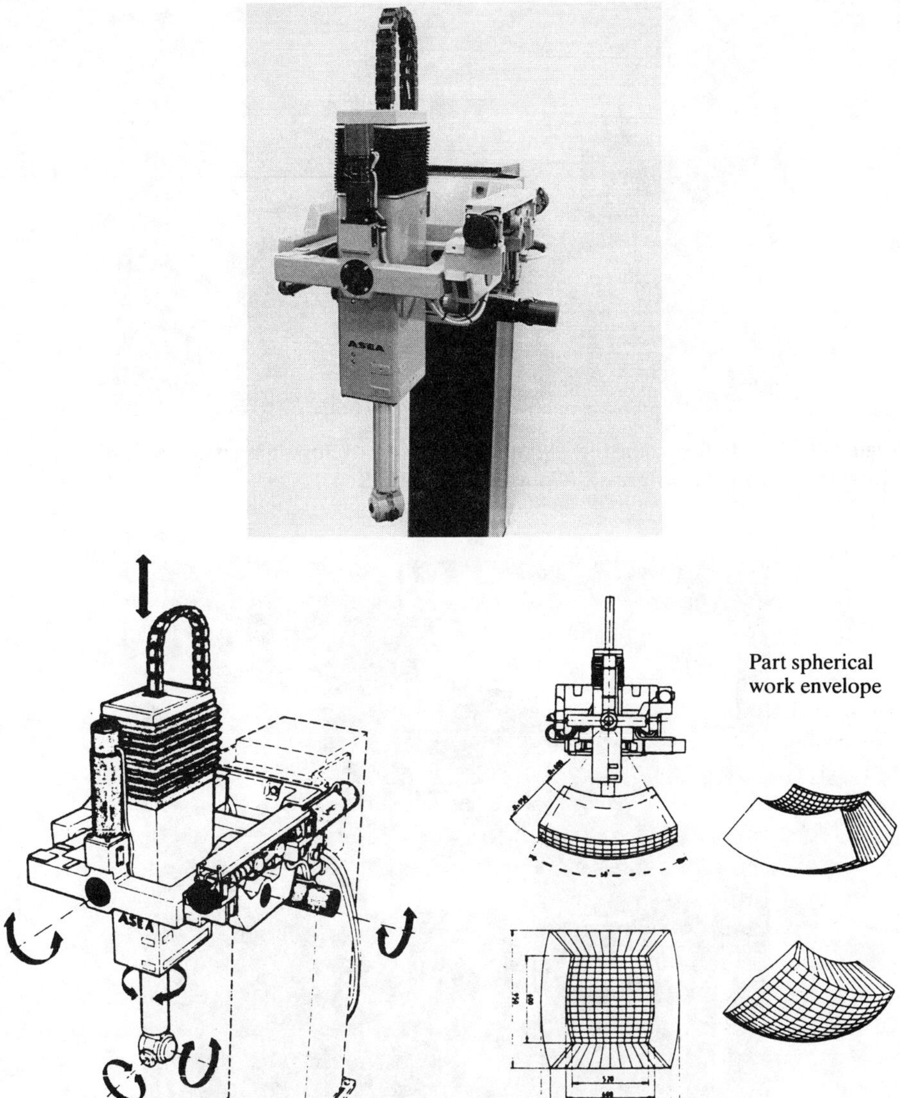

Figure 13.9 ASEA IRB 1000 'pendulum' assembly robot. (Courtesy ASEA)

ming language, Help, was specifically designed to allow parallel task programming and execution. Two arm operation allows decreased cycle times to be obtained by configuring the system to work in a manner similar to a two-handed human operator. Redundant operations such as simply holding the work, a job that can be done by a fixture, would of course be avoided. The rectangular work envelope of this type is popular with assembly robots as it provides a reasonable work area and yet allows access for parts feeding devices.

Figure 13.10 Microbo Castor cylindrical work envelope precision assembly robot (version shown has two arms). (Courtesy Microbo SA)

Figure 13.11 Pragma H3000 assembly robot with rectangular work envelope. (Courtesy DEA)

Figure 13.12 'Sysmo' assembly robot – a multi-arm gantry system with rectangular work envelope. (Courtesy Sysmo)

Other types

There are many other assembly robots but is is unnecessary in this text to pursue them further. There are quite complex systems available such as the IBM 7565 robot which is a gantry mounted, hydraulic drive, assembly system, shown previously in Figure 3.1. This utilizes the high level language AML and requires powerful computing facilities, the robot having sensory abilities and adaptive control capability. This is in contrast to the IBM SCARA types mentioned earlier which use a subset of AML and can be programmed from a personal computer. There are fewer revolute types now applied to new assembly tasks as their horizontal axis construction makes control and vertical movement more difficult, this puts them at a disadvantage compared to the SCARA, rectangular and cylindrical types. There will undoubtedly be many other assembly robots developed.

Designing products for robotic assembly

The main thrust of this section is the application of robots to assembly. However, it is at the product and component design stage that the greatest cost savings can be made. In fact, when a company decides to change from manual assembly to automation or robotics it often finds that the necessary redesign of the product and components provides the greatest savings. Thus even if the manual assembly was maintained after redesign, a reduction in costs would still be achieved. It is worth noting therefore some aspects of design to make robotic assembly more efficient.

There are two major aspects of designing for robotic assembly. First is the design of the overall product for efficient assembly by the robot. The limitations of the robot should be borne in mind here, e.g. its limited precision and the number of degrees of freedom available. Second, the design of the individual component parts to ensure efficient part feeding to the work area, orientation of the part, and suitable presentation to the robot end effector. Some points particularly relevant to robotic assembly are given below.

1. Design for 'stacking' assembly. For example, the provision of one component in the form of a stable base should be made. This should be suitable for quick location in a fixture. Each subsequent component should be capable of being added to the assembly by a downward vertical motion. No sideways or angular movements should be necessary. This may reduce the number of degrees of freedom necessary for the robot, and in fact if the whole assembly is designed in this way a four axis robot such as the SCARA type may be able to complete the entire product.
2. Standardization should be sought where possible. For example, standardization of bolts, clips and component shapes will provide a number of benefits. There will be fewer component locations for the arm to move to when picking parts. The robot gripper design is much simplified as one gripper will be able to handle a higher percentage of the total assembly. Other important benefits result from this such as a reduction in material handling problems, stores control and costs due to bulk buying.
3. Chamfers and tapers should be incorporated wherever possible to facilitate component insertion.
4. Fastening or joining operations should be the quickest and simplest possible. For example, if subsequent disassembly is not anticipated, it may be quicker for the robot to apply a bead of adhesive than to insert, turn and torque a number of screws.
5. Component design should be symmetrical where possible to assist in parts feeding and orientation. This applies when using automatic component feeders and also when using vision systems. The converse also applies if the component must be asymmetric then the asymmetry should be exaggerated. This will make its orientation simpler to achieve by automatic feeder and simpler to recognize by a vision system.
6. The component design should facilitate robot gripping.
7. The components should be designed to avoid tangling and nesting when held in bulk in feeding devices.

Economics of robotic assembly

The general economic considerations for robotics are covered more fully in Chapter 15. Here a few short notes specifically concerning assembly are provided. In existing nonrobotized assembly it is generally found that where flexibility is required manual assembly costs are high, and where high volume high speed production is necessary expensive dedicated assembly machines are used. Industrial robots are being applied in assembly in an attempt to combine the flexibility of manual labor with the consistency, high quality and uninterrupted operation of automatic machines. The costs of robotization become attractive in the intermediate volume band, especially in products with short life cycles or short batch runs.

Robots are used in assembly tasks in two ways. The first is to position them on an assembly line in series, much as in the traditional manual mass production assembly lines (see Figure 13.13(a)). Each robot carries out a specific series of tasks on the product before allowing it to pass on to the next robot in line for the next set of operations. This method may be necessary where a large number of operations involving many different components are required. In this case the amount of tooling and end effectors necessary for the robot to carry out the task will be too many for one robot to handle, also the surrounding workspace would be too restricted to include them all. This method has the drawback that

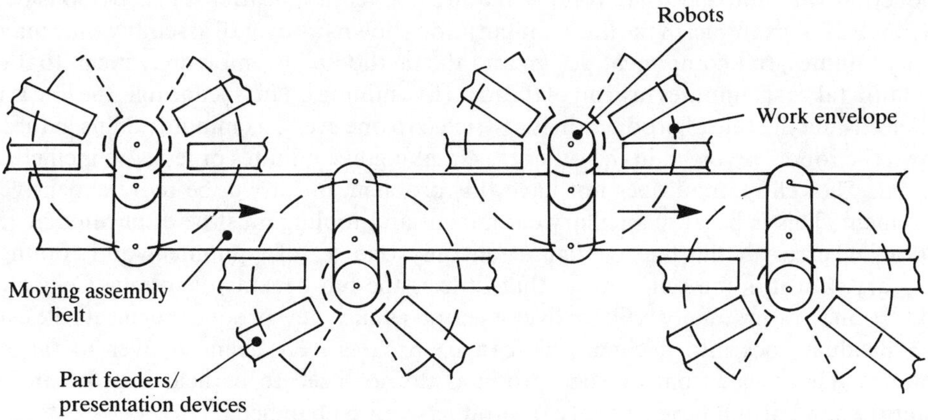

For example:
Time per operation = 5 minutes
Total time per product = 20 minutes
One assembly produced complete every 5 minutes
(a) (NB problem of line balancing)

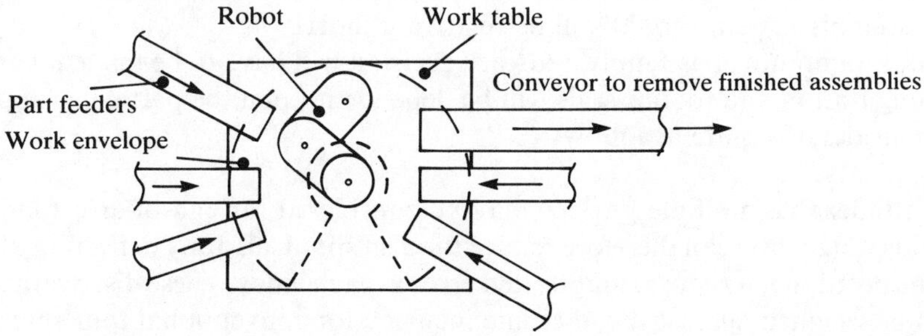

For example:
Total time per product = 20 minutes
Time per operation = 20 minutes
One assembly produced complete every 20 minutes
For equivalent production of (a) above, 4 workcells would be required.

(b) (NB Cost of ancillary equipment)

Figure 13.13 Assembly layouts for robotic assembly: (a) 'line'; (b) 'cell' if one robot breaks down then the whole line comes to a standstill while repairs are effected, or a replacement robot substituted.

The alternative method is to set up an assembly cell where one robot performs a large number of tasks. Here each robot is able to build a complete assembly independently. Thus the cycle time is much longer than for line assembly and if the same rate of production is required then additional workcells are necessary (see Figure 13.13(b)). The advantage of a cell is that if a breakdown occurs then it is only one unit that is down and the other cells are able to continue working.

The exact method chosen will depend upon the complexity of the product, the production rate and the total volume required. Each application must be considered separately. For example, in the line configuration shown the overall assembly time may be twenty minutes, but component design and the method of assembly may mean that one operation takes six minutes instead of the ideal five minutes. This means that the line is not 'balanced' and the rate of production is restricted to one every six minutes. Thus in the line shown the robots involved in the other tasks, taking five minutes or less, are inefficiently utilized. The cell method does not have this problem, but it will be more expensive to implement. This is because ancillary equipment and tooling must be duplicated at each station, whereas in the line configuration only one set of equipment and tooling is necessary. But it is the cell method that utilizes the full benefits of robots. Cells allow production to be gradually built up over a period as each new product begins its life cycle. As a product goes into decline, cells can be progressively changed over to the new products. This means that finished products do not need to be held in stock in high quantities, and also a larger variety of products can be handled.

Although the capital cost of a robot assembly system may be comparable, or even higher, than 'hard' automation (the payback period can be between two and three years), this must not be the only criterion by which a project investment is judged worthwhile. The robot system does not become redundant when the product becomes obsolete, but some specialized ancillary equipment may not be adaptable to the new product. However, if standard 'off the shelf' robots have been used, the commissioning time for the new product assembly system should still be relatively short.

For optimum rate of assembly a further decision will have to be made regarding the robot configuration and tooling, this will be dependent upon the particular application. Typical considerations are as follows:

1. An indexable multiple gripper may be considered for end-of-arm tooling. This adds weight and can therefore reduce the arm speed, also any deflection of a tool in one position may cause a duplicated error at each index. These disadvantages have to be weighed against the idle time incurred for conventional tool changing.
2. The use of tool changing facilities using a magazine and bayonet fittings may be attractive. This incurs extra time during changeover but may be simpler than a multiple gripper system. Also heavier tool and workpiece loads will be able to be accommodated.
3. A multiple arm robot system, such as the PRAGMA robot, may prove feasible. This will probably be more expensive than single arm systems.
4. For noncell layouts, as discussed above, a line layout using multiple robots may be the solution where large volumes are being produced.

Component presentation for robotic assembly

To enable a robot to carry out an assembly task the components must be brought within the work envelope and presented in such a way that they can be grasped by the robot's end effector. They can be presented in a disorganized manner, e.g. in a component 'bin', or they can be presented in a controlled orientation and position. The former method demands sensing by the robot to determine distance, orientation and position of the components, and a means of identifying single components in a crowd. However, it should be remembered that at the assembly stage the components will have already passed through various other processes. These should have provided the opportunity for placing the components in pallets, magazines, etc. so allowing them to be placed at known coordinates and in known orientations. If these are then located in fixtures around the work area the robot can move blindly to the components without the need of sensing. If the preassembly processes occur within the factory then organization of work in this manner should present few problems. If the components are delivered to the factory for assembly from an outside supplier, it may be that they are delivered loose and in cartons. There are two courses of action here. Either the vendor can be paid extra to supply them in, for example, pallets, magazines or bandoliers, or they can be dumped into a mechanical parts feeding device at the workstation. The total cost needs to be examined for each application to determine the cheapest method to adopt.

The most common piece of equipment used to hold, feed, and present small parts to the robot is the ubiquitous vibratory bowl feeder. This is shown in its basic form in Figure 13.14. It is a very versatile and popular hopper feeder and operates in the following

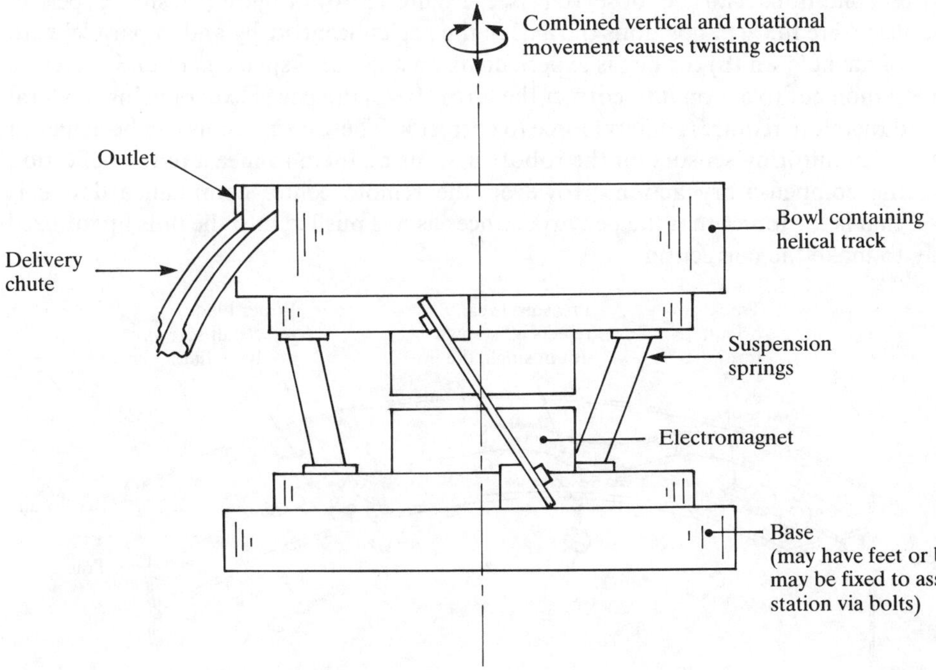

Figure 13.14 Vibratory bowl feeder

manner. The top bowl contains a helical track which curves upward from the base and around the inside wall. The bowl is supported on three inclined leaf springs secured to a massive base. The base contains a powerful electromagnet which is switched on and off at a suitable frequency. When it is switched on it pulls the bowl vertically downward with a rotary twisting action created by the leaf spring configuration. This happens at high speed so that the components are left momentarily suspended in air while their support surface drops away from them and rotates through a small angle. The components drop down due to gravity onto a portion of the support surface slightly forward of their starting position. This happens cyclically at high frequency causing the components to gradually move up the track to the outlet.

The design of the feeder track and the methods used to ensure only correctly oriented parts are fed to the robot require a detailed study in themselves. Here it will suffice to say that the component design considerations made in the previous section are also made to facilitate track design. An example of a bowl feeder track and how it can be configured to ensure only correctly oriented parts are progressed is shown in Figure 13.15.

The remote center compliance device

When attempting to use a robot to assemble products by inserting one component into another, misalignment of the mating parts can be a problem. This is caused by the robot being unable to position the end effector with enough precision over the workpiece. The problem is typified by the peg in the hole operation. Assuming the peg and/or the hole has an adequate chamfer and there is positive clearance between the peg and the hole then the following conditions can be observed (see Figure 13.16). Figure (a) shows perfect conditions where the arm positions the axis of the peg concentrically and in parallel with the axis of the hole. In (b) the peg is experiencing an angular displacement and therefore requires a moment to act on it to correct the error. In (c) the peg is experiencing a lateral error and therefore requires a lateral force to correct it. These corrections can be achieved actively by monitoring sensors on the robot wrist, using them to measure the deflection and taking compensatory action. However, the remote center compliance device is passive; that is the forces that the peg experiences as it is pushed into the hole are utilized directly to make the correction.

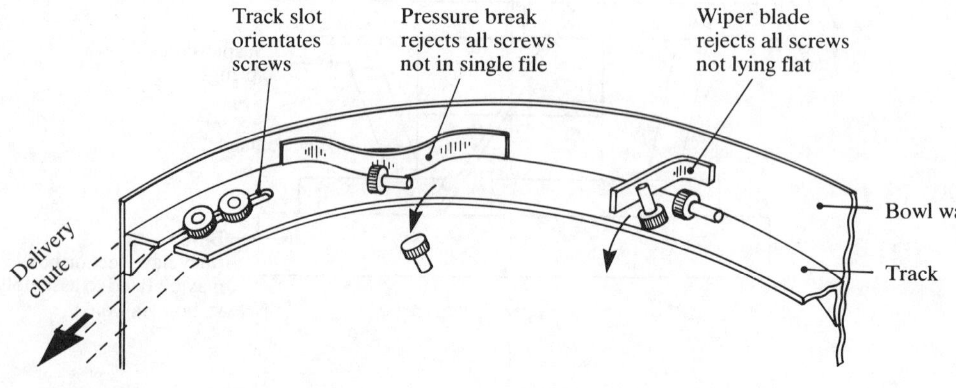

Figure 13.15 Use of track to orientate components

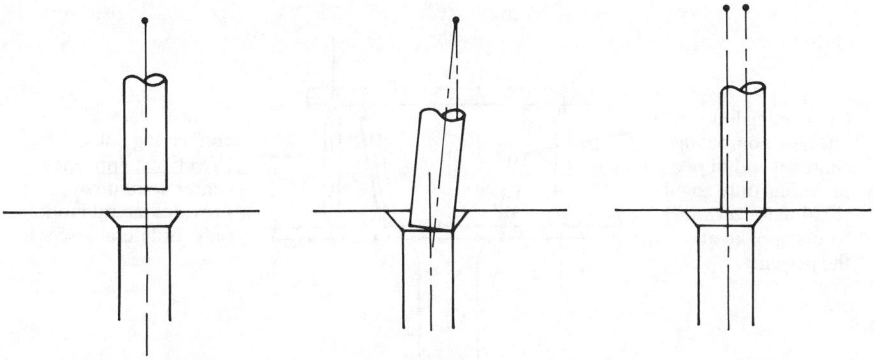

Precise alignment No movement necessary (a)

Angular misalignment Angular compliance necessary (b)

Lateral misalignment Lateral compliance necessary (c)

Figure 13.16 Peg into hole operation

The concept of the device itself is shown in Figure 13.17. Assuming the clearance ratio equals (diameter of hole − diameter of peg) ÷ diameter of hole, then the concept allows chamfered peg hole insertions with 0.0003 clearance ratios, and 1 mm or larger lateral initial position errors with a few degrees of angular misalignment. The device was originally developed at the Charles Stark Draper Laboratories in the USA but is now available commercially as an off-the-shelf unit for mounting on the robot end effector. It may be noted here that the RCC device gives angular and lateral compliance, whereas the SCARA type robots mentioned earlier have only lateral compliance. Their rigid vertical structure is designed to remove the need in most cases for any angular compliance.

There are some who would argue that if the correct robot is selected for the job and the parts are correct, then there should be no need for the extra cost of an RCC device.

Robots in mechanical assembly

A brief example is given here (information supplied by Fairey Automation Ltd) of the assembly of cylinder heads at Perkins Engines (see Figure 13.18). In this installation eight robotic assembly stations, using PRAGMA robots, assemble thirty models of cylinder head. The variations range from cylinder head type (three, four or six cylinders) to differing components in the valve train (e.g. single or double springs), and seven alternative valve types. Different models of cylinder head are assembled in completely random sequence and no changeover time is required between models. Thus one of the advantages of robotization gained here is the ability to respond immediately to changes in customer requirements. This provides short delivery times without needing large and expensive stocks of finished subassemblies. Response time on the line was cut from several days to a few hours with an estimated work in progress reduction of 30 percent.

An important consideration behind the proposed investment was the desire to eliminate costly reworking of misassembled heads. In the past this constituted between 5

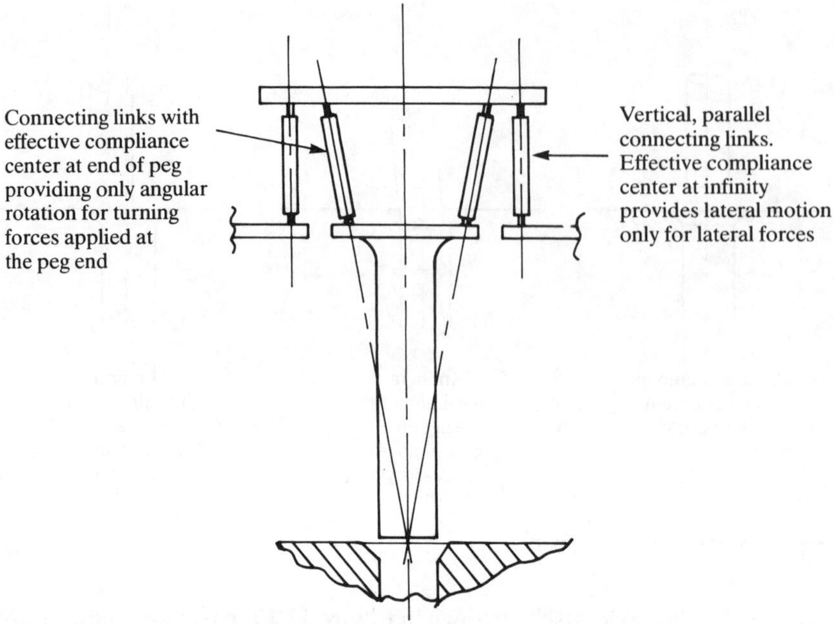

Figure 13.17 The remote center compliance concept

and 7 percent of units produced and some were not detected until the final test was carried out on the completed engine. The robotic system now performs 100 percent testing of all critical operations. Any misassembly is detected as soon as it occurs and can be rectified before additional work is done to the cylinder head. One example of this is at the valve insertion station. After inserting the valve, the robot measures the depth of the valve head below the combustion face of the cylinder head. If this falls outside a preset range it indicates a faulty assembly. The probability is that the wrong valve type has been used or the valve seat has been incorrectly machined. The robot automatically removes, discards, and replaces the offending valve. If after three attempts it still fails to complete the operation it signals to a recovery station further down the line informing the operator of the location and nature of the fault. Due to this in-line testing, the need for reworking of completed subassemblies virtually disappears.

The sequence of operations starts with washing the cylinder head casting. The valves, valve train and rocker cover studs are then assembled by the eight robots. A typical four cylinder head receiving seventy components in total. The completed subassembly is then leak tested. From insertion of the first valve to final leak test takes approximately sixteen minutes.

There are 1300 separate operations on the line and this is reflected in the sophistication of the control system. Control is distributed between four minicomputers and programmable logic controllers under the supervision of a supervisory controller. Comprehensive diagnostic and management information services are also built into the system. Integration of the system with the production control department allows the number of each model variant built in a given period to be determined by using one of the

Figure 13.18 Cylinder head assembly line at Perkins Engines Co Ltd. (Courtesy Fairey Automation)

computer terminals. Regular reports are also provided on the trends in recovery routines occurring at each station. This identifies stations where operations are drifting towards the edge of their tolerance band and enables corrective action to be taken. The comprehensive diagnostic functions report via a visual display the nature and location of any faults, as well as warning of parts feeders likely to need refilling in the near future. The overview of the system shown in the figure also highlights the extensive safety guarding necessary.

Robots in electromechanical assembly

The use of robots in this application is becoming more common. Products such as limit switches and computer keyboards, have been produced in this manner for some time now. In the IBM plant at Greenock in Scotland there is a large concentration of robots of various types engaged in the assembly of computer keyboards, control modules and monitor units. As the technology progresses and experience is disseminated this type of application should find its way into the smaller scale companies who at the moment are unable to afford the engineering time required for development.

Robots in electronic assembly

The use of robots for assembly of electronic subassemblies, mainly the population of PCBs, is still not widespread. This is because until recently robots were seen as only being viable in a very narrow volume band for this type of application. For mass produced PCBs automatic, purpose-built insertion machines are used. These machines can have insertion rates of up to 30 000 per hour and are designed to accommodate large volumes of small uniform components. They can move very fast and have the ability to carry out simultaneous insertions. For short production runs and high flexibility, manual techniques assisted by machines can achieve insertion rates of up to 500 components per hour. Robots are generally seen to be applicable in the 1800 components per hour region. They can therefore be faster than manual methods, but can also accommodate a number of component shapes and sizes, especially when fitted with multiple grippers or using interchangeable tooling.

One of the factors that directly affects insertion rate is the type of component being handled. This will also affect the ease with which the robot will be able to grasp and manipulate the component. PCBs can normally be 80 percent completed by standard component insertion machines, the remaining 20 percent of odd components are at present usually manually inserted. The cost for the manual insertion is generally much higher than that for the automatic machines. Standard components (see Figure 13.19) are classified as axial, radial, dual in line package (DIP), and surface mounted device (SMD). For automatic insertion they are not usually handled and fed using, for example, bowl feeders but are presented in their own packages (Figure 13.20). Robots normally utilize components presented in the bandolier or magazine form.

Surface mounted devices are particularly relevant to robotics. They were first used on

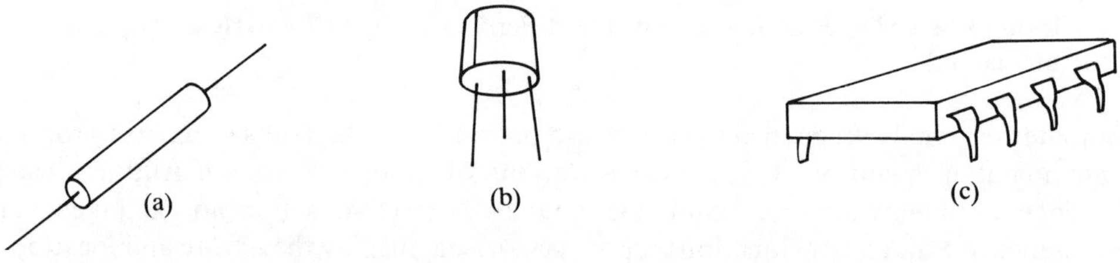

Figure 13.19 Conventional component configurations: (a) axial (e.g. resistor); (b) radial (e.g. transistor); (c) dual in-line package (e.g. integrated circuit)

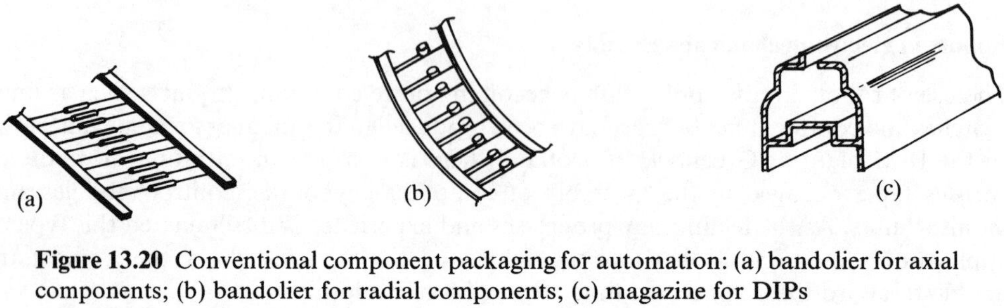

Figure 13.20 Conventional component packaging for automation: (a) bandolier for axial components; (b) bandolier for radial components; (c) magazine for DIPs

commercial products in the late 1970s, and are increasingly used for PCB population. These devices differ from the conventional electronic components in that they do not have the familiar long wire leads that connect to the PCB. The conventional component leads are inserted through drilled holes in the board and soldered to the copper as shown in Figure 13.21. Surface mounted components, however, are soldered directly onto the board thus removing the necessity for drilled holes as shown in Figure 13.22. The use of SMDs results in a more densely populated and less expensive PCB. In fact boards of three times the density, greater reliability, better electrical performance and requiring lower production costs than conventional ones are obtained. In relation to robotics their major significance is in their suitability for automatic assembly. This is due to their external similarity in appearance which enables one, or at most only a few, end effectors to handle the range of devices required to populate a PCB. This similarity also simplifies part presentation to the robot, some packaging being shown in Figure 13.23. SMDs are generally much smaller than their conventional counterparts and this therefore makes them difficult to handle manually giving a further reason for using automated equipment such as robots. Various automated soldering techniques such as wave soldering can be used in conjunction with robotic placement equipment to give fully automated board production.

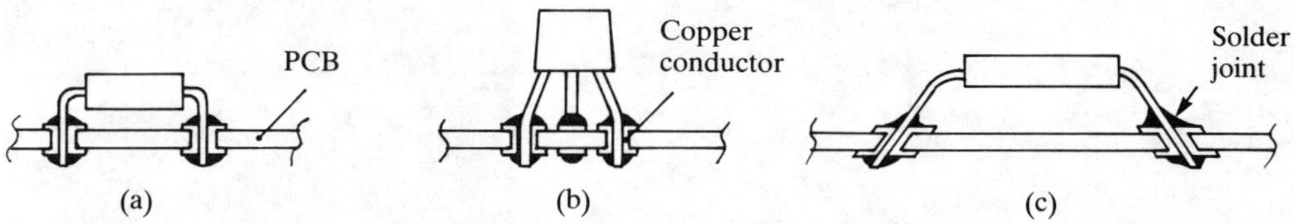

Figure 13.21 Conventional through-board mounting: (a) resistor; (b) transistor; (c) IC

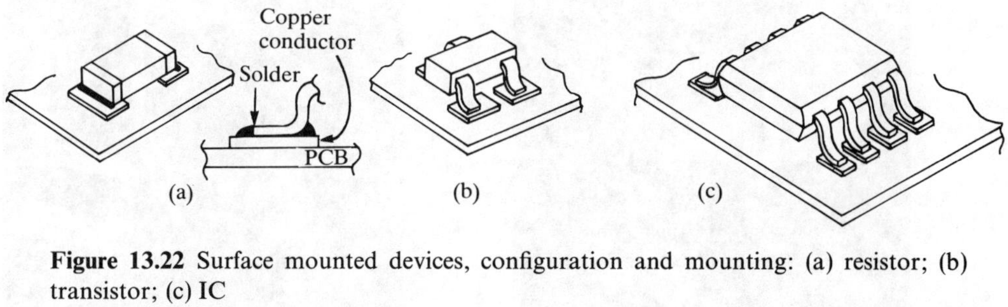

Figure 13.22 Surface mounted devices, configuration and mounting: (a) resistor; (b) transistor; (c) IC

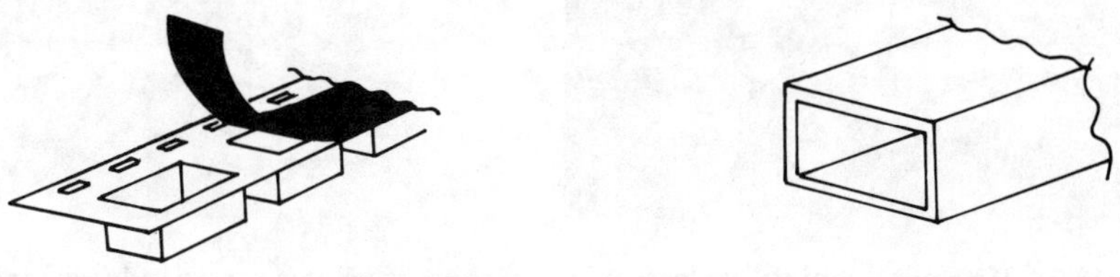

Figure 13.23 Some packaging for SMDs: (a) blister tape; (b) stick tube

An example of a robotic PCB population application is now given. Figure 13.24 illustrates the use of a multiple gripper at ICL's Kidsgrove plant (although the plant is, at the time of writing, using an exchangeable tool system in preference to this). In the installation a SCARA configuration robot assembles PCBs for a mainframe computer. The PCB is populated with five different conventional components: large and small capacitors, odd form ULAs (uncommitted logic arrays) of which there are 64 different types, and 96 pin and 64 pin edge connectors. The ULAs (which have 179 pins) and the edge connectors make the operation unsuitable for conventional PCB automation

Figure 13.24 Robotic assembly of electronic components using a rotary index gripper: (a) inserting large capacitor; (b) inserting small capacitor; (c) inserting connector; (d) inserting ULA. (Courtesy ICL)

techniques. These components could have been assembled by hand but to reduce the possibility of damage due to static and improve overall quality, automation was considered desirable. This led to the decision to construct a robotic assembly cell. Work on the cell started early in the product life cycle and so development work on the assembly process was able to be carried out in concert with product design. This approach usually optimizes the unit costs over the long term, e.g. suppliers can be approached at this stage to ensure components are provided in suitable magazines or dispensers for presentation to the robots.

Initially PCBs are manually loaded into rack magazines in a stack, one above the other. A pneumatic cylinder pushes a board out from the stack into the assembly station. A laser reader scans the bar code on the PCB and sends the information to an ICL personal computer. This computer holds all the information for each board and the relevant robot program. When the board has been identified the appropriate instructions are sent to the robot controller from the PC via an RS232 link. The board is clamped in position on a fixture which accurately locates it by means of two pins. Special component holding pallets have been designed in conjunction with the component suppliers to ensure efficient robot handling. The ULAs were supplied in pallets, the connectors in standard magazine boxes, and the capacitors in bandoliers.

The end effector for handling the components was originally in the form of a multiple rotary indexing gripper as shown in the figure, the indexing being obtained by an integral stepper motor. In this particular situation the user decided that it was more efficient to use the previously mentioned exchangeable tool system which increased accuracy but also increased cycle time. The whole system was designed for integration, for example the PC can communicate upwards with a mainframe computer and form part of a local area network, these concepts will be considered in a later chapter.

Thus robots can provide the advantage of being able to completely populate PCBs with standard and odd form components. They can also be very attractive when batches of between ten and fifty products at a time are required, e.g. for mainframe computers and defense equipment, as they can change very quickly from one batch to another. This reduces lead time for the customer and reduces the need for back-up supplies of finished boards in stock. The overall precision of the robot will probably need to be at least 0.1 mm. For example, when inserting odd shaped components into PCBs a positioning tolerance of 0.2 mm is often required. If it assumed that the PCB is held in a fixture and is located within 0.02 mm, and the components are presented to the robot gripper within 0.02 mm, the robot has a precision of 0.05 mm and the component leads are irregular to 0.1 mm the tolerance build-up approaches 0.2 mm, the maximum allowable.

ROBOTS IN MACHINE TOOL LOADING

The use of robots for machine tool loading/unloading is essential for unmanned machining cells and flexible manufacturing systems (FMSs). The concept of the numerically controlled machine tool is that the machine is capable of being programmed to produce batches of different components to order. All that is necessary to change from one product to another is to change a paper tape, floppy disk, or in a more advanced system, the instructions coming from a host computer. However, the basic machine can only cut material, not manipulate it into and out of the machine; for this a robot is needed.

Various types of robot have been used for this purpose, all have been integrated with the machine tool control system to ensure minimum cycle times are achieved. The most common types for this application are the specially designed manipulators shown in Chapter 3, or cylindrical coordinate robots produced by manufacturers such as FANUC or Cincinnati Milacron (see Figure 13.25).

Robots for this application are capable of automatic loading and unloading of workpieces, tool changing and workpiece handling. Their control systems are designed to interface with, and control, conveyors, jigs and fixtures (workholding devices), and the operation of chucks and machine guards. They can interface with a host computer and/or the machine tool controller to accept tool change requests and workpiece changes, and are also designed for easy programming of palletizing routines for tools, material and components. A variety of grippers are available such as those shown in Chapter 11.

The specially designed robots, such as the machine tool loading robot shown in Figure 3.4, are able to fit a wide range of machine tools. They are relatively small and lightweight, enabling them to be attached directly to the machine thus saving floorspace. The larger cylindrical coordinate robots can handle heavier workpieces and have a much longer reach. This enables them to be used for servicing a number of machines, possibly moving the workpieces in sequence from one to the other then palletizing the finished components. It may be noted that if the robot is only being used to service one machine tool, then the cycle time of the machine may be so long that the robot would be idle for a

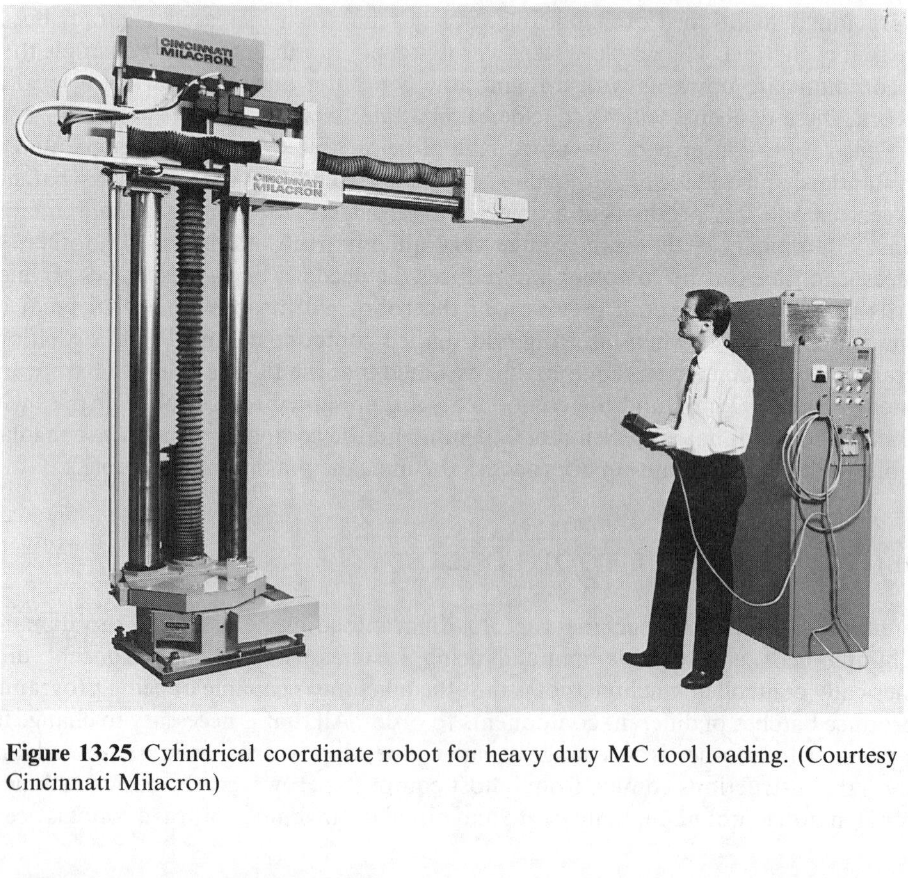

Figure 13.25 Cylindrical coordinate robot for heavy duty MC tool loading. (Courtesy Cincinnati Milacron)

considerable time. This may remove the economic feasibility of using the robot unless other work can be found for it while it is waiting. This other work may be deburring or inspection of finished components, or an entirely different operation for another component.

Ideally the machine tool used should be capable of being loaded from the back. This allows manual servicing of the machine should the robot fail or require maintenance, although these robots are essentially simple and rugged in their mechanical construction. Programming them is also usually simple using a teach pendant with point to point control.

Thus machine tool robots when linked with NC machine tools and automated material transport systems provide the final link in the concept of fully automated machining.

ROBOTS IN SURFACE COATING

Paint spraying was one of the first processes to be robotized using continuous path control. The Trallfa robot shown in Figure 13.26 is now widespread, having been specifically designed for paint spraying. Another hydraulic robot designed for this task is

Figure 13.26 Trallfa paint spraying robot in operation. (Courtesy DeVilbiss Trallfa)

the Spine robot shown in Figure 13.27. The paint spraying process is ideal for robotization due to a number of factors:

1. The paint spraying environment and working conditions are unpleasant and dangerous for humans this is due to: (a) the materials used which are often toxic and necessitate the wearing of protective clothing; (b) the noise produced from the nozzles as compressed air forces the paint out through them in a fine spray; this requires the operator to wear ear protection; and (c) the working area is often restricted and the amount of manipulation required is often complex and relatively strenuous.
2. Conventional automated paint sprayers are expensive, relatively inflexible, can be inefficient and are not usually able to fully complete the job especially in complex tasks such as spraying car bodies which requires touch-up by a human operator at a later stage.
3. Due to health and safety regulations energy costs can be high. This is due to the necessity to provide adequate ventilation and extraction of fumes to allow a constant flow of sufficient fresh air to the operators. If humans were removed from the work area then the necessity for energy intensive ventilation systems would be reduced, thus lowering costs.
4. Human operators do not produce work of consistent quality. Also overpainting often occurs increasing the overall cost of paint.

When the method the human operator uses to spray is considered it is apparent that the spray painting robot requires considerable dexterity. A human operator will skillfully hold the spray gun, connected to a paint reservoir, and move it over the area to be covered

Figure 13.27 Spine robot. (Courtesy Komatsu Ltd)

keeping the nozzle a constant distance above the surface. The robot therefore requires continuous path control and is programmed by the lead through method. The robot must also be capable of working in the highly inflammable atmosphere that exists when using some paint materials. This has meant that until recently only hydraulic robots with their low voltage signals and remotely located power packs were admissible in spray painting installations. However, the new brushless motor robots with less likelihood of sparking may mean that electric robots will become usable.

ROBOTS IN HANDLING AND PALLETIZING

This category includes a wide variety of applications. The items being handled may weigh anything from a few grammes to some tonnes. Obviously the type of robot applied will also vary considerably, from lightweight non-servo-controlled pneumatic robots, through medium sized electric types, to massive hydraulic manipulators. For many applications the cylindrical, polar and revolute types are equally suitable, and for some light duty work in restricted areas rectangular types may be used. A simple material handling task using a Unimate polar coordinate robot is shown in Figure 13.28.

Point to point programming and control is necessary. The controller should be able to store a large number of points and have the facility for easily programming horizontal and vertical increments. Thus by defining only the start point for a palletizing operation the arm can then be programmed to pick or place components at equal steps along the x, y and z axes by simply defining the length, direction and number of steps.

Gripper design is a major aspect of this application as the correct choice and design of grippers will affect the economics of the whole process. Often the robot controller will be linked to other equipment, e.g. a machine tool or a conveyor. Material for palletizing or for loading into other equipment may be presented to the robot in a random sequence.

Figure 13.28 Unimate stacking cartons. (Courtesy Unimation)

Thus the palletizing robot may be linked to, for example, a laser reader which reads a bar code on the component cartons or on the components themselves, this then indicates to the robot what sequence of movements to initiate next.

ROBOTS IN PRESS LOADING AND FORGING

In press work, components are produced from cold rolled sheets or strips of metal which are fed into a press containing a press tool. This tool is in two parts; the upper section is bolted to the press ram and the bottom section is clamped to the press table. Each time the press cycles a component is produced by shearing the metal between a punch, usually attached to the top tool, and a die, usually attached to the bottom tool. This shearing action is termed cropping, blanking or piercing, depending upon the exact operation. Components may also be produced by drawing the metal over a die and so producing a shape, such as a car door panel. Often blanking and drawing operations are combined.

In forging, components are once again produced in a press but here the metal is 'squeezed' into shape by creating high pressures to deform it. This is in contrast to the shearing, bending and drawing actions typical of press work. In forging the material takes on the form of the cavity contained between the upper and lower parts of the forging tool.

Industrial robots are usually employed in press work and forging to transfer material. Examples may be: removing a blank metal sheet from a stack using a vacuum gripper and inserting the sheet into a press; loading billets from pallets into a forging press; unloading components from presses and palletizing them. Often components are not finished in one operation and they have to be passed through a number of different presses before completion. Robots can be employed to transfer components between presses.

Press work and forging applications are ideal for robots as complex programs are not usually required, simpler cylindrical coordinate robots can be used, and the work environment is often extremely unpleasant for humans i.e. dirty, hot and very noisy. When robots are installed in these situations productivity is improved but not necessarily by reducing cycle times. In fact cycle times may even have to be increased slightly. Improvements in productivity come from reductions in nonproductive time i.e. no relaxation time is required, no lunch breaks or shift changes, and twenty-four-hour working can be achieved if necessary. Scrap rates are reduced as there are fewer 'mis-hits' due to operator error when loading the presses. This also reduces die maintenance time and costs. In situations where a line of robots is servicing a series of presses, light curtains may be used for safety purposes, these will be set to switch off all machines should anyone enter the danger area. The perimeter around the light curtains and press line could be protected from casual intrusion by a chain or guard rail system. Safety will be discussed in more detail in Chapter 16.

ROBOTS IN INSPECTION AND TESTING

There is a wide variety of inspection and testing techniques and therefore industrial robots are applied here in an equal variety of different ways. For example, a robot may be linked to a vision system to remove bad components identified by the system as they pass by a camera; a robot may pick up a series of gauges and use them to check a series of holes in a machined component; the robot wrist may be fitted with a simple probe connected to a microswitch to determine the presence or absence of a component feature; or the robot may be used to load and unload electronic components into test equipment.

ROBOTS IN FETTLING AND DEBURRING

Fettling and deburring involve the removal of unwanted metal from components by using cutting or abrasive tools. Fettling is the name given to the removal of extraneous metal, such as flash or sprues, from castings. Deburring involves the removal of the sharp or ragged edges left after machining operations. Fettling is carried out by cutting the excess metal away using a grinding disk, saw or other metal-cutting technique. The material can either be held by the robot and presented to the cutting tool or vice versa, very much depending upon the weight of the component and the load carrying capacity of the robot. Deburring is carried out by mounting a small cutting tool, which rotates at high speed, on the robot wrist. The robot then traces around the edge to be deburred. This edge following operation may also be necessary in fettling and in both instances could require continuous path control.

Robots are applied in fettling and deburring operations as these tasks can be tedious and often take place in dirty, hot, noisy environments. The cutting tools themselves constitute a further safety hazard to human operators. An example of deburring is shown in Figure 13.29.

ROBOTS IN ADHESIVE APPLICATION

Recent advances in adhesives giving improvements in performance characteristics and ease of application have led to their increased popularity. They are capable of replacing conventional methods of joining such as welding or rivetting, and are now used in the construction of aircraft, ships, cars and domestic goods. Two of the major advantages of adhesives are that they can spread stresses over a larger area, and they can be applied using relatively unsophisticated equipment.

Typical applicators suitable for use by robots are dispensers which deposit a bead of adhesive along a programmed path, or spray guns for coating larger areas. The application of the adhesive must be tightly controlled for functional and economic reasons. This therefore makes it an attractive application for robots as they can be programmed to position adhesive beads accurately and dispense precise quantities whether in bead, droplet or spray form. Robots can usually apply adhesives at a higher rate than a human therefore adhesives with shorter setting times can be used with corresponding increases in productivity. Some adhesives produce joints with better properties if they can be applied at temperatures too high for human comfort, so here again robotization is an advantage. As with spray painting robots, if programmed properly, adhesive applying robots will be more efficient in their usage of adhesive than human workers. Costs are also reduced as special clothing and ventilation are not necessary. See Figure 13.30.

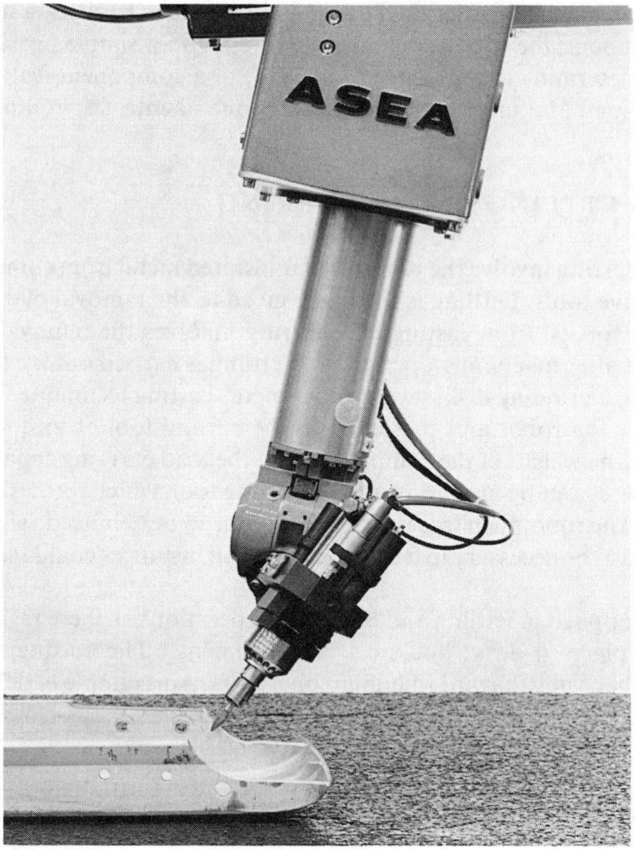

Figure 13.29 Deburring operation on a car bumper. (Courtesy ASEA)

ROBOTS IN WATERJET CUTTING

This method of cutting such materials as lightweight glass reinforced plastics and carbon fiber is gaining in popularity. It is fast, does not produce hazardous dust, and under correct conditions produces a clean cut surface. It may be used to trim excess material, or cut internal or external profiles, on three-dimensional components such as car door and body panels. The process involves the use of a waterjet gun which directs a high pressure jet of cleaned water at the material to be cut. The waterjet width can be from around 0.2 to 2.5 mm, and the pressure used up to about 3500 bar. Cutting speeds of from 50 mm per second to 125 mm per second can be achieved though material thickness is usually only a few millimeters.

The waterjet should always be maintained at an angle as close to 90° to the work surface as is possible; usually an edge will be required to be followed or a profile traced with repeatable precision; and the waterjet itself is hazardous. These factors make the process attractive for robotization.

In the robotic process the robot holds the waterjet gun and manipulates it around the

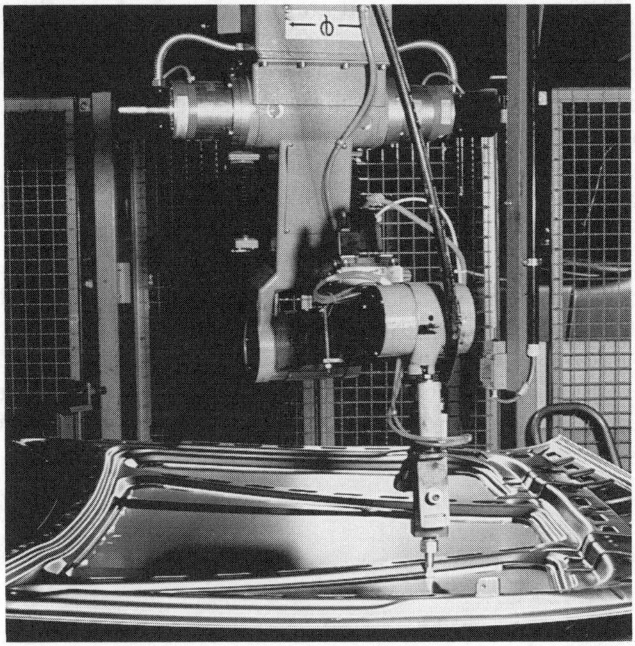

Figure 13.30 ASEA IRB 6 robot applying sealant to front and rear car hoods. (Courtesy ASEA)

Figure 13.31 Waterjet cutting of car sound absorbers. (Courtesy ASEA)

work to be cut. The movements required are thus very similar to those required in arc welding. Thus the robot may be required to work in concert with a second reprogrammable manipulating table. The robot arm and the workholding equipment must be stable and the robot movement must be smooth otherwise a ragged 'sawtooth' effect will be obtained on the cut surface. The workpiece is mounted on a fixture which may in turn be mounted on a table or on a programmable manipulator. The workpiece can be clamped mechanically to the fixture, or held in place by a vacuum. See Figure 13.31.

There are many more applications than the ones discussed here, and the range of applications will continue to increase as robots become more sophisticated and their relative costs decrease in relation to those of human labor. The following chapters highlight some further aspects needing consideration when applying robots, and also attempt to put the industrial robot into perspective within the overall automated manufacturing scene.

14

APPLYING ROBOTS

INTRODUCTION

There are three basic ways in which a firm will arrive at a robot installation decision:

1. A new plant or line is being built from scratch for a new product. Here the firm will be wanting to use the most efficient, and therefore probably the latest, technology. This will often involve the use of robots as part of a larger automated system. The type of system adopted as well as the product will largely determine the choice of robots and manner of installation.
2. An existing process or method is proving unsatisfactory. This could be for various reasons, e.g. labor cost too high, inconsistent quality, high rate of labor absenteeism or turnover. The firm in this situation will therefore have a quite specific notion of what it would like from a robotic system.
3. A desire to gain experience in robotics but not certain exactly where best to make the initial investment. This is the most general case and is looked at in more detail below.

APPLICATION SELECTION

Initially a survey working through the factory from where raw material and parts arrive to the final despatch area should be carried out. As the normal workflow is traced, the following situations should be looked for as they provide ideal conditions for robotizing.

1. Operations that are simple and repetitive; these are the ones that should be most easily robotized, e.g. unloading die casting machines or pick and place operations.
2. Jobs where cycle times are not too short, probably over three seconds and preferably greater than five seconds. This gives a larger choice of robots as cycle times less than this demand very fast moving robots probably of the high velocity and acceleration assembly types.
3. Tolerances on parts and tools such that a robot of average precision will be able to carry out the required operation.

4. Jobs where part sizes and weights will not present particular problems. For example, very heavy jobs will probably require a very robust and expensive robot, possibly not the best type for a first installation.
5. The operation being considered should not have an integral inspection element. This is because either a human will be required to carry out the inspection elsewhere anyway, or additional expense will be incurred to carry out the inspection automatically.
6. Situations where worker dissatisfaction with the job exists; this may be due to the work being tedious, dirty or socially isolated.
7. Hazardous environments such as those where toxic or flammable atmospheres may exist. These are ideal for robotization as the human workers are pleased to be relieved of the work, and savings can also be made in protective clothing and other health and safety equipment.
8. An orderly environment either exists or can be made to exist around where the robot will be installed.
9. The job should require very little in the way of intelligence or judgement.
10. The product variety should not be too great; this will minimize the tooling design and manufacture costs.
11. Situations where more than one shift can be worked thus making financial justification easier.
12. Areas with quality problems.
13. Situations where parts have been previously positioned and oriented, or the opportunity exists for this to occur.

It is also necessary to be aware of various other facts and implications. The investigator should be fully conversant with robot configurations, drive and control systems, coupled with a knowledge of the capabilities and limitations of industrial robots. The likely interfacing requirements should be noted to anticipate the potential complexity of achieving integration. Although a first installation is likely to be of a single robot, the possibility of applying more than one should be considered. This will spread development, installation and training costs over a larger number of robots. These costs, together with ancillary equipment and tooling, may easily equal that of the robot itself, and if the task is complex and novel the costs may even be a few times that of the robot. The time to get a robot installation working, from first planning through to full production, is often from three to twelve months, sometimes more if the application is complex or little previous experience in robotics is available. The job chosen must therefore allow full utilization of the robot if an economic installation is to be achieved.

Once an operation has been selected for more detailed consideration a full analysis can be carried out of the existing process. This will take the form of a method study examining what is actually being done, why, and how. Further analysis of these results, coupled with detailed information that may be useful, e.g. notes on the tendency of parts to stick to each other in a bowl feeder, and any special improvements that could be available if robotization occurs. Data on the work area within which the robot will be expected to work, and the reach, speed and load capacity required, should be gathered. All this detail will prove valuable when the time comes to robotize. Data on direct and indirect labor costs, overhead costs and production rates will also be gathered at this stage for future use in the economic appraisal. There may be many other factors to consider at this stage and these will become more apparent as the process of selection proceeds.

ROBOT SELECTION

As indicated previously, to select a robot for a specific application a good general awareness of robot capabilities and limitations is necessary. The next stage is to consider the various specifications of the robots available and check whether they match up to the task requirements. These specifications are now briefly examined; not all will be readily available from the supplier. If they are not, and it is thought necessary, they may be found by carrying out tests at the supplier's premises using test robots and prototype workstations. There may also be data available from documentation produced by other users or research establishments.

Payload

At the application analysis stage the maximum load likely to be moved will have been determined. When considering the specification of the robot it should be ascertained whether the maximum load specified is at maximum or minimum arm extension. Due to the effect of leverage the robot will obviously be able to lift heavier loads the closer the load is to the robot body. Thus a maximum load given on the specification sheet must be investigated to ensure the desired load can be manipulated at all arm extensions. In any case, it is advisable to purchase a robot with overcapacity to ensure overloading does not occur, and that precision is maintained even when working with maximum loads and full end-of-arm tooling.

Precision

This is composed of three factors in a robot: resolution, repeatability and accuracy. Many robot specification sheets show only repeatability, a few also show accuracy; probably none will show resolution. This is because although resolution contributes to the accuracy and repeatability of a robot, it is 'transparent' to the robot user.

Resolution. Resolution is the smallest increment of controlled motion that the robot can make. There are three factors relating to the robot hardware that can determine the overall resolution.

The first factor is that relating to the manner in which a positional value is held in the memory of the robot controller. For example, assume one axis of a cartesian coordinate robot is 1 m long and the position of the arm on the slide is between a maximum and a minimum specified by an 8 bit word. With an 8 bit word there are 256 control increments available. Thus the resolution of this part of the system would be $1000 \div 256 = 3.9$ mm. However, two 8 bit words or one 16 bit word would normally be used, giving a resolution in this case of $1000 \div 65\,536 = 0.015$ mm. Another approach would be to divide the axis up into a number of smaller sections, say eight. Each section could be identified by the first digit of an 8 bit word, and the remaining digits would give the resolution, $(1000 \div 8) \div 128 = 1$ mm.

The second factor contributing to resolution occurs if a digital actuator such as a stepper motor is being used to drive the arm. Assume the motor moves in steps of 0.18° and transfers this drive to the arm through a gear system which has a ratio of 40:1. With each motor increment equal to 0.18° there will be 2000 steps per revolution, and with a gear ratio of 40:1 there will be 80 000 controllable steps or increments over the whole arm movement. Thus the resolution of this component of the system will be $1000 \div 80\,000 = 0.0125$ mm.

The final factor relates to the feedback hardware; for example, shaft encoders provide a number of pulses per revolution, typically up to a maximum of about 2540. These may be mounted on the motor or joint shaft and therefore also contribute to the overal system resolution. On a 2540 pulse encoder each 0.14° rotation will transmit one pulse to the controller, any angular movements smaller than this would not be detected. Obviously once again gearing can be incorporated to increase the resolution to provide multiple revolutions of the encoder to one revolution of the robot joint axis or actuator. When analog feedback devices are used the resolution of the analog to digital conversion system will be the critical factor.

The best system performance, as regards resolution, will occur when the three factors mentioned above are all roughly equal, the worst one of the three providing the overall system resolution.

Accuracy. This term applies only to computer controlled robots and is composed of the system resolution, plus the computer control algorithms used to calculate the necessary joint coordinates, plus mechanical inaccuracies in the system. It is a measure of the ability of the robot to move to an externally taught target point. That is, a point taught by entering the world coordinates through, for example, a computer terminal, as opposed to physically moving the robot end effector to a particular point then instructing the controller to 'remember' that point. The mechanical factors arise through gear inaccuracies, backlash in the gears and joints, and bending of the arm components. Statistically the effective resolution of the robot should average out at about half of the overall system resolution. If the software is well designed, and the robot limbs made quite rigid, then it is possible in the ideal situation to get accuracies almost equalling the resolution.

Repeatability. This is a statistical term which provides a measure of the ability of the robot to return the end effector repeatedly and consistently to a previously taught point. This figure will be worse than the resolution as it is a practical manifestation of resolution plus all of the errors, inaccuracies and uncertainties in the total robot system. It will probably become worse with time as wear occurs and clearances between joints and gears, etc. increase.

As mentioned earlier only accuracy and repeatability are apparent to the user and their effect is shown schematically in Figure 14.1. Repeatability can be taken as the length of the radius of the smallest circle that can be drawn to include all tool center points achieved. Distribution curves are shown at the bottom of the figure and statistically it can be seen that repeatability can be taken as one-half the range of the curve. In the figure accuracy is represented by the distance from the center of the repeatability circle to the target point.

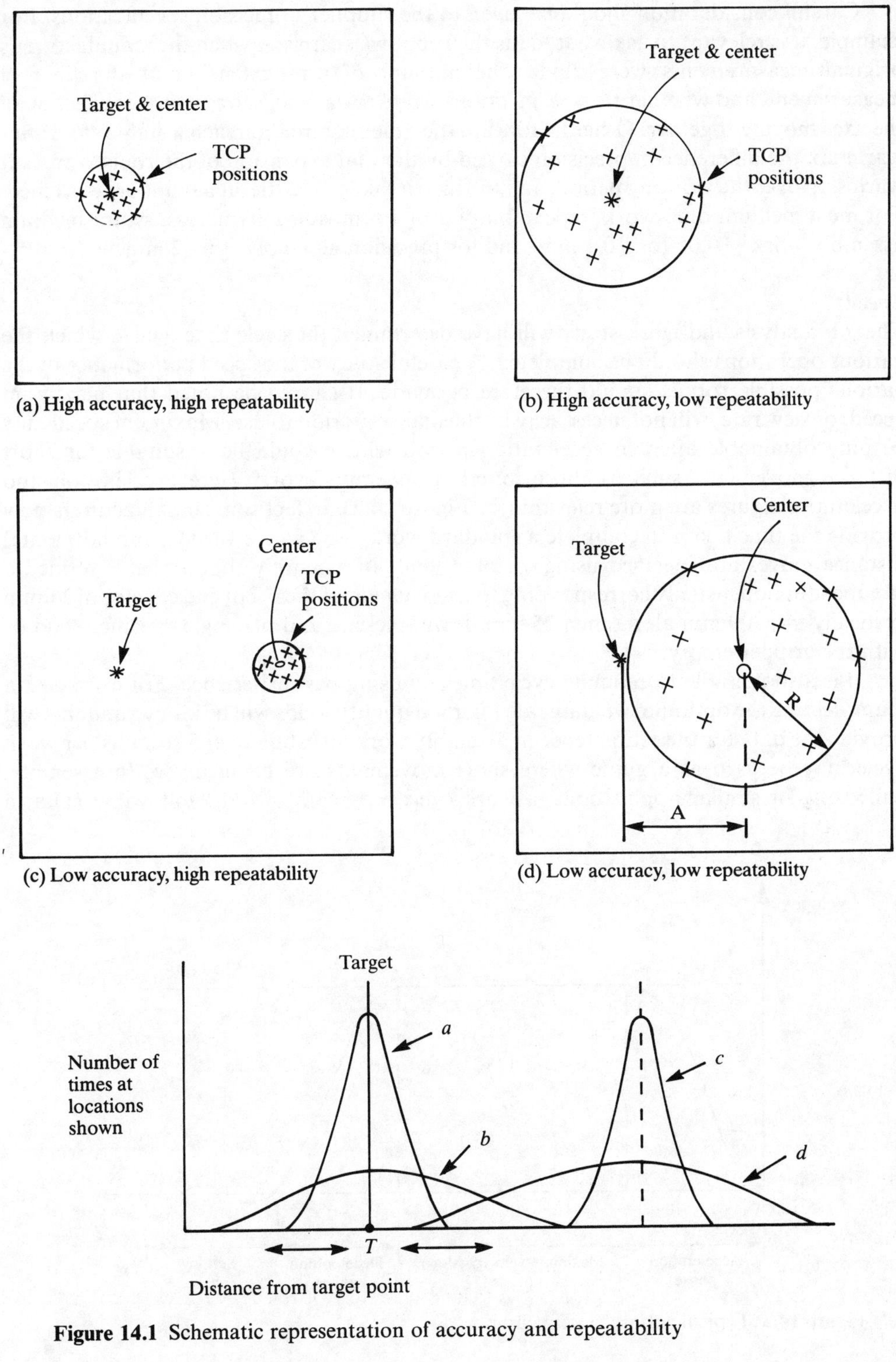

Figure 14.1 Schematic representation of accuracy and repeatability

Careful consideration should be given to the supplier's precision specifications. For example, it is relevant to ask what loads the robot was carrying when the manufacturer's original measurements were made, the amount of arm extension at the time of measurement, and whether the specifications relate to each axis, or to the net effect of all the axes moving together. Other factors are the time required to reach a stable state after warm up, the difference in precision caused by thermal expansion of the robot arm as it warms up over the working period, and the time it takes to 'settle' at a point once reached. For most medium duty work, repeatability of ± 1 mm is usually necessary, for medium assembly work ± 0.05 to ± 0.1 mm, and for precision assembly ± 0.02 mm.

Speed

The job analysis and work study will have determined the cycle time within which the various operations should be completed. A careful study of the speed performance of the various possible robots should therefore be made. It should be noted that maximum speed, or slew rate, will not necessarily be the main criterion to use. Maximum speed may be only obtainable after an acceleration period which would be unsuitable for short distance movements such as those found in assembly work. Here acceleration and deceleration figures are more relevant (see Figure 14.2). In fact some manufacturers now provide the time taken to complete a standard work cycle where vertical and horizontal distance moves are specified using a stated load. In assembly robots this is termed a 'goalpost' action, usually corresponding to a movement of the robot end effector of 25 mm vertically up, 300 mm along then 25 mm down, picking and placing a specified load or with the gripper empty.

It is also possible to estimate cycle times by using past experience. For example, in seam welding the optimum welding rate for good quality welds will be known and this will provide the data for the estimate, or in assembly work an estimate of 5 seconds per work cycle may be used as a guide where short movements are being made. In assembly, palletizing or similar applications, a work cycle may comprise the following: (1) arm

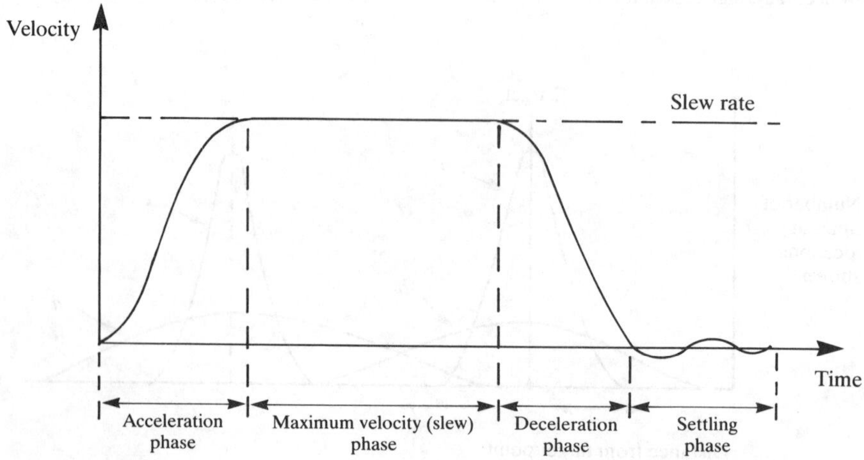

Figure 14.2 Typical robot velocity curve

moves down from home position to pick up part; (2) grasp component or material; (3) move up; (4) move across; (5) move down to insert or deposit part; (6) ungrasp; (7) move up; (8) return to home position. Each of these movements or operations can be classed as a work element and if appropriate times are available for them then quite complex tasks can be estimated. Considerable research has gone into establishing standard methods for accurately predicting robot cycle times. However, it must be noted that each robot has its own characteristics therefore data on each robot must be available for standard element times if a reasonably accurate prediction has to be made.

Other factors enhance the difficulty of anticipating cycle times, for example, similar type robots from the same manufacturer will vary in performance; as the robot heats up during the day speed variations occur; and depending upon the relative positions of the limbs within the work envelope different speeds will be obtained for similar distance movements. Thus by far the best way of anticipating final cycle times is to carry out tests on the manufacturer's equipment before purchasing, or look for previously installed identical applications elsewhere in the industry.

Work envelope and configuration

The use of scale drawings of the work area and templates of the robot work envelope will be necessary to ensure that the robot can reach all desired points and that collision and trapping points are identified. Alternatively graphical computer simulation can be used.

Control system

Point to point control will be used in applications such as spot welding, assembly and palletizing, whereas continuous path control will be chosen for arc welding, spray painting, fettling, etc.

Drive system

The most commonly used type is the electric robot due to its cleanliness, speed, quietness and ease of maintenance. Only if very heavy loads have to be manipulated should hydraulic robots be considered. If the task is a very simple point to point operation with no requirement for servo control and lightweight components are to be handled, then pneumatic robots may be the best solution.

Programming method

The job analysis will have revealed what type of programming method is desired and the supplier's specification should be examined to determine what is available for the relevant robot. In most cases a teach pendant is provided and this will be satisfactory especially for point to point work. For work such as spray painting, lead through programming will probably be necessary. If it is anticipated that the robot will be employed as part of a complex integrated system the ability to program off-line will be desired.

Interfacing capabilities

Depending upon the application, the robot will have to interface with other pieces of equipment such as NC machine tools, conveyors, process sensors, and if part of a hierarchical control system or network, other computers. Thus the capability of the robot to interface easily with other pieces of equipment is extremely important. At the simplest level, the number of input and output ports both analog and digital, and the associated hardware and software provided, must be considered. Some robots are designed with adequate interfacing capabilities within their controller cabinet, others will require the assistance of programmable controllers or microcomputers as ancillary equipment. However, if the robot is to be integrated within a larger system, compatibility of the robot with standard protocols such as MAP will also be of interest.

Number of degrees of freedom
For each robot, the more degrees of freedom required and provided, the more expensive it will be. Thus only the minimum necessary should be selected, always, of course, considering future possible applications. For example, if a robot is being selected for assembly work, probably only four degrees of freedom will be required. However, if an arc welding application is being considered a full six degrees of freedom will be necessary to ensure electrode angle is maintained when complex three-dimensional curved seams are being welded. For unusual tasks, such as in nuclear reactor maintenance and repair work, it may be necessary to have a full six degrees of freedom plus a relatively large number of extra joints, to provide high flexibility as the arm moves through pipework or around obstacles.

Self-diagnostic abilities
These can range from simple LED status indicators on PCB panels and input/output (I/O) ports, to a fully computerized diagnostic analysis with a description of faults and recommended courses of action, displayed on a monitor screen. The more sophisticated the analysis, the more expensive the system will be, therefore the system chosen will depend upon the nature of the application and the cost of downtime of the robot system. For example, a single robot in an autonomous workcell may only require simple indicators, whereas if a robot is installed as a critical part of a large manufacturing system it may be economically worthwhile to minimize downtime by providing full diagnostic abilities.

Standard end effectors
The availability of standard end effectors suitable for both the robot and the application is a further consideration. Cost and time savings can be made if the design, manufacture and development of special purpose end effectors is eliminated.

Standard sensing
Apart from internal sensing, few suppliers provide sensors as standard. However, some do provide wrist torque sensing and grippers with force and optical sensing. Another aspect is that more robots are becoming available with compatible vision systems supplied by the same vendor. This facilitates setting up robotic installations involving artificial vision as much of the implied interfacing costs are removed.

Memory size
The complexity and length of the task will determine the memory size required. With the decreasing cost of memory in relation to other hardware costs this should not present a problem.

Software capability
The application may be simple point to point but where more complex tasks are envisaged that may require, for example, circular interpolation, the ability of the software to handle the job should be verified.

Installation factors
These relate to the robot specifications that will influence the shop floor installation. Power consumption and the type of supply must be noted as this will be relevant for costing purposes and plant engineering. The robot weight must be considered (particularly important if the installation is to be on an upper floor). Should it be desired to suspend the robot from an overhead gantry, again the weight will be an important factor. The floor area occupied by the robot base will be relevant when considering the integration of the robot within a manufacturing cell, and possible interference with other equipment. The space required at maximum reach is particularly important from a safety aspect. Even if the robot is not expected ever to operate at maximum reach, it is still possible that software and hardware stops could be accidentally overridden. Thus this maximum work envelope should be considered as the volume around which guards should be installed for safety. This is applicable both at ground level and overhead where gantries and walkways may pass nearby. Also on the robot, safety interlocks and other hardware and software safety facilities should be investigated.

Cost
This is comprised of the individual robot plus all associated ancillary equipment and will be discussed more fully in the next chapter. It may be noted here that the basic robot cost can vary between 20 and 80 percent of the complete system. Also a simple, less expensive robot may require more money spent on parts presentation, orientation, and fixturing than a more expensive sophisticated robot. However, each case must be considered individually and a balance found between simplicity and flexibility. If it is decided that a cost ceiling is to be imposed then there will be a 'zero sum' situation amongst the various factors concerned. For example, if maintaining cost constant, then the interrelationship of robot precision (P), reach (R), load carrying capacity (L), and speed (S), can be shown as in Figure 14.3. That is an increase in load capacity will have to be offset against a loss in

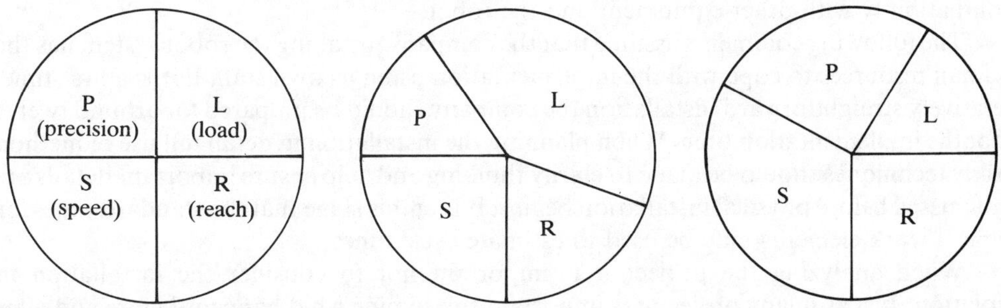

Figure 14.3 'Zero sum' representation of robot performance factors under constant cost conditions

precision, speed, or reach. Or an increase in speed will have to be offset against a loss in some other area, and so on.

At this stage a suitable application will have been found, together with probably a few possible robots able to satisfy the load carrying capacity, precision, and speed requirements, etc. It is now possible to use the criteria and methods outlined in the following chapter for financial justification. This will then allow the final decision to be made on the future of the project.

ROBOT INSTALLATION

The first point to note here is the importance of the human aspects at this stage. Safety considerations should always be paramount, also management and engineers must be sensitive to the feelings of those whose jobs are affected by the robot installation. Education of the general workforce to an awareness level of robotics may also be advantageous to remove some of the mystery and apprehension that can exist. In the case where human workers are being displaced from their jobs by robots, obviously full discussion should take place with the workers concerned, and the unions if present. During these discussions requests for suggestions concerning the project should be made and any suitable ideas included in the final installation.

Throughout the project careful planning is needed. Techniques such as critical path analysis can be put to good use here due to the large number of tasks that can be carried out in parallel. The installation may be undertaken by a 'systems house' who are hired not only to supply the robot but also to take care of the entire installation including purchasing and commissioning of ancillary equipment. The supplier may construct the system on his own premises before its final delivery to the user. The supplier may also be contracted to carry out all programming and 'debugging' to bring the system up to full production capability on the purchaser's site. At the other extreme the user may purchase a basic robot directly from the manufacturer and carry out his own design and development of end effectors, tooling, and other ancillary equipment. In between there are a variety of alternatives. For example, discrete parts of the system may be purchased as complete subsystems. Typically conveyor systems would be purchased in this way, with the supplier ensuring the system was fully operational and advising on interfacing and compatibility with other equipment and the robot.

The following comments assume that the company installing the robot system has the human resources to cope with the implementation using its own staff. If it is a first-time, relatively straightforward installation the company should be prepared for around twelve months implementation time. When planning the installation in detail full use of method study techniques should be made to clarify thinking and help ensure important details are not missed before physical installation begins. It is at this stage that the standard times for typical work elements may be used to estimate cycle times.

When analyzing the project it is important not to consider the installation in isolation. Just as in any project it is important to examine what happens before and after the specific area of concern. For example, it may be possible to ensure that parts at previous operations are immediately palletized in an ordered fashion, rather than dropped into storage bins. The palletization would not cost much more at this stage but would save much expense in devices to position and orientate the parts, or vision systems, at the robotic operation. Going even further back along the production process, it may also be possible to redesign the parts to be more suitable for robotic processing. Going forward from the installation, consideration should be given to what effect robotization will have on subsequent operations. For example, if part of a production line, the line may have to be 'rebalanced' as although the overall production of the robot may be similar to that of the previous human worker, the production rate per minute may be different.

The industrial engineer should be aware of the full potential of robotizing with respect to space utilization within the factory. For example, the robot can be mounted overhead or on vertical rails and thus require less floor space than a human. It can also reach farther and higher to lift heavier loads than a human, therefore rather than store all tools, material, and equipment at or near floor level to suit a human, different workspace configurations can be adopted, although some space should be left for human intervention in the event of a breakdown and for maintenance. If replacing a human operator it will probably be better to redesign and manufacture all associated tooling, jigs and fixtures to ensure optimum compatibility with the robot, rather than try to modify existing equipment.

Before the robot is delivered it will be necessary to ensure that the floor is suitably prepared for robot mounting, i.e. it must be flat, level, and able to take the robot weight. It may also be necessary to isolate the mounting area from adverse vibrations emanating from near-by machinery. Services such as electric power lines for drives, controllers and hydraulic power packs, and also air lines for pneumatic equipment, must be planned and provided.

There is a large amount of detail to be considered; for example, the choice of limit switches, proximity switches, safeguarding equipment, programmable controllers, conveyors, pallets, bowl feeders, etc. has to be made. Proven, rugged, equipment should be chosen where practicable and solid state components used wherever possible to ensure maximum reliability. Most problems in robot systems arise not from the robots themselves, but from the ancillary equipment. Electrical noise can present interference problems such as corruption of control signals, therefore well shielded cables should always be used for data transmission. Use of standard components where possible minimizes costs, increases the probability that they will be familiar to maintenance personnel, and makes the acquisition of replacements easier.

When the system is nearing completion individual pieces of equipment should be first tried and tested in isolation where possible. Then gradually more associated equipment can be linked together until, after modifications and development, the whole system can run in an integrated manner. The first trials of the robot should then be one step at a time. Once an initial program has been devised it should be run at slow speed moving from one programmed point to another in individual steps. When all appears to be working smoothly the speed can be gradually increased until production speeds are obtained. All steps taken and problems met during the installation should be recorded in detail for future reference. This should ensure all lessons learned at this installation are applied to future projects.

Some applications may require special development. For example, if the robot has to work in an abrasive, corrosive, or otherwise harmful atmosphere, a flexible protective cover or sleeve may have to be used to cover the robot. In some cases pressurized air can be forced into this sleeve to provide a pressure differential thus preventing harmful particle ingress. This principle can be applied in reverse in, for example, clean room work in the electronics industry. Here air is extracted from the inside of the robot and drawn through filters, decreasing the air pressure inside the robot relative to its environment and preventing particles of lubricants, bearings, or motors, etc. escaping into the clean room atmosphere.

ROBOT MAINTENANCE

The latest electric robots with brushless motors, and high quality transmission, bearing, and control components require very little maintenance. Routine maintenance will mainly be composed of checking, say every six weeks, for wear, and cleaning and lubricating where necessary, usually the manufacturer will state recommended procedures and timing. For fluid power robots additional maintenance will be required such as changing of filters, changing oil, replacement of seals and couplings. Apart from this minor routine maintenance, other major work is probably best left until the situation demands it because unnecessary disturbance of the robot is often more likely to introduce reliability problems rather than prevent them. However, full plans for maintenance action should be made in advance, to be implemented as required.

The maintenance personnel should be well trained by the robot supplier. Training courses may range from two days to two weeks depending upon the complexity of the robot system, or depth of knowledge required by the maintenance engineer. Once again the supplier may be engaged to provide maintenance under a service contract, especially if a large system has been purchased. In this case the supplier will often be obliged to provide emergency attention within a certain maximum time period from notification by the user, usually a few hours in cases of high volume, high value production. In the event of in-house maintenance personnel being trained, a variety of skills will be required for full effective cover. This can be done by having a few multiskilled technicians, each capable of handling the full spectrum of faults that may occur, or creating a maintenance team comprising individuals with particular areas of expertise in the electronic, electric, pneumatic, hydraulic, programming and mechanical fields. On return from training, full course notes and maintenance manuals should be available.

As well as comprehensive manuals and procedures to assist in maintenance and fault finding, other facilities can be provided. Some of these were listed in the section on robot specifications earlier in the chapter. These are diagnostic facilities, and even at their most basic they can considerably reduce the total downtime of the system. Estimates of average robot downtime due to faults show that it is composed of roughly 50 percent diagnosis, 25 percent repair, and 25 percent test, assuming prompt attention with no waiting time involved. The simplest type is the use of warning lights on the operator's control panel signifying a robot malfunction. One step up from this is the use of warning LEDs on the PCB units and I/O ports within the controller cabinet. These can be used for example to indicate whether data are being transmitted and/or received by each particular unit. At the highest level, full textual diagnostic information can be displayed on a monitor screen. For more complex analysis of the robot system it may be possible to connect the robot computer with the supplier's diagnostic facilities via a telephone and modem system. When carrying out maintenance or repairs on any machine, robots included, a detailed record of faults found, action taken, results of that action, and overall time taken to complete the task, should be made. This will facilitate future repairs and provide data for costing purposes.

Although many faults may occur due to simple causes, such as a switch failing to operate, or ingress of dirt into the control system, it is still advisable to hold some spare parts in stock. How much to spend on these spares can be evaluated quantitively by comparing their cost with the cost of lost production through robot downtime. For example, for a robot used in an educational establishment little or no spares need be carried. This is because if a breakdown occurs, even at a critical time such as during a demonstration, no great cost is incurred as the demonstration can often be rescheduled. However, at the other extreme in, say a car plant where a series of robots is arranged in a line and each is dependent on the performance of the previous one, the cost of downtime will be extremely high. In this case it may be advisable to have as a spare a complete replacement robot. Between these two extremes the user will decide the most suitable level of spares, with about 10 percent of the base robot price often being quoted as a rough guide to the value of spares which should be held in stock.

15

ECONOMICS AND JUSTIFICATION

INTRODUCTION

The economic aspects of robotization can be grouped under four headings as follows:

1. The macroeconomic implications of robots and other flexible manufacturing systems for national economies.
2. The strategic implications of this technology for organizations.
3. The effects of robotization at a micro or tactical level within the company.
4. The methods and techniques used for the financial appraisal and justification of robot installations.

The purpose of this section is therefore to examine these aspects, and by so doing understand the economic implications and problems of applying industrial robots, from a global view down to an individual project level.

OVERVIEW OF ECONOMIC IMPLICATIONS AT MACRO LEVEL

At the moment industrial robot sales are a very small part of the world economy. However, it is the anticipated rate of growth, assuming an improving world economy, and the worldwide inflow of funds to research, development and marketing of industrial robots that is so significant. Large international companies such as IBM and General Motors are in the field along with the more traditional robot producers like Unimation, ASEA, and Cincinatti Milacron. There have also been regular reports of licensing agreements and acquisitions between electronics and engineering companies worldwide. Appendix 1 discusses recent robot population and growth statistics. If it is assumed that the world economy is going to improve, then it can also be assumed that the robot population will grow. However, the rate of this growth should be much higher than the general economic growth due to the many advantages of robots described in previous chapters. Thus as time progresses the financial and social impact of robotics will be increasingly felt.

Optimistic views of the next few years anticipate that robots will displace only an insignificant amount of direct workers. However, the cost of this in financial terms alone must give cause for concern considering the cost of the existing world pool of unemployed. Nevertheless there is the possibility that the increased productivity resulting from the installation of robots may in fact increase employment and not reduce it. There will also be employment generated by the manufacture and design of robots and subsequent installation design, programming, maintenance and supervision.

Robots, and flexible manufacturing generally, may be the influences that stop the flow of manufacturing investment overseas. There has been a trend for the West, especially the USA, to utilize the cheap labor rates in the more recently industrialized nations. However, as the demands for higher salaries in these countries increase, the cost of transport increases, and trade barriers become more complex and expensive to overcome, it will become more attractive for a company to automate, manufacture close to the market, and avoid trade restrictions. One concern here is that the technical expertise and infrastructure, which should be better in the West, may already have become so atrophied and denuded in some countries, that it could be difficult to find enough resident engineering skills and experience.

There is certainly a desperate need for manufacturing investment if an industrialized country wishes to remain economically competitive. Since the Second World War the world economy has drastically changed. Countries whose manufacturing base was destroyed have been able to rebuild from scratch with the latest technologies and necessarily new philosophies. On the other hand take, for example, the UK which was called in Victorian times 'the workshop of the world' but is now far behind its traditional competitors in terms of manufacturing output. However, it should be noted that lack of investment is not the whole answer; some studies have in fact, shown the UK to be ahead in investment in some areas. Investment in new technology is important, but even more important is good management.

A major factor governing the attractiveness of investing in robots is the prevailing attitude of the government. Various incentives to promote investment may take the form of investment grants, tax allowances on depreciation, and assistance with the cost of feasibility studies. At the time of writing the trend in Britain has been to reduce the assistance for investment in capital equipment as part of attempts to stimulate labor employment in, for instance, the service industries. In contrast, the USA is attempting to make investment in high technology goods more attractive by organizing its tax laws to suit, and Japan's economic structure and relationship between the government, banks and industry are ideally suited to the rapid introduction of new manufacturing technology, one major reason being the longer time allowed to recover the money invested in new equipment.

There are a number of reasons given for the present low rate of investment in the more expensive capital goods; some of these are:

1. In the current world recession many factories are working at a fraction of their existing capacity, they see no obvious need to install equipment to improve output.
2. The western philosophy to date has been essentially to employ robots to reduce the

unit cost of production, through savings in direct labor. They have been seen as a means of raising output without increasing the labor force, or of maintaining that output with fewer people employed. Because of this it may be noted that in countries where wage increases are kept low, the incentive to apply robots is much less. This attitude is apparent by the prevalence of the much maligned but simple 'payback period' methods, described later, as a rule of thumb guide. This in turn leads to attempts to ensure that a robot is not employed unless it replaces two or more operators over a number of shifts.

3. Lack of business confidence in the West leads to demands for the aforementioned payback periods to be in the region of two years. Whereas in Japan, for example, the more likely payback period is five to seven years. In fact the reprogrammability of robots means that they will probably have a useful working life of at least ten years.
4. High interest rates. These cause two things to happen: (a) it becomes more expensive for the firm to borrow money which leads to higher demands being made at the justification stage before the project is deemed viable, and (b) the firm is tempted to invest its profits in savings of some kind to earn high returns. This means that the company is not developing its technology base in the form of capital equipment and experience and is now susceptible to competition that may have invested in the new technologies, including robots, and is able to offer superior goods at a lower price. Thus the company's timidity to spend money in the riskier capital investment results in a loss of sales, and if the attitude prevails throughout a country, rapid decline of that nation's manufacturing industry ensues.

THE STRATEGIC IMPLICATIONS FOR THE ORGANIZATION

When considering investment in robots the organization should be aware of their broader implications. Industrial robots properly applied will improve company productivity and quality of life of employees almost immediately. However, the *long term* benefits and implications for the *whole* organization should also be considered. If advantage is to be taken on a continuous basis of the latest ideas and equipment in, for example, computer integrated manufacture, then the company must have the structure and flexible facilities to cope with the constant changes that are required.

One way for an organization to analyze itself to determine how and where robotization could help, is to look at its overall costs by using a Pareto analysis. It could look at aspects such as costs of direct labor as a percentage of total company costs, and thus determine where maximum attention should be given. Traditionally when attacking costs it is usual to tackle labor costs first, then overheads, and lastly material costs. In fact, analysis often shows that the highest costs incurred are in material, followed by overheads, and last of all by labor. The advantage of robots is that they positively contribute to reducing costs in all categories, even if it is labor cost reductions that may initially have been aimed at.

When a company has invested in robots or other flexible manufacturing equipment, it should find that it is more able to cope with changes in market demand and product designs. It will be easier for the company to embark on new ventures and explore new

Pareto's law states that roughly 80 percent of effect is derived from 20 percent of the number of items. The Pareto curve opposite illustrates roughly how various cost sources could be grouped to analyze where attention should be concentrated to achieve cost savings.

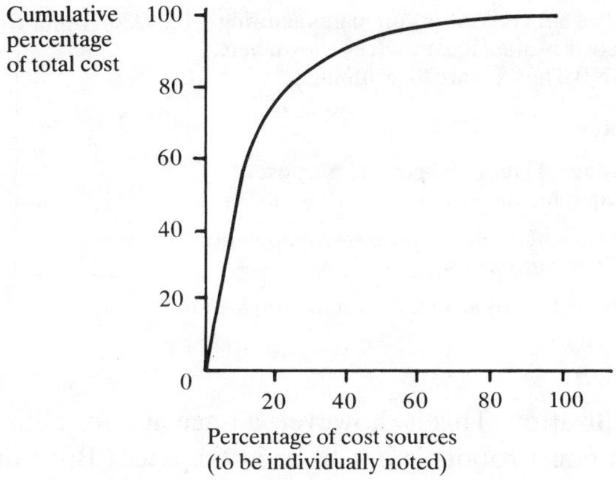

markets with a greater ability to seize new opportunities. The improvements in productivity and quality will make the firm more competitive in the world market. Also, having a strong and flexible manufacturing base it will cope more readily with innovations from the competition and the vagaries of the market.

Some other strategic implications for the organization are:

- Robotization necessitates, and therefore promotes, close cooperation with various disciplines and services within the organization.
- Industrial robots with their computer control are ideally suited for linking in with other computer systems and can form part of a hierarchical control structure.
- Reconsideration of product design is necessitated regarding suitability for production, e.g. for assembly the ideal design will allow assembly to proceed as a stacking operation with one component added on top of another, or inserted vertically downwards.
- Redesign considerations often result in material savings.
- Redesign often results in an improved product.
- The company will probably need to improve quality control by introducing automatic inspection and testing.
- A reduction in direct workers should mean a resulting reduction in support workers.
- The flexibility obtained will give fuller utilization of plant and equipment, this is because downtime between changes of product is reduced and smaller batch sizes can be handled.
- New products and modifications can be introduced more quickly.
- Due to reductions in delivery times it may be possible to deal directly with the end customer rather than use agents and stockists.

It is also worth noting that a human operator experiences a learning curve as he or she becomes competent at a particular task. This will be repeated, with the resulting increase in cost per unit, every time a new operator is required and training takes place. However, a robot once installed and operating starts immediately at the 'competent' level of the training curve. Also to be considered is the time required to achieve fully effective

The effect of robots on manufacturing costs is often illustrated as shown here. (NB The axes are logarithmic.)

Key

Manual labor and general purpose equipment ——- - ——- - ——

Dedicated, special purpose equipment (hard automation) — — — — — —

Industrial robots (soft automation)———

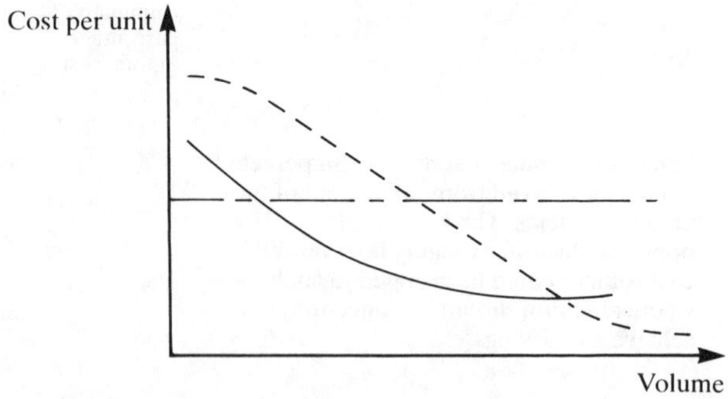

utilization. This is, however, a once and for all factor as labor turnover and absenteeism amongst robots is not to be anticipated! Both these curves are shown here.

Labor costs should be considered realistically. For example the cost of direct manual labor is much higher than is at first apparent, e.g. if an operator's wages are taken to be $10 per hour in the USA or £4 per hour in the UK then additional costs must be added on for insurance, pensions, and other benefits. This takes labor costs up to $17.5 or £7 per hour. The figures, incidentally, highlight the fact it is much easier to justify a robot installation in a country with higher labor rates as the cost of a robot does not change drastically from one country to another.

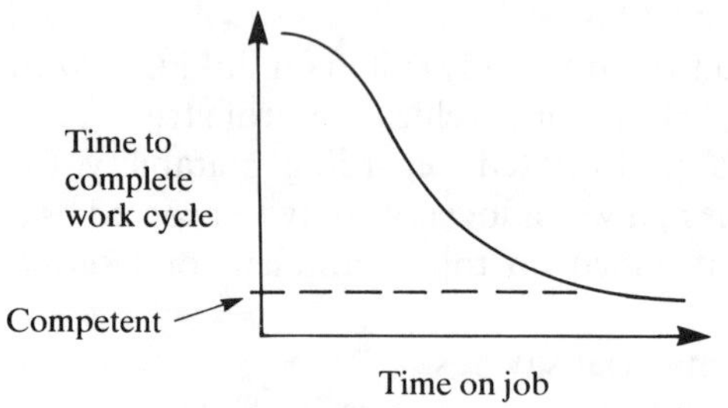

The human learning curve

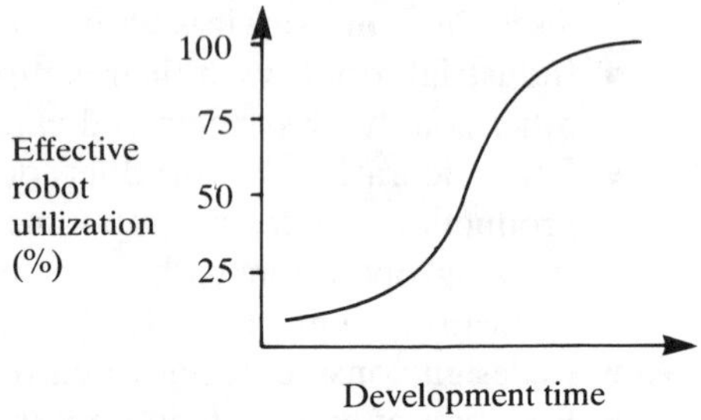

The robot utilization curve

The consistency of the robot should also be remembered. In actual operation the robot may have a longer cycle time than a human worker, but the consistent repeatability of the robot will probably ensure a higher output over the whole shift. For example, during arc welding operations a human welder, for environmental reasons, can only keep an arc burning for between 25 and 30 percent of his working time, for a robot this can rise to 90 percent. Also on die casting machine applications increases in productivity of 40 percent over manual methods may be achieved throughout one full shift. This increase in productivity, providing the volume of work is there to support it, further modifies the figures used above. Taking as a rough guide the fact that a human operator can only work effectively for 75 percent of an eight hour shift, then the equivalent labor rate that should be being used in calculations would be around $23 or £9 per hour. This compensates for the allowances that have to be added on to a human operator's basic time for relaxation, fatigue, and other personal allowances.

This leads on to the tactical considerations for the organization. The strategic effects and implications are relevant to the company's long term goals and maintenance of its own market position, but it must also be aware of the more detailed implications for its manufacturing operation. These detailed facts are also very necessary when composing a proposal to purchase an industrial robot. As well as the more obvious direct savings, there will be indirect savings such as reduced costs associated with stock and stock control due to faster throughput of work, greater machine efficiency and utilization, reduced operator training time, the elimination of secondary operations, and reduced inspection costs.

THE TACTICAL CONSIDERATIONS

Having examined the general economic environment of industrial robots, and the importance of taking a strategic view of investment for the whole organization, attention is now given to the areas that have to be quantified to enable a comprehensive capital investment appraisal to be carried out. Some areas can be quantified quite easily, for example the savings on direct labor costs. However, other areas, just as important, are difficult to quantify, e.g. indirect savings on reduced work in progress. These indirect savings *must not* be ignored as they are most certainly present, and may in fact make the difference between a nonviable and a viable project. The following is a description of costs and cost savings that should be calculated or estimated when compiling figures to use in the justification analysis.

Costs incurred

Direct capital costs

1. *The robot system.* This may be given as a consolidated figure by the supplier, or it may have to be compiled by the purchaser if separate items are being bought from various suppliers. In any case this cost will comprise the following.

(a) The industrial robot. This is the cost of the basic robot arm, controller, and power supply, and, for example, programming system. Also included, if the system was a welding one, would be interfaced turntables and associated welding equipment. Probably also included here would be basic training in programming and maintenance for the particular robot purchased.

(b) The end effector. This may be a standard off-the-shelf type gripper or, more likely, a purpose-built end effector suited to the job. This cost would be quite high if a sophisticated gripper or tooling system is required, typically between 3 and 10 per cent of the robot price.

(c) Additional equipment. Program storage facility, such as floppy disk drive and disks, although this is increasingly likely to be included in the robot cost in (a). Also a computer terminal and visual display unit, this again would be standard with a robot designed with off-line programming capability.

(d) Sensing facilities. This would relate to arm mounted sensors such as force transducers mounted on the end effector.

2. *Ancillary equipment.* This will comprise the equipment necessary to complete the

finished facility. Items included would be part feeding, orientation and holding equipment, such as bowl feeders, work fixtures, conveyor systems and turntables. Also, any additional tooling required such as motorized screwdrivers and tool holders. Vision systems used for part identification or inspection might also be included here.

3. *Perimeter guarding and other safety equipment*. This could comprise mesh fencing and gates, mechanical, electrical, or pneumatic interlocks, photo sensors and other devices. It should be pointed out at this stage that the total cost, in terms of equipment, might be composed of 80 percent robot cost and 20 percent additional equipment in simple applications; or up to 20 percent robot cost and 80 percent additional equipment in more complex applications.

4. *Cost of spares*. The optimum amount of spares necessary will depend upon the particular operation. An average value for this figure may be about 10 percent of the robot price.

5. *Residual value*. This is a negative cost and is the value of the robot at the end of the project life for which it was originally purchased. Because a robot is inherently adaptable, this value is important. The actual figure used will be an estimate and will depend very much upon the type of robot purchased and the application involved.

Labor and indirect costs

1. *Planning costs*. These are the costs of industrial engineering time spent on the feasibility studies, and detailed planning of the installation.

2. *Installation costs*. These will be mostly labor and some material costs, involved in laying on electrical and pneumatic services to the robot. Also setting up robot and installing the peripheral equipment.

3. *Development costs*. These are the costs involved in bringing the robot up to a standard adequate for production purposes. For a novel application this may be very high, say one man-year, but for well tried applications the time could be quite short, especially if a similar installation has already been attempted.

4. *Programming costs*. These will initially be part of the development costs, but after handing over the facility to production they will become a recurring cost directly related to the type of application. Programming costs for simple pick and place type operations should be quite low, whereas programming continuous path operations such as deburring or fettling could be quite high. The overall programming cost per year will depend upon the frequency of arrival of new work for which previous programs are not available.

5. *Maintenance costs*. These will be the labor costs involved in both preventive maintenance and breakdown repairs. It should be noted that these may be positive or negative as maintenance costs may actually be saved when, say, changing over from old machinery and equipment. The costs should not in any case form a very high percentage of

overall total cost. They will, of course, vary with the number of robots installed and the complexity of the installation.

6. *Training costs.* These will be borne, up to a specified limit, by the supplier. However, cost of transport to, and accommodation at, the supplier's training establishment will probably be borne by the purchaser. Also, ongoing training costs for new supervisors, maintenance personnel, and programmers will occur at intervals.

7. *Power requirements.* These should not be very high, a typical range being from 1.5 KVA for smaller robots, to 30 KVA for larger models. Compressed air at 75 psi (5 atm) should also be provided.

8. *Space requirements.* These will be a cost in a new factory, or 'green field' installation, but when replacing some other method, usually manual, it will almost certainly represent a saving, probably as much as 50 percent or even more.

9. *The cost of capital.* If a bank loan is taken to pay for the equipment, there will be interest to pay; if the money comes from shareholders they will expect a certain return for their investment. There is also the opportunity cost, i.e. what the firm would earn if it invested its money in some other way. This value would be represented as a rate of return expected by the company from the project, and would probably be included in the investment appraisal calculation.

Cost savings

Direct cost savings

These savings can be relatively easily calculated and are the ones most readily accepted by the accountants when engineers are attempting an investment justification.

1. *Direct labor savings.* These represent the human operator's annual wages, holidays and fringe benefits, and any additional payments that the government demands from the employer for each worker employed. Although this total figure is commonly used, and will be used in later calculations, it is interesting to consider what happens in reality. If it is a completely new installation then there is no problem, the robot is simply doing the job of an estimated equivalent number of human operators. But if it is an installation to replace an existing manual operation the human operators will either be made redundant, in which case redundancy payments are incurred by the company, or they will be moved to another location, in which case their wages will still have to be paid. If the redundant workers are unable to find alternative employment then they will become an extra financial burden on the state; if relocated then the direct wage savings will not really be experienced until they leave or retire, when it is probable that their jobs will simply not be filled and a general labor allocation adjustment will be made.

It is this complex interaction that suggests that labor costs are seldom actually saved, but rather redistributed within the company or the country as a whole. However, it is a recognized method of justifying capital investment in equipment that the direct costs of the labor displaced by the equipment are counted as cost savings for the purposes of the calculation.

2. *Savings on secondary operations.* When robots are introduced they often stimulate reorganization of the whole manufacturing process. This may mean elimination of manual inspection operations and removal of other secondary operations such as fettling or washing, these operations often being automated as part of the robot system.

3. *Increased output.* Although a robot may not work as fast as a human in some operations, it can be expected to produce more 'units of work' in a period of one day. This increase in productivity due to consistency of operation can be quantified as a direct cost saving, e.g. if over a shift the robot increases production by 15 percent over the manual rate, then instead of the labor costs for only one operator being saved, the savings are the equivalent of 1.15 operators.

4. *Material savings.* These are not so easily anticipated but can be quantified to a certain extent if a test facility is available. Due once again to the robot's consistency of operation, savings on material can be anticipated in, for example, spray painting applications through less overspray occurring. Also in die casting machine servicing, the constant operating cycle of the robot means fewer rejects and hence better material utilization.

Indirect cost savings

The direct savings are quantified by using data on current wage rates, work study department data, and the past experience of others using robots in similar applications. But there are other areas of cost saving just as important. These are indirect cost savings, and because of their nature they are not so easily quantifiable. Nevertheless, some estimate should be attributed, being compiled from previous experience, or if none exists, from discussion and research in conjunction with the work study, quality, accounting, purchasing and production engineering departments.

1. *Savings due to a reduction in scrap and rework.* As the robot system will only be turned over to production once it has been proved and is running satisfactorily, it follows that the proven automatic process should produce consistently good parts. This is evidenced in a reduction in the scrap costs, i.e. material and time savings, and the amount of rework, i.e. operator rectification time.

2. *Savings in training time.* Due to labor turnover and the learning curve, training and associated costs can be high for manual operation. Robitization reduces these costs considerably.

3. *Inventory or stock savings.* When robots are applied a faster turnover of stock and reduced lead time is experienced, also the ability to plan efficient schedules of work is enhanced. These factors do have a quantitative value even though only a rough estimate may be used. This saving is apparent when the following is considered. When material is being worked on value is being added to it as well as cost incurred. However, between

working operations cost is still being incurred but no value added. It has been estimated that out of the total time the material spends in the factory, value added time may be only 5 to 15 percent, the rest of the time being taken up by waiting between operations, transport, storage, and perhaps even being lost. This material is costing the company money all of this time due to, for example, interest rate charges on the money borrowed to purchase the material, and the cost of the floor space occupied for storage purposes. It follows therefore that if the material is passed more quickly through the factory and the proportion of the time spent adding value increased, then overall costs will be decreased.

4. *Other savings*. There will be a number of other areas where savings will result but these will vary from application to application. For example if wholly automatic inspection has been introduced, then a proportion of a human operator's time will be saved, the floor space occupied for the inspection operation will probably have decreased, and there will be no waiting time for inspection.

Also included here are the items previously mentioned in the costs incurred section, i.e. maintenance, training, power and space requirements. These may actually be cost savings when considered relative to the costs incurred by the old method.

Other considerations

There remain a few, extremely important factors to consider for inclusion in any robotization justification analysis. These will vary enormously from country to country and, to a lesser extent, from region to region within a country.

1. *Taxes*. This refers to the tax system in operation in relation to taxes on profits and tax allowances given to enable the cost of capital assets to be written off against profits.

2. *Government aid*

(a) Aid offered as an incentive to invest in high technology equipment such as industrial robots. This would probably be in the form of assistance with feasibility studies and development costs, and special grants towards purchase of the new robot.

(b) Regional grants offered to encourage general manufacturing investment in particular geographic areas of a country.

FINANCIAL APPRAISAL TECHNIQUES

The first item to consider is the fact that the company will have a marginal investment rate, or minimum profitability criterion. This is necessary to enable the firm to decide if the project will return any profit. As mentioned previously the firm has a certain cost of capital; thus if the firm's funds cost 15 percent and the firm employs these funds in projects that earn only 12 percent, then it will soon be out of business. The firm must therefore stipulate a minimum rate of return on its projects if it is to remain solvent. If that firm invests in projects with rates of return down to, but no lower than, for example 15 percent, then this is the firm's marginal investment rate. It is emphasized that, particularly where robots and other automation are being considered, the company should adopt a long term approach in order to turn a vision into a reality.

When financially appraising projects, two criteria must be fulfilled:

1. The appraisal method chosen should ensure that no project is chosen that gives a rate of return lower than the cost of capital or marginal investment rate.
2. When a number of mutually exclusive projects are being considered, the appraisal method should enable these projects to be ranked in order of expected profitability.

There are a number of techniques used which can to some extent fulfil one or both of these criteria, but a thorough analysis can only really be done when the full discounted cash flow techniques (explained below) are adopted, incorporating all the previously mentioned cost factors.

The payback period

Using this method the time required for the firm to recoup the initial outlay is calculated. The net cash flows resulting from the project for each year of its life are estimated. When the cumulative sum of these cash flows is equal to the original sum invested then the payback period is said to have elapsed. This method is often used in industry: it is quick to implement and convenient as a rule of thumb. It is also useful in very high risk areas where uncertainty prevails and prediction of cash flows after a few years into the future is not possible. However, it is not suitable for carrying out a thorough and reliable investment appraisal. This is mainly because it does not give any indication, or comparison, of the relative profitability of various projects, or take any account of cash flows after the payback horizon.

The accounting rate of return

This method compares the cash inflow from the project with the original outlay. It does this by taking the average yearly cash inflow over the expected project life and expressing it as a percentage of the outlay, thus giving a rate of return for the project. However, this method is again not entirely satisfactory for comparing projects as it does not properly take into consideration, the timing of the cash flows or the differing expected lifespans of the projects. In addition, the actual definition of the method varies from company to company.

The problems with these two methods are highlighted by this example. Three mutually exclusive projects are considered by both the payback and the accounting rate of return methods. The outlay for each project is estimated to be equal at \$70 000 (or £), and it is assumed that the funds the firm has available for investment amount to \$70 000. Thus the projects are all competing for the same amount of funds and the two techniques are being used in an attempt to decide which one to choose. For the sake of clarity the increases in monies coming into the company, as a result of the robotization project, are contained under the term 'cash flows'. This is the net income estimated from the project after profits, capital allowances, taxes, etc. have been taken into consideration. In reality the cash flows used in the accounting rate of return calculation may be obtained in a way that will enable the rate of return found to be compared directly with the company's marginal investment rate, so ensuring that the projects all satisfy that criterion. This may make the cash flow figures used slightly different in the two methods; however, the values used in the example are adequate for illustrative purposes.

Project	X	Y	Z
Investment	70 000	70 000	70 000
Cash flows in year			
1	35 000	7 000	7 000
2	14 000	14 000	7 000
3	14 000	14 000	7 000
4	7 000	35 000	14 000
5			17 500
6			17 500
7			35 000
8			35 000
Total	70 000	70 000	140 000
Payback period	**4 years**	**4 years**	**6 years**
Project life	4 years	4 years	8 years
Average cash flow	17 500	17 500	17 500
Rate of return	**25%**	**25%**	**25%**

It is evident from the above example that both methods are totally inadequate at providing information on which to base a decision. Projects X and Y both have a payback period of four years, whereas project Z has a payback period of six years. This would appear to exclude project Z from further consideration. But is is obvious that project Z will run for twice as long as the other two, and also that it returns twice as much in cash flow; the payback approach has not considered this. When projects X and Y are considered using the payback figure they appear equally acceptable. But project X returns most of the investment in the early years, whereas project Y has higher cash flows in the later years; the payback period has not taken this into consideration. This is important as money is worth less the further into the future it is obtainable. Looking now at the results found by the accounting rate of return method, it is apparent that it has yielded no useful information, all the projects being ranked as equally acceptable with a 25 percent rate of return. This again is not satisfactory, as the three projects have different patterns of cash flows and different life spans. One of the projects is surely more suitable than the other two, yet the accounting rate of return method gives no help whatsoever, and the payback method may at first sight be actually wrong in excluding project Z completely.

Discounted cash flow

What is required is a technique that will take into account all the tactical factors previously mentioned, to give quantifiable data for the capital tied up in the project and its residual value, and the net cash flows resulting from the project. The method should also be able to compare the investment with the income, it should include due consideration for the timing of the cash flows, and recognize the rate of return required by management before the project can be accepted.

One technique that does this is generally known by the term DCF or discounted cash flow. It essentially relates all income and expenditure to a base year, say year 0. When this is done all monies anticipated in the future are discounted back to a present value in year 0, thus ensuring all figures used in the appraisal, and possible comparison, are compatible. This is necessary due to the time value of money. An an illustration consider that if someone was offered \$100 today or \$100 in one year's time, the obvious choice would be the former, one reason being that if the \$100 was invested today at 10 percent interest, it would be worth \$110 one year from now. Conversely therefore if offered \$110 in one year's time or \$100 now, one might exhibit indifference. This is because the \$110 when discounted back to today's value is equal to \$100 assuming a 10 per cent rate of return is anticipated. Thus the discount factor used if a 10 percent RR is required is 0.909 for one year.

There are two methods of using DCF. One is the internal rate of return method in which the rate of return that will be earned on the amount of capital invested in the project over its life is calculated. This IRR will bring all future cash inflows to the same present value (PV) as the original sum invested. The second approach is the net present value (NPV) method. In this method the minimum acceptable rate of return is specified by the management, and this rate is used to calculate the present value of future cash flows resulting from the project. This total PV of future cash flows is then compared with the PV of the sums invested and the ratio calculated. This ratio is the profitability index (PI). Thus if the PI is less than unity then it is not giving a rate of return equal to that used for discounting. If a number of alternative projects are being compared then they can easily be ranked in order of profitability using this index. Both the IRR and NPV methods are often included in microcomputer programs that accept as input all the costs and cash flows mentioned previously, and output a rate of return and profitability index. An example showing the NPV calculation is now given. It must be remembered, however, that estimation of the cash flows is difficult, not only because of the uncertainty of future income, but also because of the work involved in quantifying the costs and cost savings that were listed in the section on tactical considerations.

A company in the UK is considering the purchase, in say year 0, of a simple robot installation at a total cost, for the robot and ancillary equipment, of £70 000. Due to the nature of the stocks, spares, work in progress, etc., directly associated with the project the firm will require to tie up an additional £5000 in working capital. This money will be freed at the end of the project's life. No profits are expected from the project until year 1 when they are anticipated to be £20 000. Profits in subsequent years should be as follows; £35 000 in year 2, £30 000 in year 3, £20 000 in year 4, and £15 000 in year 5. UK capital allowances will be taken advantage of by charging them against profits. The tax system in operation at the time will not allow a special high first year allowance, but will allow a writing down allowance of 25 percent of the reducing balance. The residual value is estimated to be £16 600. Corporation tax on profits will be 35 percent. The company requires a minimum rate of return of 16 percent after tax on this investment. A grant of £10 000 is available towards the cost of the robot. What is the profitability index of the project?

The example is much simplified; also certain assumptions have been made, e.g. tax is payable the year following the profits although the company is able to take advantage of the tax savings from the capital allowances immediately. However, the purpose of the

		£
Net cash investment; (Year 0)	New machine	70 000
	Additional capital	5 000
		75 000
	Less grant	10 000
		65 000

Year	Increase in profit	Tax (35%)	Capital allowances	Tax saved	Cash flow	PV factor	PV
1	20 000	—	17 500	6 125	26 125	0.862	22 520
2	35 000	7 000	13 125	4 594	32 594	0.743	24 217
3	30 000	12 250	9 844	3 445	21 195	0.641	13 568
4	20 000	10 500	7 383	2 584	12 084	0.552	6 670
5	15 000	7 000	5 537	1 938	9 938	0.476	4 730
6		5 250			(5 250)	0.410	(2 153)
						Total	69 570
Residual values			Machine = 16 600				
			Working capital = 5 000				
			21 600			0.410	8 856
						Grand total	78 426

Therefore profitability index = 78 426/65 000 = **1.21**

example is to illustrate how a number of relevant variables can be taken into account to arrive at a quantitative ratio on which to base investment decisions. As the profitability index is greater than unity, in fact 1.21, this indicates that the project should yield a rate of return greater than 16 percent. If a number of alternative projects are being considered they can be evaluated in a similar manner and then ranked according to their PI. It should be realized that although the system attempts to be very precise and thorough the figures for the cash flows are based on estimates of future performance. This means that once the PIs have been obtained other factors may also influence a decision, e.g. the relative riskiness of each of the projects, or the need for the company to gain experience with industrial robot applications.

CONCLUSION

The general economic implications of industrial robots are far-reaching and offer opportunities for companies, and indeed nations, for improvements in productivity, and stimulus for reorganization and product reappraisal. However, at a tactical level all relevant factors concerning labor, material, equipment, quality and prevailing financial conditions must be fully considered to conduct a realistic analysis of the robotization project.

The economic answer to a manufacturing problem will not always be to robotize, but it is only appropriate that the possibility should be examined and its justification proved or disproved on a completely rational basis using all the relevant data obtainable.

16

HUMAN FACTORS

SAFETY

Introduction

Industrial robots are moving machines, and the safety hazards associated with moving machinery, and their electric power supplies have long been recognized. The robot, however, presents some unique additional safety problems that have to be addressed. These arise from the following characteristics:

1. The robot arm and end effector move within a work envelope volume very much larger than the robot's own base, the only exception to this being the cartesian gantry type robot. This large danger zone within which an arm with possibly six degrees of freedom can move around in, can be difficult for the inexperienced passer-by to anticipate.
2. Due to the fact that the robot moves in response to commands from its controller it can appear to the casual observer to move both spontaneously and unpredictably. This is especially dangerous when a delay has been built into the program and the robot is immobile while awaiting a command. In this case the robot, after a period of quiescence, may suddenly burst into activity.
3. The programmability of the robot means that the integrity of the control system hardware and software is extremely important. Faults in hardware or errors in programming can cause erratic, unexpected behavior.
4. The fact that the heavier duty robots are powerful, massive and fast makes them more dangerous. When a robot is designed for moving heavy loads with precision it is currently constructed with strong inelastic materials to prevent deflection. Also the actuators, drives and transmission components will often be heavy. When this is considered in conjunction with what may be at the end of the arm, e.g. a spot welding gun or large gripper holding a steel billet, it is evident that a moving robot is a force to be reckoned with.

It is ironic that robots are often introduced into factories to eliminate hazardous jobs for humans, yet the robots themselves contribute their own hazards. It is still unusual to find present-day commercial industrial robots with integral safety sensors. This means that they are oblivious to the effect they have on their environment and vice versa. They could therefore cause injury to a human, damage another machine or robot, or damage themselves, without knowing it. However, there are some suitable sensors now available for attaching to robots and these will be considered later.

This section is intended to provide a general overview of the areas of most singificance. There is still considerable activity in the area of legislation with much liaison between safety authorities in various countries. However, there have been guidelines and standards produced; e.g. in the UK there is the Health and Safety Executive (HSE) guidance booklet *Industrial Robot Safety*, and *Safeguarding Industrial Robots, Parts 1 and 2* produced by the Machine Tool Trades Association (MTTA); in the USA there is the American National Standards Institute and Robot Industries Association *ANSI/RIA R15.06-86 Industrial Robots and Robot Systems – Safety Requirements;* and in Germany there is *VDI Guideline 2853 Safety Requirements relating to the Construction, Equipment, and Operation of Industrial Robots*. There is also basic legislation in existence in most countries for industry and industrial equipment and this must be adhered to in all cases, e.g. in the UK there is the Factories Act 1961, and the Health and Safety at Work Act 1974. There are also government regulations, many of which are applicable to robotic installations.

Throughout this section it will be attempted to follow through the safety aspects as they are encountered in chronological order from the basic robot design through to final production operating conditions.

Robotic accidents: causes, effects and prevention

The dangers of industrial robots have their sources in a number of areas. As the general statement, 'accidents do not just happen, they are caused', implies, most accidents can have their source traced back to human failure at some stage and can be prevented if sufficient foresight is applied by all concerned.

The causes can be grouped under seven main headings:

1. Poor design. This should be uncommon but could give problems due, for example, to an inadequate material being used for a seal or a coupling, or mechanical and electrical components not being adequately protected from their working environment and allowing dirt ingress or electrical interference. The former will give rise to problems with the robot hardware, e.g. leaks, motor failure, loss of power, the latter will cause apparent software faults such as unexpected movements and missed operations or cycles.
2. Poor quality components. Examples here would be accidents caused by premature failure of components such as servo valves, electronic components or safety equipment hardware.
3. Programming errors. These are most likely to occur when a new program is being tried out, or when the robot meets unanticipated or unusual conditions that the software has not properly allowed for.
4. Inadequate installation. If, at the installation stage, due regard has not been given to the workplace layout, robot guarding and other appropriate safety hardware, then accidents can easily result.

5. Insufficient training. All concerned with the robot must be adequately trained. A lack of appreciation for the robot capabilities and requirements can cause serious problems for any personnel involved in programming, operating, maintaining, or simply observing the robot.
6. Human carelessness. When dealing with industrial robots, the phrase 'familiarity breeds contempt' needs to be changed to, 'familiarity needs respect'. Most accidents seem to have occurred at the stage where those familiar with the robot have been working either on it or within its vicinity. This familiarity made them complacent and therefore susceptible to lapses of vigilance and lack of adherence to the proper safety procedures. The opposite situation is of course when the human is totally unaware of the hazards expected. The presence of this person in a hazardous area will be due to unauthorized entry coupled with a failure of safety hardware or procedures.
7. Other human failings. Lack of proper maintenance can cause premature failure of components and unexpected behavior by the robot. Also deliberate interference with safety hardware and deliberate avoidance of safety procedures may occur when an operator wishes to take 'short cuts' to make life easier or to improve bonus rates. These and others are all potential causes of accidents.

The above accident causes will produce the following effects if the necessary conditions arise:

1. Accidents effected by impact. This happens when the robot arm, the end effector, or the workpiece, impacts a human in the robot work area. It is usually the result of an unanticipated movement of the robot. The impact may also occur if a workpiece is not properly gripped by the robot and is thrown by the arm as it moves or comes suddenly to rest.
2. Accidents effected by trapping or crushing. These are effected by the robot arm trapping the human between itself and ancillary equipment, machines, stanchions, pillars, safety barriers, or even between the arm and its own support system. It is also possible for associated equipment such as automatic guided vehicles or other work handling devices to create trapping points within the robot system.
3. Accidents effected due to miscellaneous factors. These may arise due for example to electrical faults on the robot drive or control system creating an electrocution danger, or leakage of hydraulic oil creating a slippage hazard. They may also arise due to the application of the robot. For example, the flash from a welding robot as an arc is struck, spillage of molten metal if a robot is handling a ladle at a die casting machine, or a fragmenting cutting wheel if a robot is carrying out a fettling operation.

The keyword as far as prevention of robot accidents is concerned is 'awareness'. The manufacturer, vendor and user must be fully aware of all potential accident causes, hazards and methods of prevention. The methods of prevention are considered in detail below, where the manner in which safety can be tackled by each of the parties mentioned above is examined. However, at the moment it should be said that safety requires that attention be paid to the hardware of the robot and safety devices, control software, training of all personnel involved, and adherence to such company and government standards, regulations, procedures and law as is applicable.

Safety at the robot design and manufacturing stage

If safety is given sufficient attention at this stage, considerable cost savings can be made in the final installation. This is because it is much cheaper to 'design in' safety to the robot system rather than to leave safety considerations until after the robot has been installed in the factory.

Considering first the mechanical, electrical and electronic hardware. The more massive the robot is, the more difficult it will be to stop in an emergency, and of course the more damage it can inflict in an impact or crushing accident. Thus a light structure is preferable for safety reasons though this has to be balanced against payload and rigidity considerations. The materials used should be rugged, resistant to corrosion and attack by chemicals and oils, and nondegradable. Normal good design practice should prevail, e.g. no moving parts should be exposed, no trapping points for human fingers and limbs, no sharp edges or protruding elements that could cause abrasion, and no trailing leads or hoses that could cause tripping. There should also be fixed mechanical stops and shock absorbers fitted where possible to prevent overtravel of the arm members.

Easily seen warning lights should be fitted to signify 'power on'. The whole robot system and how it interfaces with the user, whether it be operator, technician or maintenance engineer, should be designed ergonomically. For example, emergency stop buttons should be easily accessible at all times, one of which should be on the operator's teach pendant. This teach pendant should require both hands to operate it to ensure that the user does not implement a robot movement while his other hand is in a dangerous position. The 'start' button should always be close to a 'stop' button. There may be many emergency stop buttons distributed throughout the system and installation, but there must be only one start button. The stop buttons should protrude clearly from any mounting surfaces and will probably be of an easily hit mushroom shape. The start button, however, should be recessed or shrouded in such a way, that only a deliberate action on the part of the operator could cause it to be depressed. Emergency stop buttons should be hardwired into the system and override all software commands. It may also be possible to fit a 'dead man's handle' which would shut the robot down if pressure was no longer applied to it by the programmer, this would only be necessary when programming.

The electrical and electronic hardware should be suitable for the environment intended. For instance, environments that are explosive or potentially explosive demand that intrinsically safe electrical systems be used. This will mean that only a certain maximum electrical energy level can be tolerated, thus necessitating low voltage controls and hydraulic power. Other areas may simply require that there be no opportunity for sparking or arcing to occur, therefore brushless electric motors may be used. The power cables and electrical enclosures must all be adequately protected and they usually have to be flameproof. The signal carrying wires, especially in the umbilical cord between the controller and the robot arm, should be adequately screened against electrical noise. This 'noise' is electrical energy produced by various sources in the vicinity of the robot system which affects the data wires, and so corrupts the control and data signals to and from the arm and controller. It may also be possible for the manufacturer to fit safety sensors onto the robot arm (see Figure 16.1) although this is uncommon.

The software in a modern robot can contain many safety features. For instance different maximum speeds can be set for operation under programming and production conditions. There should be a range of speeds available for use when programming, with a maximum speed of say 200 to 250 mm per second. This will prevent the robot from moving too far too fast when a command is mistakenly given to it. However, even slower speeds are advisable when programming is first being learned, as at the higher teaching

Figure 16.1 This robot has safe edges on the moving arm. If the arm comes into contact with people or objects, the sensing safe edge stops the robot arm, protecting the individual or object from contact damage. (Courtesy Herga)

speeds it is still possible for the robot to move fast enough to hit itself against a worktable or to impact a human before the programmer's reflexes can push the emergency stop button. The software should also be designed to allow software stops to be created for each axis. These stops should be able to be arranged to cause each axis to stop before coming against the mechanical stops. It should also be possible to create a 'virtual' work envelope which will include prohibited volumes, e.g. once the environment around the robot is known it should be possible to program the arm to stop before infringing the volume that could be occupied by a human, if say standing beside a stanchion or other piece of equipment. This will help avoid possible trapping points. The software can also be designed to cope with sensory data from external sensors mounted on the robot arm, e.g. proximity sensors and process sensors such as light curtains mounted within the work area.

Safety at the robot supply stage

It is the responsibility of the robot supplier (be it the manufacturer or their agent) to provide full and comprehensive literature on the operation, service and maintenance of the robot, and to provide sufficient training to ensure the operator and programmer can use the robot safely and that maintenance and service engineers can maintain the robot in a safe working condition. A full awareness of the limitations of the robot and particular hazards should also be presented. The vendor may simply be providing a discrete robot, but more likely a full system will be supplied and in this case advice will be given on the overall safety measures that should be taken. The training courses should be thorough; and specialized courses for each of the personnel involved may be offered. When the robot is being installed the supplier must also assume responsibility for ensuring it is working properly with all fine tuning adjustments made, and ensuring that all of the purchaser's personnel concerned are conversant with all relevant details and safe working practices regarding the robot.

Safety at the robot planning and installation stages

Attention should be given to layout details by drawing up plans of the area using templates to signify the robot work volume. These can help ascertain the optimum safe position of the robot with respect to trapping points such as roof support pillars, and to adjacent manual work areas or passageways. The robot chosen should be capable of lifting all anticipated loads. Selection of too light a robot can cause safety problems due to premature failure of robot components or unexpected release of workpieces.

The preliminary work on the creation of full safety procedures to be followed by staff can also be done at the planning stage. Possibly a short safety manual could be compiled for all those in contact with the robot, and key personnel identified to be authorized for access to the work area by using a 'permit to work'. Identification of training needs by the purchaser should also be made here, in conjunction with the supplier or manufacturer, or both.

The main focus of attention at this stage will be on the safeguarding of the robot. As mentioned earlier, most developed countries have standards to act as guides for industry. In the UK the British Standard 5304 *Code of Practice – Safeguarding of Machinery* gives the basic principle of safeguarding: that unless a danger point or area is safe by virtue of its position, the machinery should be provided with an appropriate safeguard which eliminates or reduces danger before access to the danger point or area can be achieved. The cost of the safeguards has to be balanced against the possible severity of accident that may occur. Whenever a hazard is such that it may cause personal injury then it should be safeguarded. The problem arises when the complexity of safeguarding the situation with full integrity increases the cost of the installation to a prohibitive level. It is here that careful consideration of the situation, the hazards involved, and the potential cost of safeguarding must be given by the user, supplier and government inspection authorities in order to arrive at a working, economic and safe solution.

To help in deciding on the level of safeguarding it is advisable to carry out some form of risk assessment. The UK Health and Safety Executive has written a framework for assessing the risks involved in the use of robots. Within this, three modes of operation of the robot are identified. These are normal working, programming and maintenance. Each of these modes is then examined under two types of condition. The first is normal designed behavior, i.e. the robot is working as it was designed to do when properly programmed and used. The second condition is where it exhibits aberrant behavior, i.e. when the robot makes unanticipated moves due to a malfunction of the hardware or a programming error. A summary of an early HSE safety framework published in the Machine Tool Trades Association booklet *Safeguarding Industrial Robots, Part 1* is shown in Table 16.1. The detailed version of the HSE checklist separated out the areas of concern into eight categories:

1. The environment within which the robot will work.
2. The location of the robot.
3. A hazard assessment.
4. A consideration of the robot specification.
5. Examination of the programming carried out by the manufacturer.
6. Examination of the programming carried out by the user.
7. The causes and implications of aberrant behavior.
8. General considerations including safeguarding.

The points themselves are self-descriptive, and their range highlights the fact that safeguarding is only one aspect of all that has to be considered for safe use of industrial robots. More information can be found from the HSE, MTTA, ANSI/RIA and VDI publications mentioned at the beginning of this chapter.

Safety hardware

An important principle to maintain is that all safety devices should fail to safety, i.e. they should be 'fail-safe'. There are many types of device that can be utilized as safety equipment on and around industrial robots. Three physical regions can be identified around the robot. The first is the perimeter, that region bordering the work area of the robot. This is usually guarded by some type of perimeter fence or guard. The second is the volume of space within the perimeter and around the robot. Safety mats are typical of the safety devices used here. The third is the contact surface of the robot arm itself. Safety edge switching or other types of contact sensors may be used in this case.

The first consideration usually given is to some method of caging the robot. That is, some barrier is erected between the general factory environment and the industrial robot's local work area to ensure the robot is totally isolated from personnel while it is powered. This will only apply during normal production conditions since during installation, programming and maintenance, personnel will often be inside the work area while the robot is live. Depending upon the environment and the application, the physical barriers used may vary considerably. For instance, in a research establishment using light duty assembly robots it may be possible to simply use a cord or belt suspended between pillars. The cord or belt can also be linked to a small microswitch contained within one of the pillars. In this case if a change in tension is experienced due to a person pushing against the barrier, then the microswitch will be activated and the robot will stop. However, under normal production conditions more robust guards are necessary.

If access to the general factory shopfloor is restricted to personnel aware of the dangers of the robot, then a low level fixed fence or handrail may be all that is necessary to indicate the danger area (see Figure 16.2). This, of course, assumes there are no application hazards such as welding flash, molten metal, or the possibility of a heavy

Table 16.1 Summary of UK HSE safety framework. (Courtesy MTTA)

(a) Determine the mode of operation i.e. normal working, programming, maintenance.

(b) Carry out a hazard analysis to determine potential areas for doing harm.

(c) Determine whether 'designed' or 'aberrant' behaviour is to be considered.

(d) Determine if hazards are liable to lead to injury.

(e) If so, then consider whether there are any recognised methods of guarding the particular machine concerned. At present, such standards may well be available for the associated machinery but probably not for the robot.

(f) Consider whether such standards are appropriate particularly in the context of machines being used in conjunction with robots. For example, the risk assessment could be different for a machine with a human operator than one which is associated with a robot. One factor that will affect this risk is whether or not the human operator will take over from the robot during, for example, robot failure.

(g) If no standard is available, consider what the logical steps should be in order to establish a reasonable standard for the particular application.

(h) Determine whether a fixed guard can be used.

(i) If fixed guards cannot be used then consider the use of interlocked guards. Determine the type of interlocking system appropriate to the circumstances.
The interlocking system should give a reasonable level of integrity appropriate to the risk in question and should enable regular effective maintenance checks to be made. Any 'monitoring' system should be carefully examined to assess its effectiveness.

(j) When analysing the system under 'aberrant' behaviour, a similar process of examination of the hazards and then a risk assessment is carried out. The hazards may be more difficult to determine because they may only exist on failure of part of the machine system such as a control system malfunction on the robot. An alternative to guarding in such circumstances might be to improve the safety integrity of the control system in question, retaining the interlocking safeguards proposed for 'designed' behaviour.

(k) After particular measures have been taken a re-assessment of the system integrity will be necessary. If the hazards are minimised/prevented re-assessment of risks will also be necessary. In most cases and particularly where the risks are high it is preferable to assume 'worst case' when designing the safeguarding system and work on the premise that specific malfunctions will occur. It is not prudent to rely solely on the digital programmable electronic system of, say, the robot for all safeguarding features unless a very detailed assessment has been carried out which may be beyond the competence of the average user.

(l) After the analysis has been carried out for normal working, programming, and maintenance – any safeguarding interlocks considered necessary for any one of these modes must be compatible with the requirements of the other from both a functional and a safety integrity point of view.
Consideration should also be given to emergency stop controls and whether adequate integrity is achieved.

(m) The need for documentation concerned with the analysis, decisions and systems of work etc. relating to hazards analysis, risk assessment, safety integrity assessments, maintenance requirements etc. cannot be over-emphasised.

component being thrown. However, it will normally be necessary to fully cage the robot off, and in this case the fence will need to be at least 2 m high with access via interlocked gates. Interlocking is the term used for interfacing the mechanical operation of the guard with the robot so that if the guard is opened the robot cannot operate, or if the guard is closed it will remain locked shut until dangerous conditions within the enclosure cease. The interlocking system itself should be failsafe. The associated electrical circuits should be directly connected to, and operate through, the robot electrical or power circuit system, rather than the software and internal logic system. This principle of hardwiring applies to other safety components as well as to the emergency stop buttons mentioned earlier.

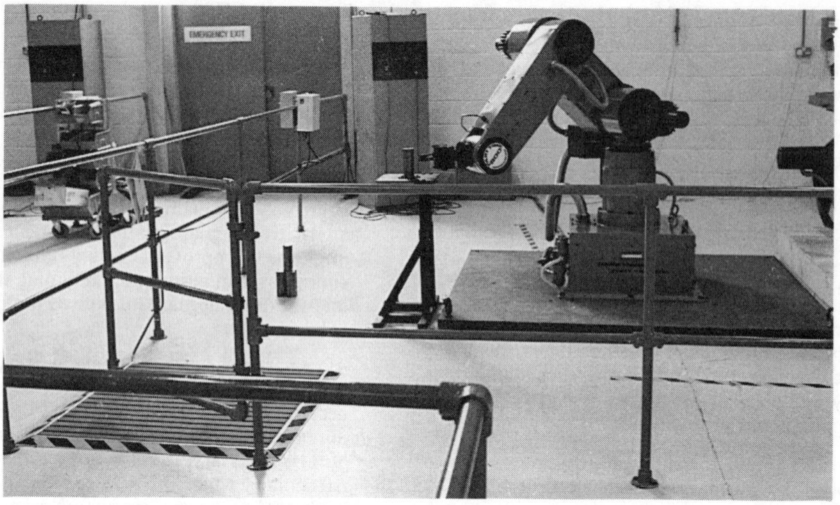

Figure 16.2 Heavy duty safety mat and barrier rail guarding access to a robot. Anyone treading on the air operated mat will stop the robot. It can only be started again by pressing the restart button on the control box mounted away from the robot. (Courtesy Herga Ltd)

Figure 16.3 shows an industrial robot safeguarded with fixed and interlocked guards. Figure 16.4 shows a typical switch used for interlocking. This switch can be directly connected to the hinge pin of the gate so that as soon as it is opened the robot is halted. The circuitry should be configured so that the robot cannot be activated again without the use of a start switch outside the robot area. It is also possible to fit the gate with a lock and captive key. When the gate is opened by use of the key, the key and handle can be removed from the lock by the operator entering the work area. This prevents the robot from being activated until the operator leaves the area and relocks the gate using his key.

A popular piece of safety equipment is the pressure sensitive safety mat. This is placed either across the entrance to the robot work area or around the robot in such a manner that anyone approaching the danger zone will stand on the mat. The pressure of the person on the mat operates a switch which deactivates the robot. There are two common types of safety mat, one electrical and the other pneumatic, both having failsafe control systems. The former comprises a matrix of low voltage electrical switches, the latter a web of pneumatic tubes and valves. The matrix or web is sandwiched between two rugged surfaces to form a mat which is then connected to the robot controller. In the case of the pneumatic mat, air is blown through the pneumatic tubes and the air pressure measured at each end. This is sensed and the signal produced is used to switch off the robot.

Another technique used is photoelectric guarding or light barriers. These are devices utilizing an arrangement of photoelectric emitter and receiver elements to detect the presence of a person entering or present in the robot work area. A typical commercial system is shown in Figure 16.5 together with an example of a robotic application. They can utilize visible light or infrared radiation, the latter being used to overcome problems of sensitivity and ambient light. Arrays of light emitting diodes can be used to provide light

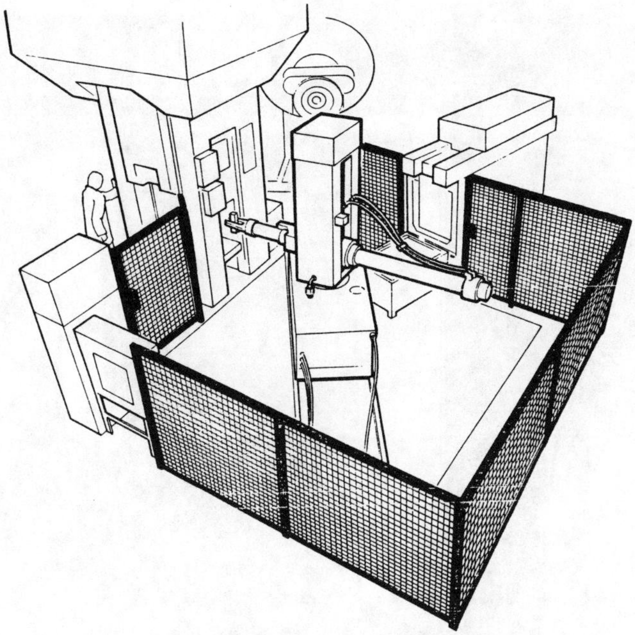

Figure 16.3 A robot safeguarded with fixed and interlocked guards. The robot is removing material from an induction furnace and loading an upsetting press. (Courtesy MTTA)

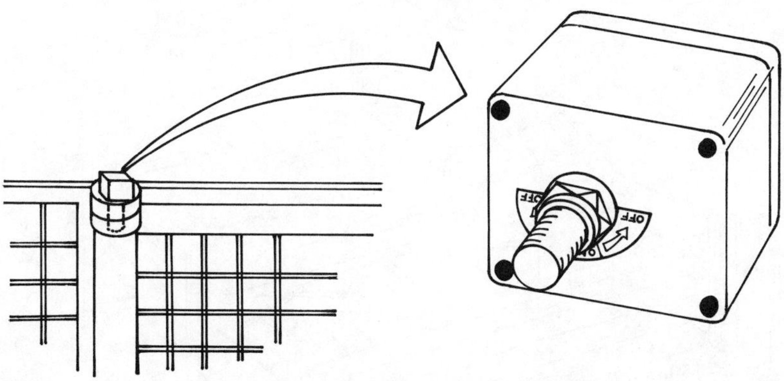

Figure 16.4 For hinged gates, a shaft operated switch can be connected directly to the hinge pin

curtains. Any person or part of a person passing through the light curtain will be sensed and a signal sent to the robot to stop moving.

For contact sensing by the robot, safety edge switching can be mounted on the arm of the robot as shown in Figure 16.1. Once again this may operate electrically or pneumatically. In either case, when the switches are depressed by an obstruction or a person they will immediately cause the robot to be deactivated. The electrical type can be composed of normally open electrical contacts which will complete a circuit when the

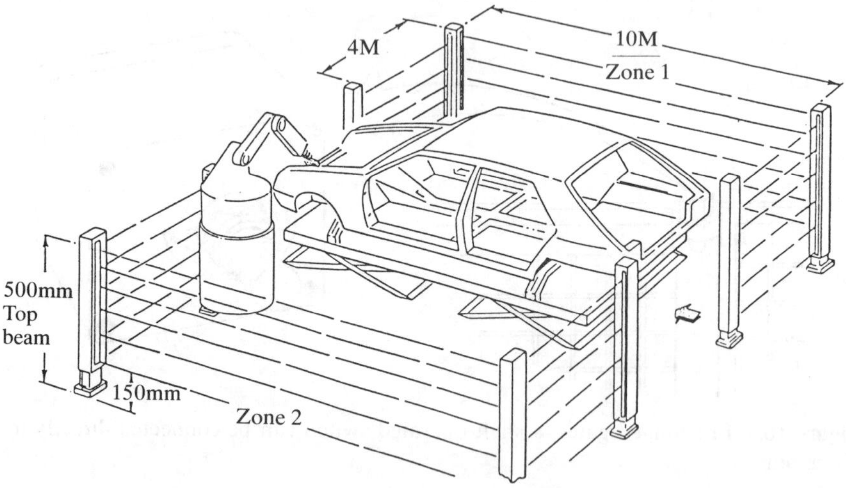

This solid state light curtain can operate in total darkness or full sunlight. The system uses a linear array of diodes sequentially projecting pulses of infrared energy at 4000 pulses per second across the danger area. A typical layout is also shown.

Figure 16.5 The light curtain. (Illustrations courtesy Lightguards Ltd) outer sleeve, within which they are encased, is depressed thus switching off the robot. The pneumatic type contains a long pneumatic tube, when depressed a difference in air pressure within the tube is sensed and so the robot is deactivated.

EDUCATION AND TRAINING

Education and training for robotics can occur at many different levels. To simplify this section the subject has been split into two. Firstly, education in its broadest sense is considered, this is where the fundamental teaching is given to provide a knowledge base on which a robotized industry can build and prosper. Secondly, the training necessary for those within industry who are adopting, or considering adopting, industrial robots is examined.

Education

For all students coming to industrial robotics there is primarily a need to first grasp the fundamental concepts of the subject. These fundamental concepts may be said to include the following elements:

1. An appreciation of the reasons why industrial robots are useful.
2. A basic understanding of how they operate, i.e. what components are used and how they relate to each other.
3. Understanding how robots integrate within the total manufacturing system.
4. A knowledge of their capabilities and limitations.
5. An awareness of their implications for individuals, industry, and society in general.

From the contents of this book it should be apparent that to progress from here, particularly in the areas of robot design and control, a fuller understanding of each area is necessary. This in turn implies a much deeper background knowledge on which to base a study of the specific subjects concerned. It has been said that the last man to live who was apparently conversant with the whole body of human knowledge was Leonardo da Vinci. From his time onwards, man's knowledge seems to have increased exponentially, so that now one man can no longer hope to be fully conversant with even one of the sciences. Consider now the spectrum of disciplines that contribute to robotics. Within the robot design area alone there is a need for mechanics, kinematics, dynamics, materials science, metallurgy, electrical power theory, hydraulic theory, a good knowledge of mathematics, control theory, programming, microprocessor control and design, interfacing, electronics, instrumentation, artificial intelligence, vision systems . . . . No one person can be expert in all of these and even this does not consider the practical knowledge and experience required for effective application and management of the robots.

In fact it is still unusual to find universities or colleges offering degrees in industrial robotics. This is really due to two factors: the interdisciplinary nature of the subject, and the different interests and viewpoints held by those coming to robotics. It is often a subject provided as part of a traditional degree or as a supplementary subject. For example, students who have elected to pursue a career in manufacturing engineering and management will be interested in gaining a sound knowledge of robots to the extent that they will be able to apply them effectively. They will know their capabilities and limitations and will be able to appreciate fully how they will integrate into a manufacturing system. On the other hand, students intent on following a career in research and development, possibly specifically in robotics, may be concerned more with control theory and gaining a good knowledge of mathematics, computing, and electronics, to enable them to innovate and improve on existing designs. Between these two extremes there will be students from various disciplines who may be able to bring to bear their own specializations to the subject, e.g. materials science or artificial intelligence.

Training

Training has been distinguished here as applying to those personnel who are already in industry and require knowledge to work with robots in their occupation. Consider the types of people who will require training: programmers, technicians, operators, maintenance and service personnel, and general personnel (e.g. management, personnel officers, union representatives, safety officers). Adequate training is essential for a successful and safe robot installation and must be given to all those who work with the robot in some way.

All the categories of personnel mentioned above will benefit from a one day general awareness course on robotics before the first robot is installed. This is beneficial from an industrial relations aspect as it will tend to put the subject in perspective and allay any suspicions as to what the machine may look like or how it may behave. The course will also introduce an awareness of safety considerations and possible dangers. Those directly involved with the robot will require more specialized training. The robot supplier or manufacturer usually provides training courses for the programmers and operators. (In fact this is essential to check when purchasing an industrial robot system.) The cost of the training will usually be included in the system price, and will often cover one or two personnel for possibly one week, excluding accommodation, at the supplier's premises. This type of training is important not only for the programmer and operator, but also for the technician and maintenance personnel. Thus three approaches can be taken: either the extra cost is paid to send all the personnel for training, or the supplier runs his course at the purchaser's factory, or the one or two personnel that have been to the supplier's course return and run their own course for all those interested.

Initially, the type of knowledge needed by the user will be programming skills, knowledge of the robot limitations, safety considerations, and basics on how the robot operates. For detailed knowledge on servicing, repair and maintenance courses will probably be necessary. It may be, however, that a service contract will be made to ensure expert attention is given when necessary by the supplier and appropriate maintenance carried out. Technicians will require a more thorough training, especially if the firm is considering engineering its own installation, i.e. interfacing conveyors, sensors and computers with the robot. This will require training by the robot supplier and also possibly attendance at relevant courses in colleges or other training centers.

SOCIAL EFFECTS OF INDUSTRIAL ROBOTS

Introduction

One of the major advantages of industrial robots is that they can efficiently carry out work previously necessitating human operators. It is therefore obvious that robots are having, and will continue to have, some impact on the human aspects of industry and society in general. This effect has already been considerable in the areas where robots have been applied in large numbers, e.g. spot welding in the automobile industry. One spot welding line that was robotized was recorded as having reduced the number of human workers from 138 to 13. This is certainly an extreme example but various reports from different countries estimate that each robot installed in industry will do the work of between two and five humans. From pure economics it is unlikely that a robot could be financially justified under present conditions unless it could do the work of two or more operators, or one operator for each of two shifts. Of course, as the relative costs of robots decrease and the human worker's wages increase, it may become the case that a complete robot system to carry out a simple task will be less than a year's costs for a single human. In this case the number of jobs taken over by robots will escalate rapidly. The use of more intelligent robots with advanced sensing will also increase the range of jobs to which robots can be applied.

Estimates of the number of workers anticipated in the future to be displaced by robots are often given in the press, journals and conference proceedings. However, they fluctuate considerably in relation to the predicted state of the economy. At present it is sufficient to say that the social influence of robots within industry and society merits some consideration. It should be remembered that industrial robots are only a part of the microprocessor-based technology which is changing the employment patterns in society. However, it is attempted here to look at the impact of robots in as much isolation as is practically possible.

Implications for the employee and employer

Robots displace human workers; this is an apparent fact. What is not so apparent is what happens to those displaced workers. In general terms the improvements in the competitiveness and profitability of the firm may mean that work is created elsewhere in the company, or that as the economy of the country improves then jobs will become available in the service sector. These implications have already been considered in Chapter 15. At the moment unemployment is a general problem in the West and it could be said that this is a result of the present trough in the world economy. The concern is that if recovery takes place the new technology involving robotics, automation and computers instead of humans may prevent any reduction in unemployment. Take, for example, the introduction of one robot into a job in a factory, and the alternative routes for the management to follow with respect to a worker whose job has been robotized. These routes are shown in chart form in Figure 16.6.

When the worker's job is robotized the first possibility is that the worker may be retained to work with the robot. This situation may occur if the company is installing the

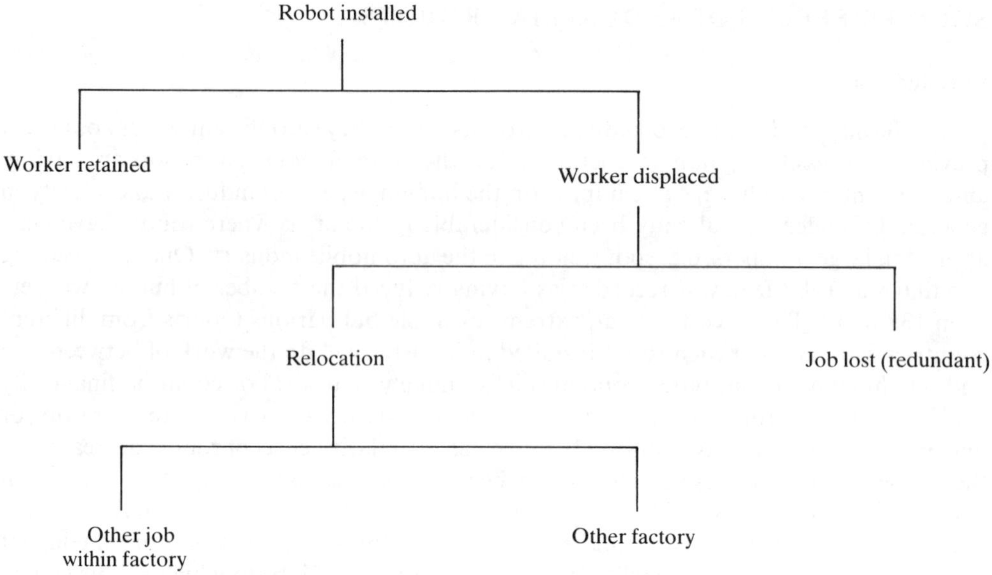

Figure 16.6 Possible effects of robot installation on worker

robot for health and safety reasons, or justifying it financially by, for example, improvements in quality. Or it may be that in fact three workers are being replaced by one robot, in which case it could be possible to retain one of them in this manner and leaving the other alternatives open for the remaining workers. It could also be possible that a number of robots are being installed and the worker can distribute his time between them. In any event it is highly unlikely that if a robot is installed to do the work of one worker then this worker will still be needed full time in that job.

Suppose, however, that the worker is retained for one of the reasons given above, what are the implications? There are three possible ways in which the worker can be utilized: (a) he may be required to program the robot and set up any jigs and fixtures necessary; (b) he may simply need to be alongside the robot, monitoring and tending it to ensure satisfactory operation; (c) he may have to work with the robot on an associated task.

In case (a) the skill of the worker is being changed or added to. For example, the worker may have been a highly skilled arc welder. Robotization of his job with him as programmer implies that he will require training to allow him to acquire the necessary programming skills. On the other hand, the job concerned may be a palletizing task and the worker, previously used to a relatively simple though physically demanding job, will now require to have his skills upgraded to program the robot. Programming for palletizing is much easier than that for arc welding but in each case training must be given. This training is important not only for the overall efficiency and safety of the work as mentioned earlier, but also for the job satisfaction of the worker.

In (b), simply observing or tending robot, the skills required are normally reduced or removed. This deskilling of work should be implemented carefully, especially when it is remembered that one of the reasons for installing robots is to remove tedious work from humans. A human worker can be relegated to being a servant of the machine if the operation is not properly planned. In some cases workers have complained that the introduction of a robot made them feel much less satisfied with their work, they felt increased stress, and their work was more boring. Also, if robotization is occurring over a large area of the factory then there will probably be a much lower population density of human workers. This removes the opportunity for casual conversation, sharing of ideas, and general social interaction. Psychologically this can produce worker alienation, and therefore give rise to other problems through job dissatisfaction.

Thirdly, in category (c) many of the factors in (b) also apply. Additionally, care has to be taken that the worker is not paced by the robot. For example, if the robot is working as part of an assembly line, the human worker may be positioned after the robot. Here the worker will have to work at the same rate as the robot. In itself this would be simple enough for the human, but the robot has the ability to carry on working at a constant speed without a break for relaxation, etc. This type of work organization can therefore again add stress to the worker as he feels compelled to perform at least as well as the robot. The other problem experienced here is, of course, worker safety, which was covered earlier.

These then are the considerations to be made when retaining workers whose jobs have been robotized; now those who are completely displaced by the robot are considered. The most attractive option to all concerned is to relocate the worker, the simplest option being to relocate him within the same factory. This will probably mean retraining for the new job, and may once again mean a change in skill levels. If the company is large it may be possible to relocate the worker to a similar job in another factory. Individual cases would be judged on their own merits, but a move to another factory will usually prove unsatisfactory to the worker and cause high expenditure in relocation costs to the firm.

It will be noted that in all of these cases the worker is required to adapt to a new situation. It may be simply learning to work with the robot, or it may require the learning of new skills with considerable training involved. Should the worker have been employed in the same job for a number of years and he has passed the age where adaptation comes easily, then great care is obviously necessary on the part of management to ensure more problems are not created than solved.

The final option, job lost, is the least acceptable to all concerned. The term 'redundancy' has so many connotations in today's society that it is usually avoided where possible and euphemisms used in its place. However, should this occur there are economic as well as social costs incurred. The company usually has to make redundancy payments, the individual loses his earning potential for the length of time he is redundant, and the country has to pay the cost of benefits and allowances. In actual fact redundancy due to robotization on its own is unlikely. Overall reduction in labor forces are more likely to occur due to major organizational changes and large new technology installations where robots form only a part of a new system. In single robot installations, if the firm is affluent enough to purchase a robot system, it is usually solvent enough to retain the worker in some capacity within its organization. What does happen of course is that although some firms may operate a policy of no redundancies through automation, they will not replace people who retire or leave for other reasons. This means that the jobs available for particular skill levels within an industry will gradually decrease due to automation. However, in a broader sense when work at higher skill levels is considered the net amount of jobs available may actually increase. This is due to the new skills required to install, program, maintain, and develop robot installations.

Implications for the vendor

Those designing, manufacturing and installing robots are obviously providing a new source of employment for individuals. These firms are not very large in number as yet, nor do they contain a large staff. However, other spin offs occur such as systems houses that will take the robots from the supplier, gather from various sources the necessary ancillary equipment, and then install and commission complete systems on the customer's premises. These new job opportunities will demand a higher level of education and training than the jobs that are being lost. They will also be fewer in number.

General implications

When the benefits of robotizing, and automation generally, are considered the end result should be an improvement in the economy. This in turn should mean more money for people to, for example, invest in banks, or spend on entertainment and travel. Thus the service industries should prosper so providing jobs in this sector. The move then is away from 'blue collar' workers to 'white collar' workers. This trend can be self-perpetuating for one unattractive reason which is perhaps most evident in the UK. Due to the poor economic climate that has existed for some time, companies have been reluctant to start young people as craft or technician apprentices. This means that there is now a shortage in many areas of the country of skilled craftsmen such as toolmakers and welders. Therefore when the need for skills arises firms find it difficult to recruit the appropriate people. The new solution will be to find one person with the skills who will then be used to train robots to carry out the work. Thus gradually the infrastructure of skilled human workers is being eroded away by yet another factor. Care should obviously be exercised here on the widest possible scale, to ensure the country maintains a credible knowledge, skill, and experience base on which to build when the economy improves.

Workforce response

Usually a workforce will be amenable to new technology providing it is not seen as a threat. Unions say that they realize investment in new production methods must be made to enable their company to remain competitive. This approach includes robotics, which unions increasingly see as being just one other tool to aid survival. Careful discussion between unions and management is necessary to ensure a smooth transition from manual to robotic operations. It is worthwhile to ensure that the unions or workforce representatives are brought in at an early stage, possibly even at the initial planning before the robot is purchased. Sometimes protective agreements are made and signed by management to state that no redundancies will arise as a direct result of robotization. There are therefore no modern-day Luddites, but there are practical and powerful bargaining bodies to ensure unnecessary redundancies are not implemented, that early retirement is an option where possible, and that training and retraining is made freely available.

Working conditions

It should be remembered that industrial robots can improve working conditions considerably. For example, although they have the potential to be dangerous if safety precautions are not taken, they commonly reduce accidents due to their application in hazardous environments and tasks. Many occupational diseases and injuries are caused by poor environments and dangerous and heavy labor. Robots can be put into these jobs. As mentioned elsewhere they are also ideal for the monotonous, and boring jobs that can cause both accidents through complacency and industrial unrest.

This section has highlighted some of the implications of industrial robots for society. It has shown that these implications can be beneficial to individuals if robots are carefully and compassionately introduced. It has also shown that problems can arise should the full implications of robotization not be responsibly considered. Should a company, or country, wish to remain competitive in the world marketplace then the adoption of industrial robots and other types of new technology is a necessity. However, this adoption must progress quickly, humanely, and with proper concern for the engineering infrastructure of the economy.

17

THE AUTOMATED FACTORY

INTRODUCTION

In previous chapters an overview of industrial robotics has been given, the individual components that go to make up a robot have been examined, and the manner in which they are organized to create and control the robot has been discussed. Examples of applications have been presented, economic factors explained, and some implications of robots for the workforce and society in general considered. However, the 'natural environment' of an industrial robot, i.e. the modern factory, has yet to be looked at. Industrial robots form only one part of a manufacturing system and their true relevance can only be grasped by putting them in perspective with the rest of the production organization. To do this we will first accept that the industrial robot is one element of the system. This chapter will firstly examine some other elements of the modern factory before going on to consider some contemporary manufacturing concepts and how these elements and concepts can be combined to form the automated factory.

An industrial robot may exist in a factory as a single unit built into a small system. For example, a machine tool loading robot together with a numerically controlled lathe, and material transfer and automated inspection systems, may constitute a discrete work cell within a factory. However, in major installations robots often form part of a large, integrated manufacturing system. If this integration could be totally controlled by computers at all levels this would be termed a CIM (computer integrated manufacturing) system. No full CIM systems exist as yet although many large companies, such as Boeing in the USA, are moving towards it. The cost of a computer integrated system for a large company is extremely high; however, it should be noted that the maximum effect comes not from the use of computers, but from the efficient 'integration' of all the functions of the organization. Thus careful planning creates the benefits and economies; the use of computers can only enhance them.

Acronyms have become very popular, for example CIM, CAD (computer aided design), CAM (computer aided manufacture), FMS (flexible manufacturing system), MIS (management information systems). Their proliferation can also be confusing e.g. the programmable logic controller described later is sometimes referred to as a 'PC' or programmable controller, while the personal computer is also referred to as a 'PC'. Here 'PLC' will be used for the former and 'PC' for the latter. As this chapter progresses the meaning of these terms, and others should become clear. Their interrelationship should also become apparent as should the place of the industrial robot within the total system. Some elements of the automated factory are now considered.

NC MACHINE TOOLS

For a manufacturing facility, as opposed to one simply engaged in assembly, machining of metal is almost certainly required. This machining mostly involves metal cutting by a chip removal process, i.e. metal is removed from a bar, block, or casting to form a component. This is done by using machine tools such as lathes, milling machines, and machining centers. The principle of operation of these machines and a full description is beyond the scope of this text, although many of their components and controls are the same as those used in industrial robots. Most of this equipment is now numerically controlled (NC), and indeed this is a prerequisite if the machine has to become part of a computerized system. A modern NC machine tool is microprocessor controlled and programmed using its own programming language. Various terms are used to describe the type of control such as NC, CNC, DNC.

In basic NC the part program containing all the information necessary for the machine to produce a workpiece is contained on a punched paper, or magnetic, tape. This information is fed to the machine's controller which translates the part program instructions to machine operations on the workpiece. Each time a workpiece is made the tape has to be fed through the control system again. In CNC (computerized numerical control) the machine has its own dedicated microcomputer. An example of this type of machine was shown in Chapter 1, Figure 1.2. This performs many of the basic NC functions with its own stored program and provides greater flexibility and more powerful computational ability than basic NC. Part programs are usually entered via tape as before. However, once the program has been entered, the tape need not be rerun for each part as the computer will have stored all the necessary information in memory. DNC (direct numerical control) is of most interest in this text. The paper or magnetic tape program storage medium is eliminated and the machine tools are controlled directly by a supervisory computer which downloads part programs to individual machine controllers. Thus a central computer can hold programs for many machine tools. The communication between the machines and the computer is two way and the system has to operate in real time, i.e. the response time must be so short as to appear instantaneous for each machine.

PROGRAMMABLE LOGIC CONTROLLERS

With the advent of the microprocessor these devices have become very powerful in recent years. Both personal computers (PCs) and PLCs are now used in manufacturing systems, both are microprocessor-based, and both are used for similar type tasks on the shop floor. In fact the distinction between them is now blurred, especially when the more sophisticated PLCs are considered. It could be said that personal computers, using a high level language such as Basic, are not specifically designed for the type of application that a PLC is prepared for. Personal computers are usually designed for business use with control a secondary function, whereas PLCs are general purpose devices designed for control of equipment and processes in a factory environment with say a high temperature and humidity range. Thus PLCs are designed for easy direct connection to industrial plant and machines, whereas personal computers usually require additional interfacing hardware and software. However, many stand alone PCs are becoming available that do have the necessary construction to withstand the rigors and demands of the industrial environment.

The National Electrical Manufacturers Association in the USA defines a PLC as, 'a digitally operating electrical apparatus which uses a programmable memory for the internal storage of instructions for implementing specific functions such as logic, sequencing, timing, counting and arithmetic to control machines or processes through digital or analog input or output modules. A digital computer which is used to perform the functions of a programmable (logic) controller is considered to be within this scope. Excluded are drum and similar mechanical type sequencing controllers.' The sequencing function in a PLC is used to control the sequence and timing of switches, motors, solenoids, lights or other sensors. The PLC can be linked in with instrumentation, and using its ability to calculate results from inputs, it can then take appropriate action and display or print any necessary information. It can produce analog outputs and provide position control functions when used as part of a closed loop system. Industrual robots are commonly connected to PLCs to enable interfacing with process sensors and safety equipment.

The basic components (see Figure 17.1) of a PLC are:

- The input and output interfaces to allow simple connection to and communication with the industrial equipment being controlled. Interfacing requirements have been covered in an earlier chapter but typical signals the PLC would have to handle would be: various AC voltage levels, DC voltage, binary coded decimal signals, and low level analog signals. The voltages used within the PLC are usually at low levels, e.g. 5 V DC.
- The processor or CPU. This is a microprocessor system capable of carrying out all

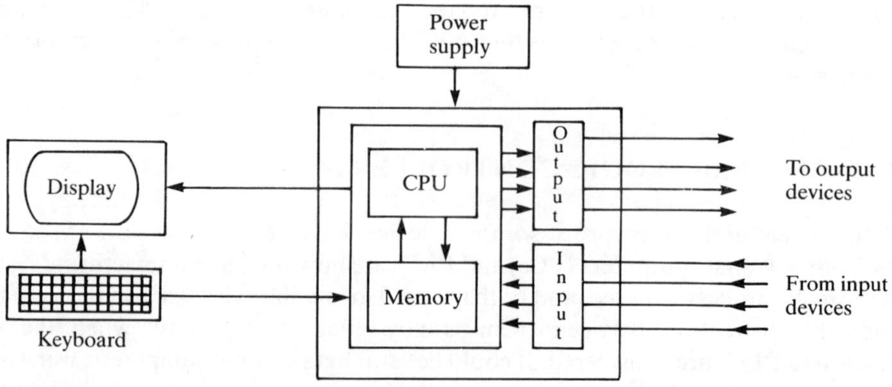

Figure 17.1 Programmable controller major components

the decision making, logic, and mathematical functions, and also supervision of all the input and output signals.
- The memory. Just as in computers this contains the program instructions on how the PLC is to control the external equipment.
- Power supply to provide noise-free power to the controller and for the output signals.
- Program input device, e.g. a keyboard.

The PLC is programmed via pushbuttons or a keyboard with a visual display unit. Depending upon the PLC type, the programmer may use simple symbols to represent relays, to set timers and counters, or to carry out mathematical functions. The program may be developed by first constructing an electromechanical diagram of inputs and outputs, this is then organized into what is called a 'ladder logic' diagram for use by the PLC. These ladder diagrams are relatively simple to learn and use, even by personnel that have had no programming training. Becoming more popular, however, are systems with the ability to be programmed using a high level language such as is normally used in a personal computer. This facility may be integral with the PLC or it may be offered by the supplier as an 'add on, plug in' module. Where PLCs are used for complex tasks in process and manufacturing systems then the greater sophistication of a high level language is attractive.

Programmable logic controllers and industrial quality PCs can therefore be used to control individual machines and processes. They can be used to monitor process sensors, status signals from conveyor systems, machine tools, automatic inspection systems, and robots. They can also be linked together to form communications systems with supervisory computers, or a central control system. One possible system is shown in Figure 17.2 although this concept will be discussed more fully later. The front panel view of a typical PLC is shown in Figure 17.3.

MATERIAL HANDLING

Consider the need to transport parts and materials within the automated factory. Transport and material handling is required within the raw material and bought-in components stores, between these stores and the work areas, between the carriers that take the work to the work areas and the points of work, between different work areas and back to the finished goods store. For full automation all of this handling equipment will need to be compatible with computer control, i.e. it must have a control system capable of being interfaced with a supervisory computer. The equipment should also be capable of transporting parts and material in any order desired to any location desired, whether it be a machining cell, an assembly station, or a store.

Access to all machinery within the factory for maintenance and repair should not be hindered by the handling equipment, and facilities should allow for queuing of parts in the event of a breakdown. If the work involves machine tool loading there will probably be a primary handling device, such as a transporter of some kind, to take material from the store to the machine, and a secondary handling device, such as a robot, to transfer the work from the primary carrier into the machine tool. This principle will also apply to

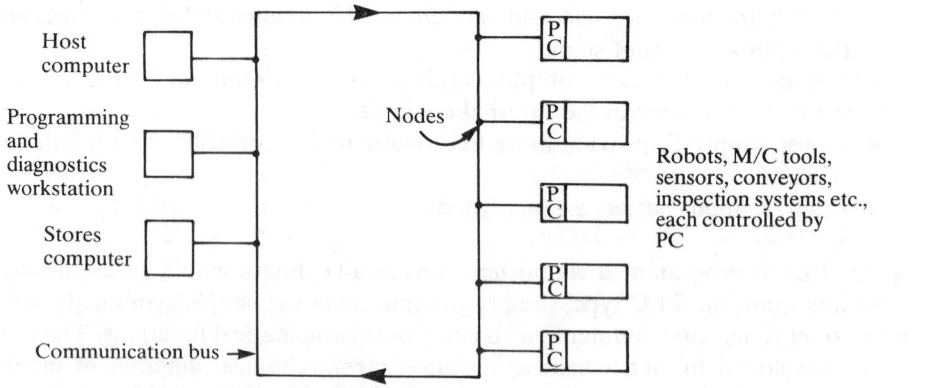

Figure 17.2 Using PLCs in a local area network

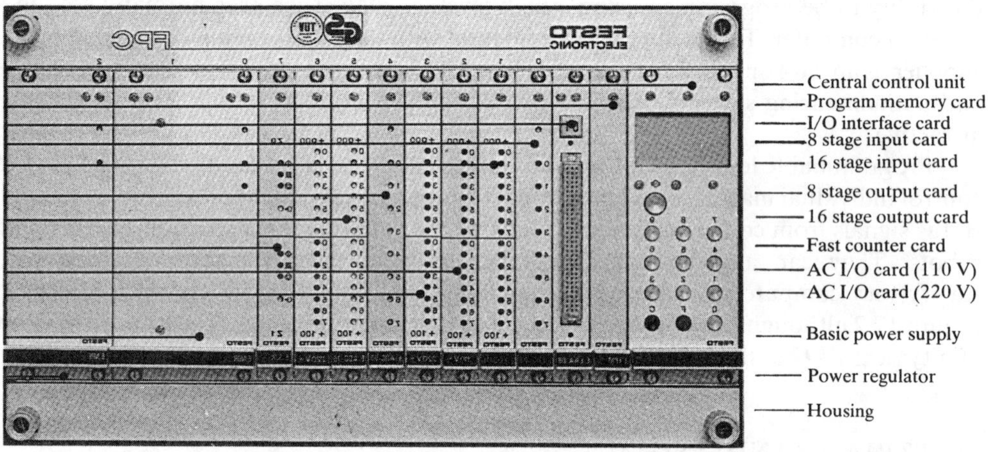

Figure 17.3 A PLC front panel view. (Courtesy Festo)

assembly tasks. Work holders such as pallets, trays, boxes, or racks may be used. There is a large variety of handling devices used to carry these work holders, but the two primary handling devices mentioned here are conveyors and automated guided vehicles (AGVs).

Conveyors

Conveyor systems are quite ubiquitous in factories, having been used in the roller form for centuries. Modern conveyors come in a wide range of types designed to suit all kinds of load. They can have mechanisms for changing the speed and direction of their loads and these together with process sensors can be controlled and monitored through a host computer. Free running roller, or wheeled, gravity conveyors are used to carry material between powered roller sections. The conveyor units are often available as bolt together modules. For fragile loads, inclined tracks, or accurate positioning, belt conveyors are used. A conveyor system carrying a work pallet and workpiece for robotic assembly is shown in Figure 17.4. For secondary handling, i.e. transferring pallets from the main conveyor line to a machine loading bay, assembly cell, or storage area, pneumatic pushers such as those shown in Figure 17.5 can be used.

In a fully automated conveyor system, PLCs are used to direct movement of material along the conveyors and they can be interfaced with a supervisory computer. This may be a PC which also may be a part of a larger hierarchical control system responsible for other functions within the factory. Mimic boards as shown in Figure 17.6 are used to provide

Figure 17.4 Roller conveyor for robotic assembly. (Courtesy Dexion)

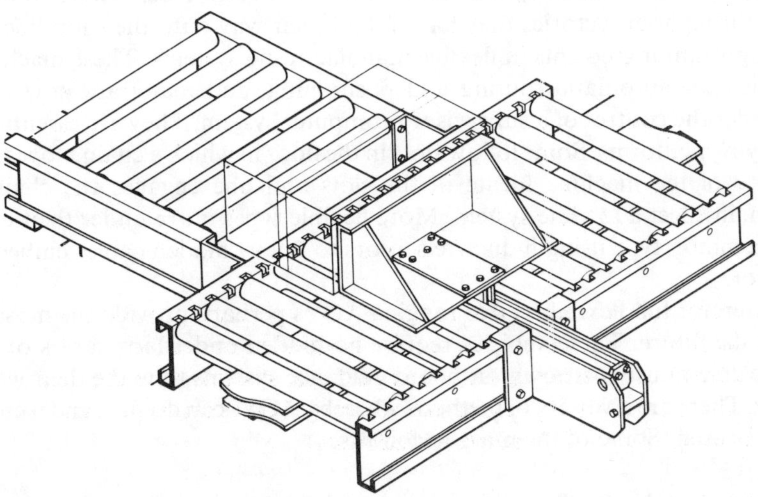

Figure 17.5 Pneumatic pusher for secondary handling. (Courtesy Dexion)

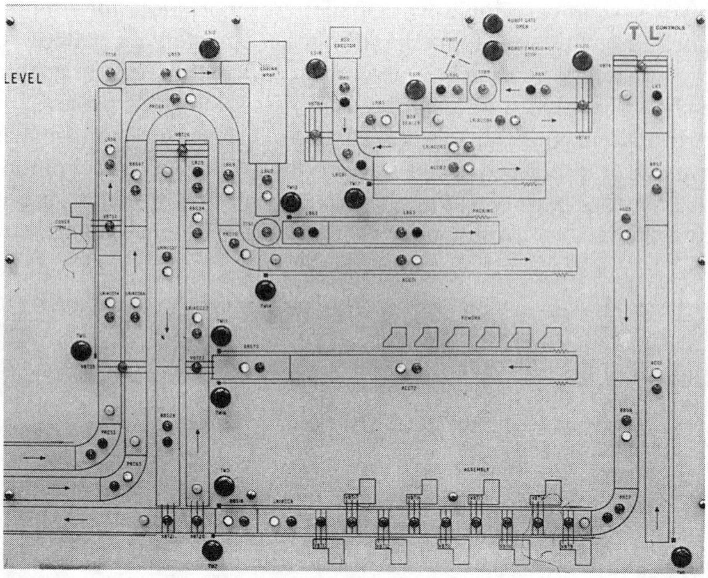

Figure 17.6 Conveyor system mimic board. (Courtesy Dexion)

representative displays of the system status at any time, alternatively interactive computer control consoles can be used.

Automatic guided vehicles

Figure 17.7 shows an AGV in a pallet handling operation. They are often also used in FMS systems for servicing CNC machine tools. AGVs have been used for a considerable time especially in automated distribution centers or warehouses. With automation now spreading throughout factories they have found their way onto the shop floor and have become important ingredients in flexible manufacturing systems. These machines can be programmed for automatic routing and positioning, and many are now designed to operate under the control of a supervisory computer system. They are essentially mobile pallet carrying platforms. Some follow slots in the floor in which a chain moves, guide pins on the base of the machine locate in the slots and one engages the chain to allow propulsion, in effect a towline system. More flexible types move under their own power, the most common type using inductive sensors to follow hidden cables embedded in the factory floor.

However, for full flexibility, free ranging AGVs perhaps provide the most attractive option for the future. These vehicles require no hidden underfloor cables or tracks but instead are able to use their own on-board guidance systems to make their way through the factory. There are a variety of methods whereby AGVs can do this, and a combination is frequently used. Some of them are as follows:

(a) Inertial guidance. This utilizes an inertial gyroscope which is set up parallel to the required direction of movement. When the AGV deviates from the desired path

Figure 17.7 An AGV being used to lift and transport loaded pallets. (Courtesy Jungheinrich (GB) Ltd)

an acceleration is created at right angles to the direction of motion and this is detected by the gyroscope. This acceleration is used to determine the deviation. The information is then fed into a control system to modify the AGV position accordingly.

(b) Position reference beacons. These are set up at specific positions and used by the AGV as it moves around the factory. Various methods may be adopted by the vehicle to pick up its bearing from each beacon as it passes. For example, laser beams may be used to scan for the beacons which could have bar coded information on them. The vehicle will have on board sensors to pick up the reflected signals and so take appropriate action to ensure the proper route is being followed.

(c) Imaging and vision. Ultrasonics or stereo vision may be used to allow the AGV to 'see' its way along its desired route. The vehicle will hold a map of the factory in its memory and will constantly compare the information from its sensors with that map and associated instructions.

(d) Internal sensor monitoring. Here the vehicle operates in a similar way to a robot in that wheel movement, speed and direction are continuously monitored. This allows the AGV's position to be known at any point in time assuming a 'home' position was first fed into memory as a reference point.

Method (d) is probably the simplest to implement but is not 'absolute' in the sense that errors can occur due to wheels slipping or skidding. Method (c) is probably the most attractive for true autonomy but no commercial system of this type exists at present for factory use. Indeed, free ranging AGVs are still uncommon even in highly automated factories. It is also possible to mount robots on AGVs and thereby create a mobile robot. At the time of writing at least one commercial manufacturer is producing a robot of this type. This allows the robot to service more than one machine as well as transport materials and components; in fact, it generally increases the opportunities for exploiting the advantages of a robot's flexibility. One particular advantage of a mobile robot would be in the case where the workload in a manufacturing facility did not justify mounting fixed robots at each machine. For safety reasons mobile robots and AGVs have a flexible touch sensitive 'bumper' at front and rear which if touched, say by contact with a human, will remove power thus stopping the vehicle.

AUTOMATIC IDENTIFICATION

As parts move through the computerized factory some means of automatically tracking and identifying them is necessary. This will ensure, for example, that a machine tool knows what component is being loaded into it by a robot, that an automatic inspection machine checks appropriate dimensions, that the correct route for a component at a conveyor junction is chosen, or simply that the production control system knows where each item is at any point in time. It is possible to put identification onto the component itself, although this is unusual as operations such as machining or painting will probably remove or occlude it as some stage. More commonly the identification is on the pallet carrying the work, this pallet usually staying with the job during its progress throughout the factory.

There are various systems used. The simplest is to use a pattern of holes or cams on the pallet which are pushed over an array of microswitches as the pallet moves. As each pallet has its own unique pattern the sequence of signals produced identifies the work. At the other extreme, human alpha numeric characters can be read by a vision system with character recognition capability. However, the most common methods used today are magnetic labels, and bar coded labels for laser reading. The magnetic identification labels contain magnetic strips in which are encoded data items such as part numbers, these are read by appropriate scanners. Bar coding is also very popular and a typical code is shown in Figure 17.8. A series of bars of varying thickness and separations represent the numeric data. Each digit is represented by two black bars and two white bars in alternating sequence. The codes are read by a laser scanner as the pallet with the label attached passes by. The scanner emits a beam of laser light which is reflected back onto a sensor. The series of light and dark bars produce a corresponding series of binary voltage levels, which are then decoded to provide the identification data. These can then be relayed to the

000189 171848

Figure 17.8 Typical bar code

Figure 17.9 Bar code laser scanner in use on robotic assembly line (Courtesy Dexion)

appropriate control system. A scanner in operation is shown in Figure 17.9.

Another identification device which can be attached to the product itself is shown in Figure 17.10. This is a small programmed tag which contains a custom integrated circuit and a long life lithium battery. The IC can hold as many as one hundred alpha numeric characters in its memory, to form a unique identity code of large capacity. When the tag comes within one meter range of a sensor containing a transmitter and receiver it is automatically activated. This allows it to transmit the information held in its memory thus identifying the host item to which it is attached. The tags are reprogrammable either by proximity methods or remotely and can operate in extreme industrial environments. Unlike bar coding which depends upon a direct line of sight, these sensors can be interrogated around or even through obstructions. They are driven and controlled by intelligent, solid state decoders which format the data for subsequent transmission via a range of serial data links to the next level of computer control. The system is shown in Figure 17.11.

COMBINING THE ELEMENTS

Industrial robots, and other elements of the automated factory such as those mentioned above, have been combined in many installations throughout the world to form large automatic facilities within factories. The two most common types of integration are flexible manufacturing systems and automatic storage and retrieval systems.

Flexible manufacturing systems

Flexible manufacturing systems (FMS) are arrangements of reprogrammable automation equipment to allow manufacture of components in batches, or even single quantities on demand. They comprise NC machines, robots, automated material handling systems, PLCs, etc., and also supervisory computer control. Most FMS in existence are flexible machining systems, but flexible assembly systems (FAS) are also beginning to appear as

Figure 17.10 Programmable identification tag with IK memory. (Courtesy Eureka Systems)

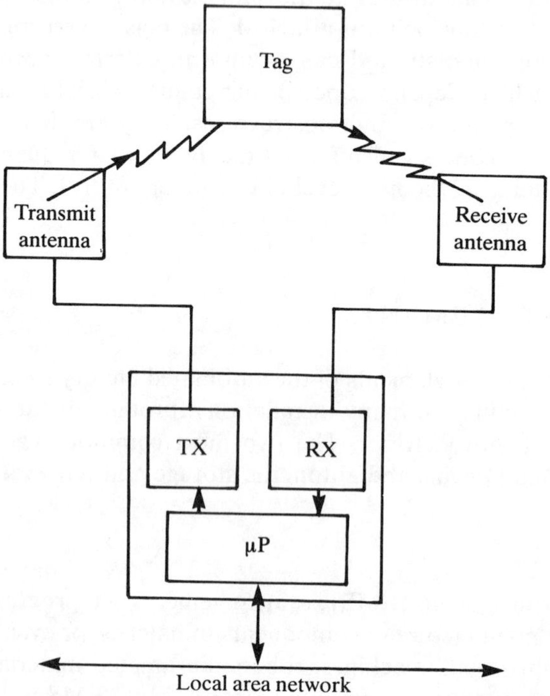

Figure 17.11 Tag communication system

assembly robots become better designed and cheaper. The domain of application of FMS is within the reprogrammable automation block shown in Chapter 1, Figure 1.6. They grew in an evolutionary process from NC and DNC systems and now would be an essential part of a computer integrated manufacturing system. Although only one part of a full CIM system, massive savings in stock and work in progress (WIP) and manpower reductions of, for example, sixty down to five have been claimed as a result of the adoption of FMS.

Automatic storage and retrieval systems (AS/RS)

One of the greatest financial burdens on a manufacturing company is the cost of holding materials and parts in storage, and as work in progress. The concept of 'just in time' manufacture (JIT), where material arrives at the factory just in time to be processed to meet customer orders, is an attempt to reduce this cost. The principle seems to work well in a country like Japan with short communication lines between densely packed, closely interrelated, manufacturing centers. However, in other countries, that perhaps rely more on imported materials and components, this luxury does not exist, although JIT can still be applied to some extent. There is, therefore, still a need to store material before usage in quantities that ensure an even production rate can be achieved on the shopfloor. In the computerized factory the storage area can be automated just as thoroughly as the rest of the plant, to improve efficiency and reduce the time that material is lying unused. The terms 'storeroom' and 'warehouse' are not now used by those who possess an AS/RS. These words have, for them, connotations of huge volumes of static materials simply lying gathering dust and incurring cost. Terms like automated materials distribution center (AMDC) are much more popular. Thus the AS/RS in the automated factory is a totally automated system using conveyors, AGVs, automatic cranes, and sensors, etc., all under computer control. Each element has its own controller, usually microprocessor-based, which is interfaced to a higher level supervisory computer. There may be a number of levels of computer before the final host computer is reached.

OTHER COMPONENTS OF THE AUTOMATED FACTORY

Up to this point the main concern has been the hardware used on the shopfloor of the automated factory. However, if full computerized integration of a company is to be achieved the other essential functions of the organization should be considered. For example, if the quality of production is to be monitored then some form of computer aided quality control must be adopted. Also at the initial design stage of a product computers have now been established as an essential part of the design process in many companies. In the computer integrated manufacturing concept it is the data produced at this initial design stage that are used as the basis for all subsequent activities. Thirdly there are many general management services which will need to be computerized to provide management, or supervisory computers, with the necessary information on the state of the factory to enable efficient control of the manufacturing process. These three areas are now briefly considered.

Computer aided quality control

Contributing to CAQC are the elements of computer aided inspection (CAI) and computer aided testing (CAT). CAI involves the linking of devices such as NC inspection machines, vision systems, and other inspection facilities with a computer controller to log the data and monitor the process. CAI therefore involves inspection of components and products to ensure they are within laid-down tolerance limits. CAT, on the other hand, involves testing the performance of products, such as the performance of an internal combustion engine or the correct functioning of a printed circuit board under environmental changes. The computer monitors the data from sensors attached to the products, logs them and analyzes the results. The information from both CAI and CAT systems can be further analyzed by computer to check for trends, so that the process can be stopped when necessary before any scrap or unsatisfactory products are manufactured.

Computer aided design

CAD as the name suggests involves the use of a computer to assist the engineer to create and develop a design for a product or system. By the use of interactive graphics the designer can analyze the performance of his design and its response to simulated influences. He can also use it to increase his efficiency by drawing upon libraries of standard shapes, routines, and programs. This increases productivity by aiding the designer in visualizing, thus reducing both design time and overall project time. Another advantage is that if using a draughting facility the drawings produced are clearly legible and of standardized format thus improving communication.

However, in the final analysis it is the ability of CAD systems to integrate with the automated factory that is perhaps their most intriguing aspect. In a full CIM system there would be no necessity for paper drawings as all data would be transferred as digital signals from the CAD system to the other components, such as material ordering, or to postprocessing to produce programs for machine tools and robots. The CAD process will in fact be the source of most of the information necessary to create a database from which to manufacture a product.

The use of a CAD system for simulation to facilitate robot programming has already been mentioned in Chapter 12, but it is worth noting here the essential components of a CAD system. Firstly, there is the designer, the human source of the original design concepts. Secondly, there is the hardware, this comprises a CAD terminal or workstation composed of a monitor screen, keyboard, and programming tablet, a printer, the computer, and the mass storage system. The third component is the software; this allows the designer access to the standard packages specially designed to speed up certain tasks, e.g. finite element analysis, PCB design, plant layout, or robot programming. A typical system is shown in Figure 17.12. A wide range of CAD systems is now available. The subject is vast and no attempt can be made to embark on it here. As well as engineering design, which is our concern, CAD techniques are used to produce artwork, film for video programs, and special film effects. Wire frame and solid images can be produced with shading, hatching, sections, color coding, and animation all included.

Management information systems

To ensure efficient functioning of the factory there are many other factors that require

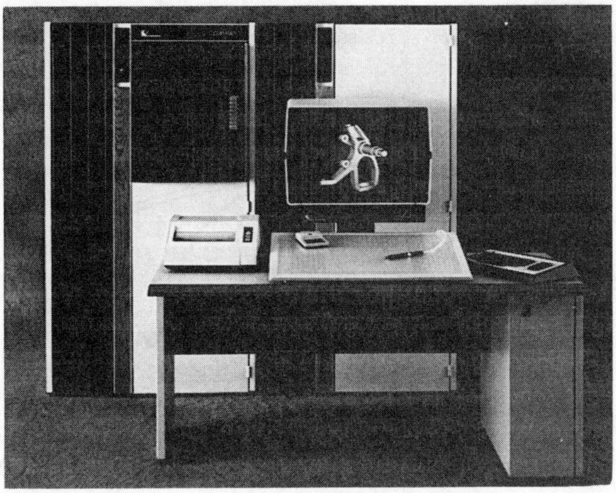

Figure 17.12 A CAD system. (Courtesy Computervision)

attention. For example, the planning necessary to ensure material is available at the correct time and at the right place can be a complex task in a large production system. Assuming again we are considering the automated factory, then a computerized system is required to do this. Various techniques and methods have been devised to handle the problem, the main solution sought being to reduce stores, work in progress, and throughput times. Two typical approaches to this are MRP (materials requirements planning) and OPT (optimized production technology). Essentially, whatever the approach adopted, material, components, job numbers, etc. must be tracked through the factory from goods inward to despatch. Orders for material must be placed at the proper time and progress monitored. Also the whole production operation must be planned to prevent bottlenecks and ensure a smooth flow of work through the factory with all machines and equipment utilized at optimum levels. The computer system that handles this will, in a CIM system, interface with the other factory computer systems, and in a truly integrated fashion will simply form part of one large homogeneous unit.

COMPUTER INTEGRATED MANUFACTURING

This is the ultimate manifestation of the computerized manufacturing concept. The totally unmanned manufacturing organization is still really only worth considering as a potential future event, but CIM utilizes contemporary technology to take us almost to the 'unmanned factory' stage. The components of the automated factory so far described are in themselves now quite commonplace. FMS cells, for example, often exist as 'islands' of automation within a large factory. The linking of all computer controlled equipment and functions within the factory is now seen as very desirable, and it is this linking, or integration, that is referred to as CIM. Figure 17.13 gives an illustration of this concept.

One of the main advantages of full computerization is that a common database can be used. Exactly the same information on, for example, a component will be available to design, manufacturing, quality assurance, and purchasing departments, and it would be

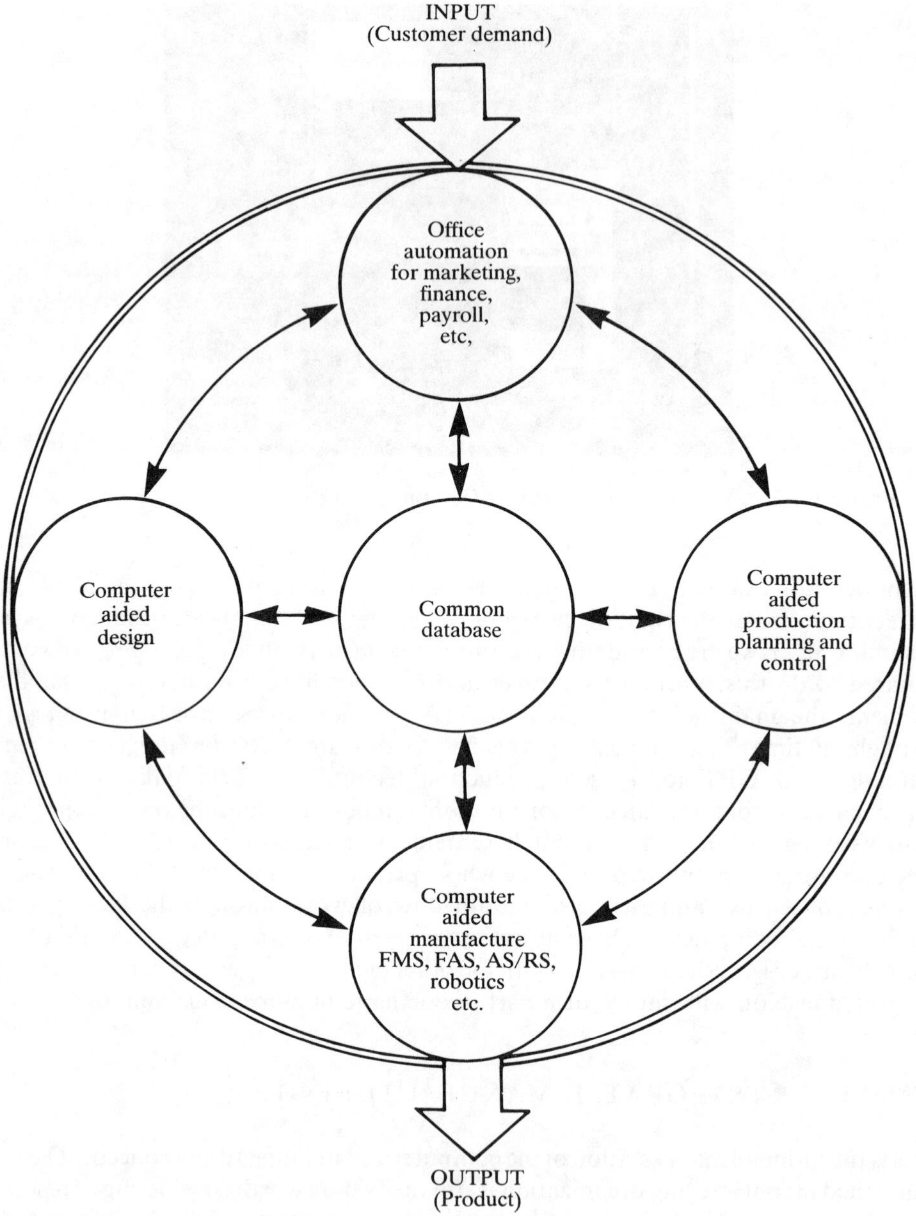

Figure 17.13 The CIM concept

available virtually instantaneously to all users. All functions within the company would be able to access data from various departments simply by consulting a computer display screen.

The integration of the components of a CIM system is a complex problem. Within the factory, computer- or microprocessor-based control exists at various levels. A hierarchy of levels can be identified as shown in Figure 17.14. Complete integrity of information must be maintained as it is transferred from one machine to the next. Interfacing different machines and computers can be a problem but compatibility of equipment can ameliorate this. As computers have developed, so also have methods of using them. Initially the one computer, probably the only one in a very large organization, was used by one operator.

Factory level Management information coordination	CAD; MIS { Order entry, Purchasing, Production control, Quality control, Costing and finance, Other data processing }
Cell/Center level Multistation coordination	Production control
Station level Device coordination	Industrial robots; NC machine tools; Instrumentation; Process control
Process level Device control	Servomotors; Sensors; Switches etc.

Figure 17.14 Levels of electronic control systems that must interconnect to form a CIM system

As computers became more powerful a number of users were able to utilize the machine by having their work processed in batches. This was very slow, therefore the next stage of time sharing arose where many users had direct access to the same computer simultaneously from remote terminals. However, now that computers have proliferated and all large companies and organizations have large numbers of computers, some dedicated to specific tasks, the fourth stage has arrived. This is evidenced by the desirability for various computers to talk to each other.

This communication may be required over a large geographical area, say between company divisions or universities at various parts of a country. To implement this a network of computers is established typically using the telephone system as a communications link; these are termed WANs (wide area networks). Today networks are required within places such as factories, hospitals and large offices. These networks do not span such a large area and therefore communications links in the form of coaxial, or fiber optic, cables are used. These are termed LANs (local area networks). Thus a computer integrated manufacturing system will, by definition, be composed of such LANs. Some typical network configurations, or topologies, are shown in Figure 17.15. The bus system, in which all computers or controllers are connected to a single communication line or bus, is very common. The loop system is similar to the bus concept, except that the common line is looped round to join up with itself. The star system is similar to a telephone system in which each device communicates through a master controller. In the ring configuration information is passed through each device which checks to see if it is addressed to itself, if not it passes it on to the next device. Many other networks or combinations are available but the bus and ring are the most likely to become standard for LANs on the factory shopfloor.

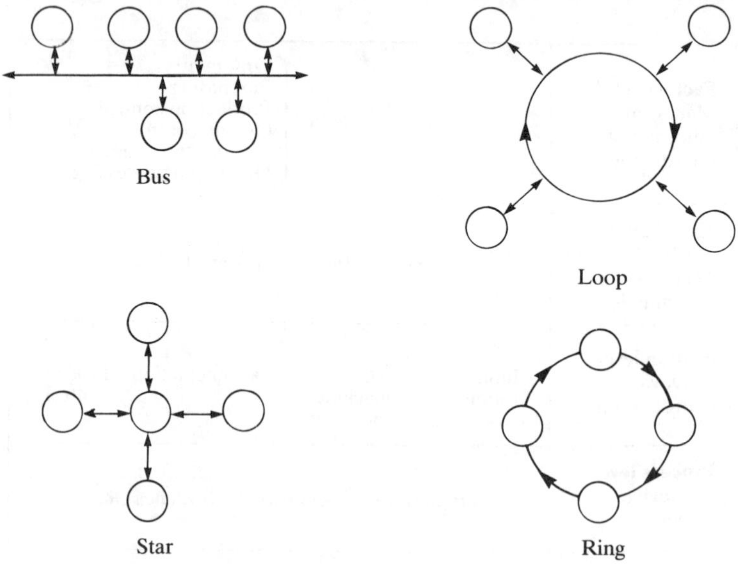

Figure 17.15 Computer network topologies

One of the main problems in setting up a system in which all the computerized elements communicate with each other is the compatability of equipment and the communications protocol used. That is, not only should the hardware for communication be standardized, but so too should the communication signals used by that hardware. Thus some sort of standardization is necessary ultimately leading to plug-in compatibility between computers. At the time of writing the most likely contenders for worldwide acceptance are the General Motors' Manufacturing Automation Protocol (MAP) for the factory floor, and the Boeing Technical Office Protocol (TOP) for CAD, engineering management and other administrative areas. Both MAP and TOP are very similar and facilities using these protocols will be easily interconnected.

Considering MAP a little further, it is based on the International Standards Organisation's (ISO) Reference Model for Open Systems Interconnection (OSI) which was originally developed in 1979. This standard was devised to allow computers to be interconnected in a network in such a way that there would be full compatibility within the system. There were seven layers identified (see Figure 17.16). The lowest layer at level 1 is

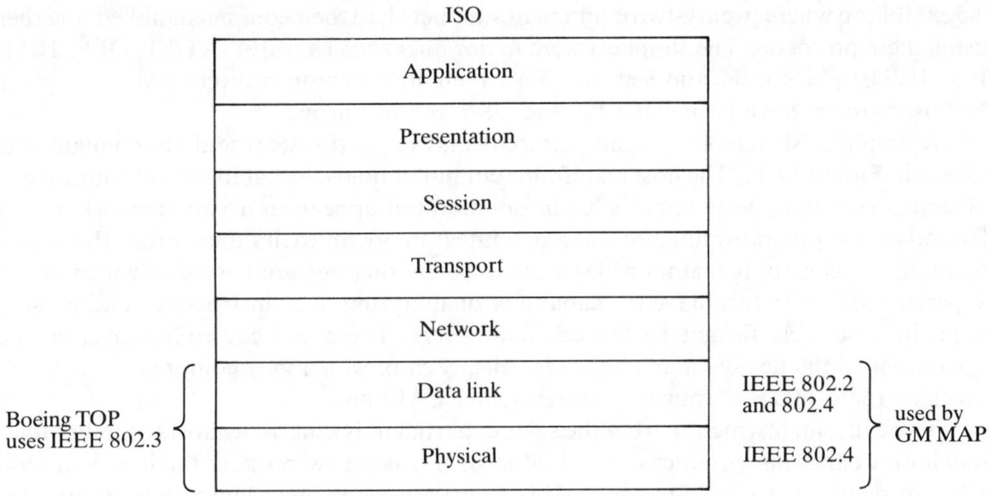

Figure 17.16 Layers of ISO reference model for OSI

the physical layer which defines the mechanical, electrical, and physical interconnections. Various functions are defined by the intermediate layers until level 7, the applications level, is reached. This provides the user facilities such as file transfer and message services. MAP takes each of these layers, which are only functional, and applies a protocol to them. At the time of writing, the American Institute of Electrical and Electronic Engineers (IEEE) 802 committee has worked out standard formats for communication at the bottom two layers of the OSI model. These are 802.2, 'logical link control'; 802.3, 'carrier sense multiple access with collision detection' or CSMA/CD; 802.4, token bus; and 802.5, token ring. The 802.3 is patterned after an established system developed by the Xerox Corporation called Ethernet, this is used for the two lower levels of TOP. MAP uses 802.4 for the physical link, i.e. a coaxial cable, and 802.2 and 802.4 for the data link.

Data can be transmitted in LANs as 'baseband', 'carrierband', or 'broadband'. In baseband transmissions information is sent digitally as ones and zeros. It uses the entire bandwidth of the cable without modulating the signal. In carrierband the entire bandwidth of the cable is used and the signal is modulated. In broadband systems carrier waves are used which allow the available frequency spectrum to be split into a number of channels which can be used for different purposes, i.e. a portion of the cable's bandwidth is used and the signal is modulated.

In 'token' passing systems the only member, e.g. a computer, in a network that can transmit is the one holding the 'token' or 'permission to transmit'. When it has completed transmission it then passes its token to the next station. Each member is only allowed to hold the token for a limited period thus ensuring all members get a chance to transmit. In the CSMA/CD method each member is able to listen constantly for messages meant for it on the bus. However, if it wishes to transmit it must wait until current transmissions are finished. As collisions between transmissions can often occur, methods have been devised for recovery and resumption of communications.

General Motors and Boeing first exhibited MAP and TOP together at the Autofact

'85 exhibition where twenty-two equipment suppliers had their equipment linked together using these protocols. The suppliers were major ones such as ASEA, AT&T, DEC, IBM, ICL, INTEL, Motorola, and Siemens. Since then other demonstrations have taken place and user groups have been started in the USA and in Europe.

A simplified sketch of how some components of a CIM system could be configured is shown in Figure 17.17. The host mainframe computer holds the database and coordinates all activities within the system. Taken in isolation this appears as a 'star' network, but in fact other computerized functions are also integrated with each other. From the host a hierarchy of control is maintained. In this case instructions are passed down to group supervisors, they in turn have the capability of analyzing these instructions and passing them in a suitable format to the cell controllers. These cell controllers oversee the operations of the flexible manufacturing cells, each of which may comprise robots, NC machine tools, material transport systems, and CAI units.

Typically, an instruction from the sales department is sent to the host computer for a batch of a particular product to be produced. If it is a new product the host will send relevant details to the CAD facility. This will use the common database to ensure, for example, standard components are used wherever possible. In the ideal system the CAD function would be able to get feedback from purchasing and manufacturing on the likely effect of different design proposals on ease of production and overall costs. An example of

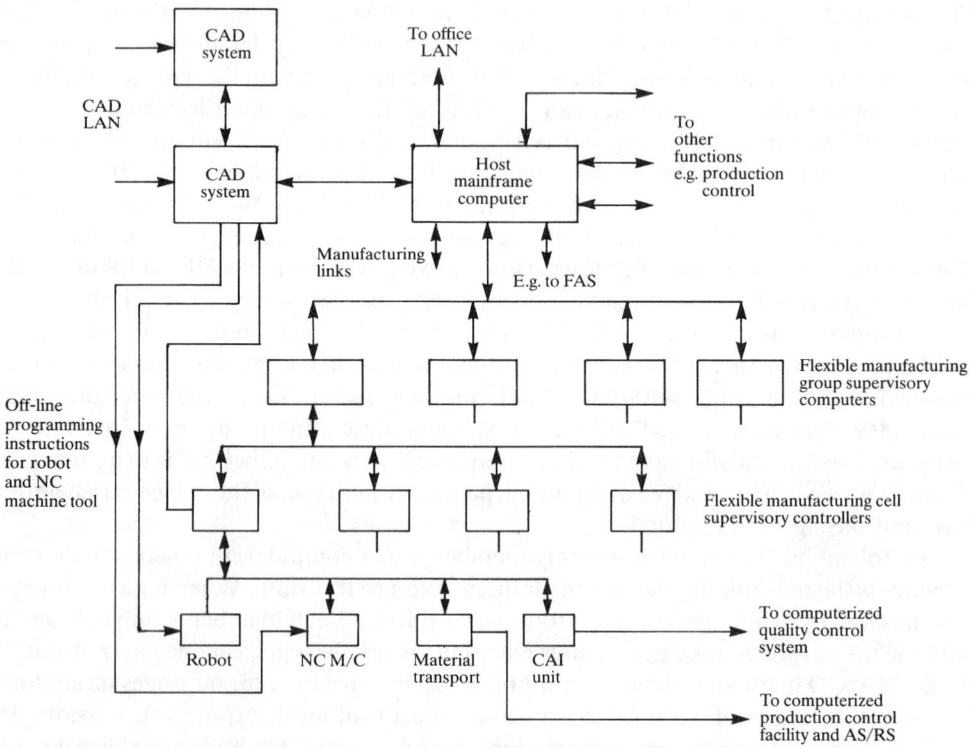

Figure 17.17 Simplified illustration of hierarchical control and integration of various local area networks to form one CIM system

one advantage of off-line programming for robots and other equipment is apparent here in the making of a prototype. In the figure a link from a cell supervisor is shown going to the CAD system. This could be sending a message signifying readiness for communication. The CAD system would then download the relevant programs to the robot and the NC machine tool.

Once the design has been finalized, and the relevant information passed to the database, the purchasing facility can order the material and components necessary from outside vendors, and the production control function can plan the manufacturing operation. When production is ready to start the order details are fed down through the various levels in the hierarchical structure.

Thus the fully computer integrated manufacturing system has been considered and the place of the industrial robot within this system should now be apparent. The robot is only one component of the automated factory. However, it is an essential component and it is also particularly interesting because a study of the robot leads to an understanding of many of the principles and problems encountered both in computer controlled machines and in modern manufacturing systems.

18

THE FUTURE OF ROBOTICS

INTRODUCTION

This book has developed over a period of about five years. When the original list of contents was devised it was anticipated that there would be a sizable chapter on 'the state of the art' and a similar one on the 'future' of robotics. However, during this period it has become obvious that the speed of diffusion of ideas and designs from the research stage to commercial production is very short; concepts that were being researched when the book was started are now commercially available products, in production and on the market. It is therefore hoped that as the book was continually updated while writing the main body of the text itself reflects the 'state of the art'. The future is therefore considered as beginning at least a decade ahead of publication. In this way it is anticipated that this relatively brief chapter will not date too rapidly.

Many aspects of robotics will of course change in the short term. For example, only a few years ago 8 bit microprocessor control systems were predominant, now 32 bit systems are quite common. However, the basic principles of robot control should remain valid for some years.

The format of this section is in two parts. Firstly the possible future uses of robots are examined, and secondly the areas requiring further research and development to achieve these uses are identified.

FUTURE USES OF ROBOTS

Up to this point in the text the 'industrial' robot alone has been considered. However, as technology develops the knowledge gained from industrial robot research, development and application will be utilized not only in the robotic devices used in factories but also in many other applications.

Factory robots

These are the 'industrial robots' with which this book has been mainly concerned. These robots will continue to utilize the latest developments in hardware and software, and as the equipment improves so the range of applications will increase. So too will the number of human workers displaced increase as the costs of the robots continue to decrease in relation to the cost of human labor.

As mentioned in the previous chapter, industrial robots are an essential part of the computer integrated manufacturing system. Factories will be so organized and automated that the industrial robot of the future will not require to be universally adaptable. That is, the 'specialization' of robots will continue so that classification by application will be quite appropriate, e.g. 'assembly robots' or 'seam welding robots'. They will be made with only the sensory abilities necessary with which to carry out their designated tasks and many may require no external sensors. This will be in contrast to the robots used outside the factory in disorganized environments that will demand the full range of sensors available.

Assembly
This is one of the areas of greatest potential for industrial robot development. The speed and precision of robots will continue to improve, facilitated by the use of new materials to allow lightweight arm constructions. Some robots may retain rigid structures suitable for heavy lifting, others may use very lightweight flexible arms coupled with special sensors and control to allow extremely precise high speed operations for lighter assembly work. Multiarm robot systems may become more common, as these would remove the necessity of robot 'assembly lines' and allow truly flexible robot 'cells' to be constructed. Parts feeding and presentation will also be improved to make cells cost effective.

Artifical intelligence, which will be explained later, in conjunction with advanced world model languages, and vision and tactile sensing, will allow robots to carry out complex assembly work based on simple instructions. Commercially available universal grippers will also enhance robot flexibility within the assembly function. As well as these sophisticated robots there will also be a proliferation of the simpler, low cost robots with a limited number of degrees of freedom and simple grippers. Both types of assembly robot will be part of flexible assembly systems, receiving instructions on what to assemble and when through a computer hierarchy as in the CIM systems discussed earlier.

Materials handling
Taking this in its broadest sense, i.e. encompassing machine tool loading, unloading, palletizing, stacking, etc., there will be no major changes in the types of robot used. Their performance characteristics should of course improve, but there should be no need for significant improvements in control or sensing. However, where mobile material handling robots are concerned there will be many improvements, but these are considered separately.

Material joining
A successful prediction of future developments of robots in this area is dependent upon the progress of materials joining technology. For example, spot welding and seam, or fusion, welding are the most popular techniques. But as new materials are introduced and subsequently new joining methods, then the utilization of robots will also change. The use of plastics, composites and other materials in products may make adhesive application one of the more important joining tasks. Generally speaking, robots engaged in any

material joining techniques will experience improvements in 'adaptive control' techniques, which will allow them to monitor and control their movements and process characteristics as necessary. In adaptive control generally, the parameters used by the computer for control are constantly adjusted automatically in response to measurements of process variables, with the purpose of achieving better performance.

The other industrial robot applications will benefit too from new developments, but perhaps more interestingly there will also be emerging new applications within the factory. The advanced free ranging AGVs mentioned in the previous chapter will be more fully developed and become commonplace in large factories. As well as being used for material and tool transport they may be modified for other tasks. Examples could be security guards, or even in a trivial sense as novelty factory tour guides. The technology applied to industrial robots will also find application outside the factory environment, some of these applications are now considered.

Robots for use outside the factory environment

Mobile robots

The AGVs that have already been mentioned worked in the highly structured and controlled environment of the factory and its surrounding area. However, outside the factory these vehicles will require more autonomy and rely more on preprogrammed instructions and sensors. When communication with a central controller is required radio, sonic, infrared or laser communication may be used. Mobile robots will be used in a large variety of tasks and some of these will be considered later. Generally speaking these robots will have internal navigation systems, 'maps', and sensors. They may use stereo vision, ranging devices and touch sensors linked into artificial intelligence technology in its 'on board' computer control system. As well as wheeled robots the attractiveness of having robots that can climb stairs or cross very rough terrain makes the commercial availability of quadruped and hexaped robots a strong possibility. Four and six legged robots have already been constructed in a number of research establishments. The control of the limbs is one major problem. Another problem is the power to weight ratio of the final machine considering it will probably be required to carry its own power supply, fuel and a payload. Practical biped robots in human form are certainly a long way from commercial viability. Most of the robots now considered will be mobile and capable through their sensors and programming to operate autonomously in unstructured environments.

Robots in the nuclear industry

Devices similar to robots have already been used here for inspection, maintenance and manipulation of radioactive material. However, although very sophisticated, these devices are usually under human control. Intelligent robots may in the future be used in radioactive environments, but due to the dire consequences of an autonomous robot making a mistake or running out of control it is likely that there will always be some degree of human supervision. Conventional closed circuit television systems and umbilical cords to the robot will still be used to achieve this safety feature. In highly radioactive areas once the robots are installed they will probably be sealed in, never to be taken out except as waste material when their work is finished or they break down. In the event of nuclear accidents mobile robots could be invaluable for investigation and preliminary decontamination of 'hot' areas.

Fire fighting and rescue

Mobile robots equipped with appropriate sensors and impervious to smoke and fire will be ideal for this task. Their mobility will have to enable them to climb stairs and navigate narrow passages and sharp corners. This will require tracked or legged robots possibly with articulated bodies and arms. By mounting hoses on them they could get into areas to extinguish fires previously impossible to humans. By using infrared sensors to see through smoke they could be used to search for people in smoke-filled buildings. They could then offer oxygen masks and lead the humans out retracing their own memorized paths.

Agriculture

The use of robotic devices for farming is interesting as it was here that the majority of the human population was employed before the first industrial revolution. Robot tractors for ploughing, fertilizing, sowing and spraying may be possible, and research has started in these areas. As well as removing much human toil, a major advantage would be that daylight hours would no longer be important. Agricultural tasks such as those mentioned could be carried out by robots twenty-four hours per day, no matter what weather conditions prevailed. Robots could also be employed in forestry in cultivating, planting, tending, and cutting down trees for timber.

Military robots

This is probably the most dramatic future application of robots. Machines that can engage in combat in an autonomous manner conjure up frightening images reminiscent of science fiction. However, by combining mobility, artificial intelligence, sensory abilities and armaments, an efficient and deadly robot war machine can be produced. These machines could use wheels, tracks or legs for locomotion depending upon the terrain they are expected to cover. Their artificial intelligence and sensors could guide them rapidly to their targets which they would then, to use a military euphemism, 'neutralize' with their weapon systems. Robot aircraft are also likely, being simply one step on from present day remotely controlled surveillance drones. The existing Cruise missiles already come into this category as they have internal navigational systems and carry their own 'map' of the terrain over which they fly. Although combat machines are the most threatening type, other robots used for reconnaisance will possibly be the first to be used. These machines would be relatively small and quiet, they would also be expendable, which is a distinct advantage when the condition of the destination is unknown. Other robots could be sent behind enemy lines to emit 'dummy' radio signals to confuse enemy listening systems. Already satellites have been used to provide a Digital Terrain Database in which the terrain of almost the whole world has been digitized. These data can be used by the Cruise missiles and future mobile robots to provide an internal digital map of their area of operations. In combination with vision systems and AI the data will allow them to plot the best routes by which to accomplish their tasks.

Medicine and health care
Robots may be seen in hospitals tending to patients, transporting materials and drugs, and even being involved in surgery. Outside the hospital, in domestic life, robots could bring aid to the disabled and aged, but this will be considered later. In hospitals where patient isolation is necessary, robots could be used as nurses, servicing the patient in bed much in the same way a robot tends a machine tool in a factory. Food, drink and magazines, etc. could be retrieved from a storage cabinet and presented to the patient as necessary. More complex tasks could possibly be undertaken under remote human supervision. Generally speaking, however, humans will always prefer human contact if possible and robots will only be applicable in extreme circumstances. In some surgical tasks robots coupled with vision systems may prove to be useful due to their speed and precision, although supervision by a human surgeon would be mandatory. It is also worth noting here that a great deal of the research into robots may have spin-off effects in prosthetics. Motorized limbs, hands and fingers with tactile sensors will provide the disabled person with more control over his or her own surroundings.

Domestic
This is potentially the most lucrative market for robot technology. In western society almost every family has their own car and must now claim it to be a necessity rather than a luxury. In the future the same may be said for the family robot. It can be easily seen that a machine that could do the tidying, vacuuming, dusting, setting and cleaning tables, carrying heavy loads, and acting as an information source would soon also become a necessity. One chore not mentioned was 'the ironing' as the author believes this will resist even the most strenuous attempts of engineers and scientists to robotize; future fabrics, however, may be designed so as not to require ironing. Many devices in the home are already automated and reprogrammable, e.g. the automatic washing machine and the electronically controlled cooker. However, these devices are analogous to the NC machines in a factory, they do their own specific job but do not manipulate the material or products they are concerned with. For example, the washing machine does not put the clothes out to dry, and the cooker does not serve the food. In these examples robots will be able to collect, load, and unload the washing from the washing machine, and prepare, load, unload and serve the food from the cooker.

For disabled people a domestic robot would change their lives considerably. To be able to reduce the burden on relatives and helpers by using a robot would also remove some of the mental stress on the invalid. Naturally human company would still be absolutely necessary from a social aspect, but the obvious advantages of a robot servant would make it a very desirable possession. Depending upon the degree of disability, the robot could be made to respond to a variety of commands from voice control to simple whistling or clicking noises, etc. A specially constructed environment and ancillary equipment would be practical to reduce the amount of sophistication required by the robot. That is, the robot would still be working in a relatively structured environment.

However, for the general purpose domestic robot operating in a disorganized environment such as a family home, a great deal of sensory ability and adaptability will be necessary. Although the robot may contain a 'map' of the home in its memory, and know the location of all the major items likely to be used, there will no doubt be many random obstacles and unexpected movement of items. For example, a baby may be crawling across the floor, a table may be moved temporarily, or a valuable and delicate ornament may be placed on a shelf that the robot is about to dust. A domestic robot would have to cope with all of this and at the moment would be prohibitively expensive; but in the future it is quite possible.

Mining

This is an extremely hazardous job for humans and a number of deaths occur every year. Using robots in place of human miners will remove people from underground work completely. Already mining is highly automated but there is still a need for humans to undertake a number of tasks. If these tasks can be robotized then there could be a reduction in safety precautions and the size of underground galleries, which would give financial savings in addition to the reduction in accident statistics. This would possibly require the use of intelligent robots capable of following the mineral seams through the ground, cutting them and ensuring the cut material was transported to the surface.

Marine

Remotely operated vehicles for use in marine conditions are well established in off-shore engineering. They are used for surveillance, inspection, repair, and retrieval of materials and devices from the sea bed. By using them an operator can extend his influence into the subsea environment which has conditions that would be extremely hazardous to a human diver. However, there are problems with control. For example, if an umbilical cable is used for communication it can become twisted and trapped in debris, weeds or manmade objects such as pipelines. If this cable is broken then control is lost completely, and in some conditions this may result in loss of the submersible. Other methods of communication such as sonar present difficulties due to noise, corruption of signals, and time delay between transmission and reception.

Thus a robotic device, able to operate autonomously and following preprogrammed instructions, would be ideal for remote undersea work. The robot could be programmed to seek out pipelines and follow them for inspection purposes. It could be provided with sensors, manipulators and tools, to identify faults and repair them. Many other tasks could be given to these automatic robot submersibles. They could explore the ocean bed seeking valuable minerals, they could construct pipelines and erect other artefacts such as underwater mining equipment. As the world in the future turns to the oceans for food, these devices could be used for fish farming and marine plant growing, etc. The power sources, drive systems and manipulators all exist today. Once again these robots will be made practical by the increased computing power, AI techniques and sensor technology that will be available in the near future.

Outer space

Space may be the ideal environment for the robot, offering many opportunities for applications. Due to the extremely high cost of sending humans safely to space, maintaining them there with all their encumbent life support systems, then bringing them back to Earth, robots will be economically very attractive for space applications. Reduced safety considerations for the inanimate robots will mean lower operational costs for transport to space, and of course no life support systems will be necessary. Once up in space there will be no need to bring the robots back as repairs could be effected by other robots, or the occasional human engineer. The robots will be weightless, require no air, food or water, and if properly constructed, be impervious to radiation and extremes of temperature. Power requirements for manipulation could be met by solar cells, possibly the robots could 'plug in' at regular intervals to central solar power stations to charge their batteries. Conventional reaction fuel could be used for propulsion. In space the robots could be employed in the construction of space platforms, solar power stations, zero gravity manufacturing, etc. Many of the tasks undertaken by conventional robots today could be undertaken in space, although the design of the robot and associated tooling would have to be different to cope with weightless conditions.

Moving further 'out' in both distance and time, robots may be used on the other planets and asteroids for mining purposes. 'Robot' explorers have been used in the past to explore the moon and Mars. However, in the future more sophisticated and powerful robots will probably be employed in the search for, and exploitation of, ore deposits in the extraterrestrial environment. It has been postulated by NASA that these robots could be self-replicating, that is they will use the material they mine to build processing plants and eventually manufacturing plants to build more robots and so on; this is of course highly speculative.

There are many other possible applications but those mentioned should be sufficient to indicate to the reader the potential impact of robots on everyday life in the future. Briefly examined now are some of the areas for research and development required to achieve these applications.

AREAS REQUIRING RESEARCH

The various chapters of this book have themselves indicated the important areas of robot technology and management. These areas will continue to be developed to enable the future applications just described to be achieved. Some specific research areas can be identified that are critical and these are now outlined.

Artificial intelligence (AI)

A subset of this important technology is now available commercially in the form of expert systems and other software packages. However, much research has still to be done as it has only recently come onto the marketplace, its domain having been the research laboratory for the past forty odd years. As it expands and improves it will have a powerful effect on future robot developments.

During the Second World War a mathematician called Allan Turing was working on the code breaking techniques at Bletchley Hall in England as mentioned in Chapter 2. After the war, in 1950 Turing wrote an article, 'Computing machinery and intelligence', which posed the question, 'Can machines think?' To achieve an answer to this he proposed a test whereby a human, termed the 'interrogator', in one room would communicate with either a computer or a person in another room. It is not necessary to go into the details of the test here, but if the interrogator could not ascertain whether he was communicating with a computer or a human, and indeed it was a computer that was being interrogated, then the computer could be said to be 'intelligent'. This test has been named the 'Turing Test'.

A precise definition of AI has not been made as many people in the field view it from different perspectives. It can be said that AI is the ability of an inanimate machine, such as a computer or a computer controlled robot, to exhibit the characteristics associated with human intelligence. Robots may utilize AI techniques in many ways. For example, expert systems, i.e. programs that contain factual data and procedural knowledge and are able to use this in conjunction with a set of rules to enable them to copy human experts, may be integrated into a robot memory. Also artificial vision, high level language programming, and speech recognition and synthesis will all be combined with AI techniques to produce extremely flexible and intelligent robots.

Control
To enable large amounts of sensory information to be handled efficiently and at a high enough speed for practical applications, control systems must be improved. This will involve research into computational methods to model the robot even more accurately in software. Another area for research is the attempt to enable a robot with AI, as it carries out a task, to progressively reduce its data processing times as it undergoes a 'learning curve' similar to a human being.

Sensing
Research into sensor technology should be carried out in three broad areas. Firstly, broadening the scope of the different types of sensor available to allow new applications to be attempted. Secondly, the resolution of both internal and external sensors could be improved; this will, for example, increase robot precision, allow component identification by touch, and decrease inspection costs as this would be able to be carried out 'in process'. Thirdly, improving the speed, reducing the size, and lowering the cost of sensors will make them more attractive for use in robots not only in industry but also in other applications such as domestic use.

Vision research should continue to develop methods of identifying components or items in a jumbled or overlapping background. Research in this area will aid the development of mobile robots attempting to find their way in unstructured environments. 3D vision is being researched but commercial availability of such potentially useful systems is still some distance in the future.

Speech recognition still requires considerable development to be made to ensure consistent and fast operation. This will enable voice programming and two way communication using a voice synthesizer.

Sensor–controller integration
Integrating and interfacing sensor technology with the control function mentioned above is also important. For the control system to be able to handle input data from a number of sensors, analyze them, and make appropriate coordinated movements in real time is a difficult task, and although implemented to some extent in present-day robots, it could still be tackled more efficiently. Progress in this area will be evidenced by many forms of adaptive control, two examples being collision avoidance of the robot arm with moving objects in the work environment, and error recovery. A fixed industrial robot will have to

avoid collision with humans that inadvertently stray into the work envelope if other safeguards fail. Mobile robots will also need the ability to quickly see an object, identify it, then take appropriate action (such as avoidance) all in real time. Error recovery would also be a desirable robot attribute as it would enable considerable cost savings to be made in the factory. Here an industrial robot could quickly check a workpiece, component, or product, as it arrives and leaves its own workstation, and take corrective action to remedy any faults found or mistakes made.

Autonomous mobile robots
Research in the above areas will provide the basis for the full development of the truly autonomous mobile robots. For legged robots many areas have still to be developed, how to keep the machine balanced, how to maintain a sensible 'walking gait' to ensure payloads are not damaged, and to obtain an adequate speed. Semi autonomous wheeled vehicles are in use today, but independent machines capable of rescue, farming, construction, and domestic work are still quite a distance in the future.

Arm design
Improvements in the physical arm design are important to allow speeds and precision to be increased. Research into the use of materials such as fiber graphite, special high strength low alloy steels, and composite materials for arms will improve lightness and rigidity. Also developments in motor design to increase power to weight ratios and reduce heating effects are still required. Continuing research into lightweight direct drive motors and rigid arm materials, in conjunction with the previously mentioned improvements in control resolution of software and hardware, will all aid precision and speed. In fact, it may be possible to produce robots for lightweight applications of high precision capable of accuracies and repeatabilities of less than one-thousandth of a millimeter. These would find their application in microsurgery and electronics component manufacture.

Standards
Much research needs to be carried out here for standards applicable to both hardware and software. In hardware, standard end effector connections, for example, would probably cheapen the cost of end of arm tooling. Also standards for peripheral connections and sensor interfacing would decrease installation development times and costs. Progress towards this could already be underway in the form of the protocols such as MAP mentioned in the previous chapter. If a standard robotic language could be developed this would improve the ability to integrate robots with CAD/CAM systems and simulation packages. Training of robot users would also be much simplified if a standard language could be taught. It would also make it easier for suppliers of software to justify creating standard packages for particular applications, e.g. as with present-day PCs a robot could be purchased and the control software appropriate for a particular task purchased separately.

Social implications
Due to the many human problems associated with unemployment and other types of displacement from work, the social implications of robotization should be thoroughly examined. This is obviously a difficult task as it is not possible to predict the state of the

world economy for any reasonable time into the future. It may be, as with so many other technological developments, that the effects of robotization will simply be observed rather than planned. This is usually not the ideal situation.

CONCLUSION

This text has attempted to describe industrial robots, how they work, their applications, and their implications. It has also tried to place them in perspective, initially within their ideal environment, i.e. the automated factory, and in a wider context by looking at possible future applications. Robots will change as time progresses but the broad principles outlined in this book should not become obsolete for some time. One principle that will never lose its importance is that robots must be applied responsibly with due consideration always being given to the implications for human beings. The chief end of robots should always be to improve the lot of the human individual.

Appendix 1

INDUSTRIAL ROBOT POPULATION STATISTICS

Although the first industrial robots appeared on the market around a quarter of a century ago, it is only in recent years that their use has reached significant levels. The USA were the first to manufacture and install industrial robots, but were overtaken by Japan just as the world robot population began to increase. Due to the definition problems mentioned in the earlier chapters, comparisons of populations in the early years are unrealistic. More compatible estimates are now possible, though as indicated below some problems continue to exist.

From estimates compiled by the British Robot Association and other sources, the tables and figures below have been constructed. These figures became available in 1987 and refer to statistics for the year ending December 1986. The reader may obtain up-to-date information by contacting the robot associations mentioned at the end of this appendix.

Table 1 compares the world robot distribution for the countries from which population figures were available for the period 1982 to 1986 inclusive. These figures give an indication as to the total numbers of robots in use worldwide and the numbers in use in each country. They also highlight the interesting fact that the number of robots claimed to be existing in Japan took a very unusual leap in

Table 1 World industrial robot population

	1982	1983	1984	1985	1986
Japan	13 000	16 500	64 600		90 000
USA	6 250	8 000	13 000	20 000	27 000
West Germany	3 500	4 800	6 000	8 800	12 400
France	950	1 500	3 380		7 500
Italy	700	1 800	2 700		5 500
Sweden					3 800
UK	1 152	1 753	2 263	3 208	3 683
Belgium	350	500	860		
Spain		400	516		
Australia		300			
Finland		120			
Estimated total	c. 30 000	38 000	97 000	110 000	155 000

Notes
1. All figures are approximate and the totals are only rough estimates.
2. When comparing figures and compiling totals the problems of compatibility of definitions should always be remembered.

1984. In the author's opinion this is a reflection on the difficulty of acquiring compatible information on robot population considering the definitions and classifications adopted by each country; this applies especially to Japan. Having said this, it is probably true that Japan still has the largest industrial robot population in the world, this being an indication of the Japanese approach to industrial investment which differs considerably from western countries.

Figure 1 illustrates the UK robot application distribution. Injection moulding, at the time of writing, is the most popular application, though this is because most of these are non servo, inexpensive robots that are simple to install. Spot welding has the second highest number of robots employed; this is as expected due to the large number of robots so applied in the car industry. The almost equally high number of robots in arc welding reflects the increasing availability of robot systems designed specifically for that task. An interesting figure to note is the number of robots employed in assembly. This figure is increasing at a relatively high rate as the number, and sophistication, of robots designed for assembly increases and their costs decrease. From Table 2 it can be seen that the number of robots applied in spot welding, as a percentage of the total, is decreasing, whereas the percentage of assembly applications is increasing. The number of car installations is finite and the reducing robot application percentage here indicates a trend towards

Injection moulding	674
Spot welding	534
Arc welding	454
Machine loading	356
Assembly	348
Education/research	211
Surface coating	210
Other applications	191
Not specified	191
Handling/palletizing	145
Press loading	76
Grinding/deburring	75
Glueing/sealing	68
Inspection/test	62
Die casting	40
Investment casting	15
Laser cutting	14
Forging	11
Water jet cutting	8
Total	3683

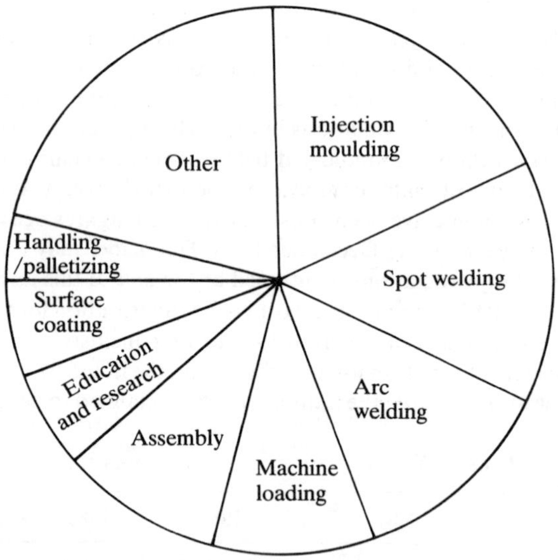

Figure 1 UK robot application and distribution

Table 2 UK spot welding and assembly application trends

Year	Total robots	Assembly	Assembly as percentage of total	Spot welding	Spot welding percentage of total
1982	1152	32	2.8	249	21.6
1983	1753	103	5.9	349	19.9
1984	2623	199	7.6	471	18.0
1985	3208	294	9.2	511	15.9
1986	3683	348	9.5	534	14.5

saturation. Assembly, however, continues to be seen as a major growth area over the next few years due to improvements in robot design, increased speed, more powerful languages and use of advanced sensing techniques.

The growth in industrial robot application is perhaps more important than their existing distribution, as it gives an indication as to how rapidly they will impact on industry and society in the future. The values shown in Figure 2 are estimates based on data obtained from a number of sources. Despite a recent slow down the trend remains quite remarkable, and if the world economy

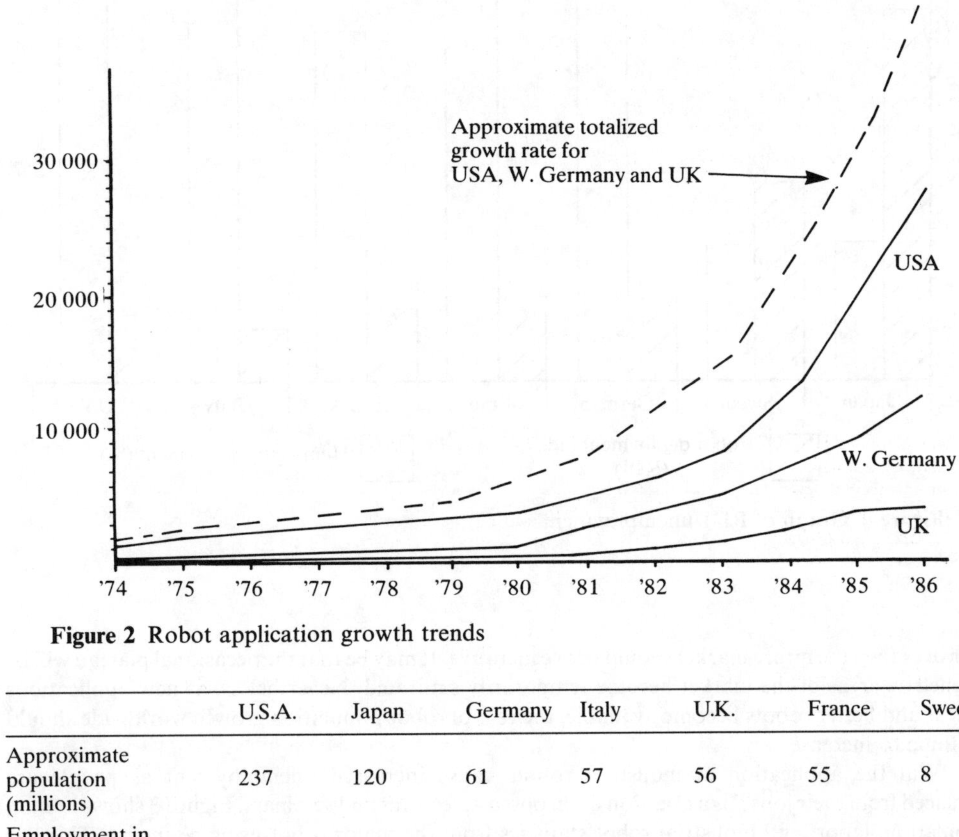

Figure 2 Robot application growth trends

	U.S.A.	Japan	Germany	Italy	U.K.	France	Sweden
Approximate population (millions)	237	120	61	57	56	55	8
Employment in manufacturing (millions)	19.2	12.3	7.8	5.7	5.4	4.8	1.0
Unemployment %	6.9	2.8	9.0	11.0	11.5	10.3	2.7
Industrial robot population (000's)	27.0	90.0	12.4	5.5	3.7	7.5	3.8
Industrial robots as percentage of manufacturing workforce	0.14	0.73	0.16	0.10	0.07	0.16	0.4
Robot deployment index	1.4	7.3	1.6	1.0	0.7	1.6	4.0

Figure 3 Human and robot statistics

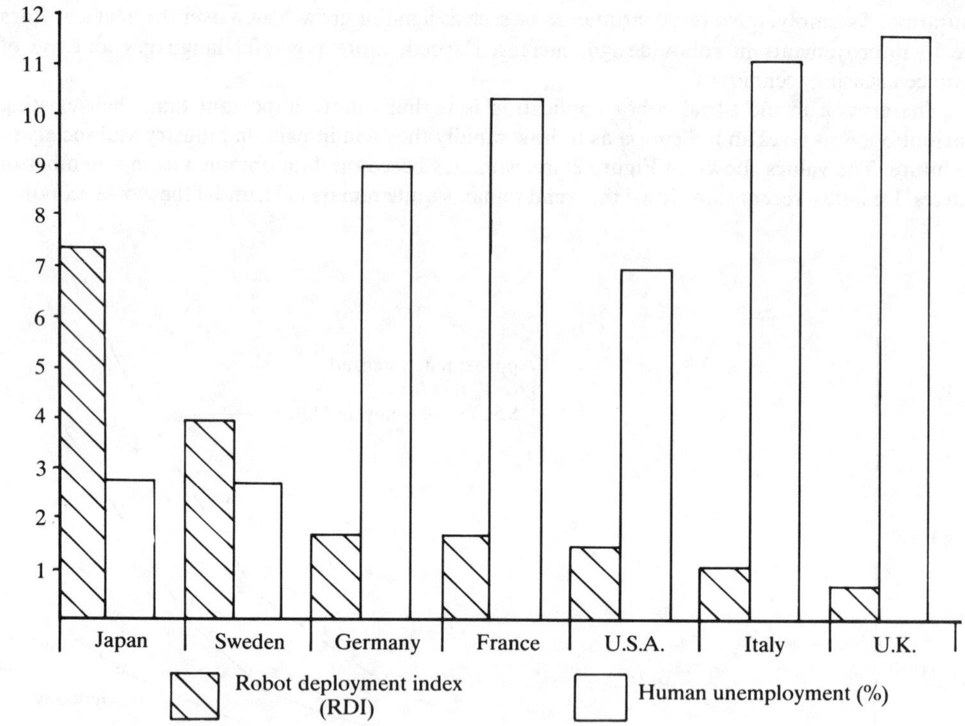

Figure 4 Graph of RDI/unemployment

improves then the future market should prove lucrative. It may be that the occasional plateau will be reached as areas of the market become temporarily saturated. Nevertheless, as new applications appear and better robots become available, the rate of robot population growth worldwide should continue to increase.

'Will the application of industrial robots cause increased unemployment as people are displaced from their jobs?' is a question often posed at lectures and seminars. Figure 3 shows human population, labor, and industrial robot statistics from the major robot-using countries. This has been compiled from data obtained from the *International Labour Office Bulletin of Labour Statistics*, the *OECD Labour Force Statistics*, and the various robot associations.

A 'robot deployment index' has been constructed by taking, for each country, the industrial robots in use as a percentage of the manufacturing workforce and multiplying by ten. Thus the higher the index, the stronger the deployment of robots. This is compared with the percentage human unemployment figures and the results are shown in Figure 4. There is apparently a negative correlation, i.e. the countries that employ a higher ratio of robots have lower unemployment levels. This at least shows that robots are not causing massive unemployment and in fact their application appears to be indicative of high levels of employment.

For up-to-date figures on industrial robot statistics the reader is advised to contact the robot associations shown below. Statistics are usually published early in the year following that concerned. Also the BRA, for example, provide regular newsletters, magazines, and other information to association members.

UK

British Robot Association (BRA)
Aston Science Park
Love Lane
Aston Triangle
Birmingham B7 4BJ (Tel: 021 359 0981)

USA

Robotic Industries Association (RIA)
900 Victors Way
PO Box 3724
Ann Arbor
MI 48106 (Tel: 313 994 6088)

FRANCE

Association Française de Robotique Industrielle (AFRI)
89 rue Falguière
Paris 75015 (Tel: 13 27 92 02)

GERMANY

Institute for Production Automation (IPA) Stuttgart
Nobelstrasse 12
D-7000 Stuttgart 80 (Tel: 0711 781602)

ITALY

Società Italiana per la Robotica Industriale (SIRI)
c/o ETAS Kompass PT
via Muova Rivoltana 95
20090 Limito (Tel: 02 92771)

JAPAN

Japan Industrial Robot Association (JIRA)
Kikai-Shinko Building
3-5-8 Shiba Koen
Minato-ku
Tokyo 105 (Tel: 03 434 2910)

SWEDEN

Swedish Industrial Robot Association (SWIRA)
Industrirobotgruppen inom Sveriges
Mekanforbund
Box 5506
Storgaten 19
S-114 85 Stockholm (Tel: 08 22 70 00)

Appendix 2

FURTHER READING

BOOKS

There is now an extremely large number of textbooks available on the subjects of robots, industrial robots, and robotics associated topics. General introductory books are not included here. New books are continually appearing and the reader should contact the publishers listed here to find out the latest publications.

Ayres and Miller. *Robotics Applications and Social Implications*. Cambridge, Mass.: Ballinger. 1983.

Bonney and Yong (editors). *Robot Safety*. Bedford, UK: IFS Publications. 1985.

Hartley, J. *FMS at Work*. Bedford, UK: IFS Publications. 1984.

Hollinghum, J. *Machine Vision*. Bedford, UK: IFS Publications. 1984.

Husband, T. (editor). *Education and Training for Robotics and Automation*. Bedford, UK: IFS Publications. 1986.

Morgan, C. *Robots: planning and implementation*. Bedford, UK; IFS Publications. 1984.

Nof, S. Y. (editor). *Handbook of Robotics*. New York: Wiley. 1985.

Owen, T. *Assembly with Robots*. London: Kogan Page. 1985.

Pham and Heginbotham (editors). *Robot Grippers*. Bedford, UK: IFS Publications. 1986.

Pugh, A. (editor). *Robot Sensors. Vol. 1 Vision*. Bedford, UK: IFS Publications. 1986.

Pugh, A. (editor). *Robot Sensors, Vol. 2 Tactile and non vision*. Bedford, UK: IFS Publications. 1986.

Ránky, P. *Computer Integrated Manufacturing*. Hemel Hempstead: Prentice Hall. 1986.

Rathmill, K. (editor). *Robotic Assembly*. Bedford, UK: IFS Publications. 1985.

Robot Technology Series. London: Kogan Page, 1983–87:
- Vol. 1 Modelling and Control
- Vol. 2 Interaction with the environment
- Vols 3A & 3B Teleoperations and robotics
- Vol. 4 Robot components and systems
- Vol. 5 Logic and programming
- Vol. 6 Decision and intelligence
- Vol. 7 Robot performance and evaluation and CAD
- Vol. 8 Index and bibliography

Scott, P. (editor). *The World Yearbook of Robots R&D*. London: Kogan Page. 1987.

Snyder, W. E. *Industrial Robots: computer interfacing and control*. Englewood Cliffs, NJ: Prentice Hall. 1985.

MAGAZINES AND JOURNALS

The following magazines and journals (in the English language) provide up-to-date information on the state of commercial robotics and R&D. The magazines will also contain advertisements by robot and robot system suppliers. Most magazines contain 'tear off' cards to facilitate contact with the advertisers. There are also many other magazines that regularly carry articles on robotics. These may be found in libraries listed under categories including: Automation, Control, Design, Electrical, Electronic, Mechanical, Production, Robotics, and Industrial Engineering.

UK

Automated Assembly. IFS Publications. (Quarterly)
Automation and Robotics Times. Gower. (Monthly)
BRA Newsletter. British Robot Association. (Quarterly)
CIM Review. IFS Publications. (Quarterly)
FMS. IFS Publications. (Quarterly)
The Industrial Robot. IFS Publications. (Quarterly)
The Production Engineer. Institute of Production Engineers. (Monthly)
Robotica. Cambridge University Press. (Quarterly)
Robot News International. IFS Publications. (Monthly)
Robotics and CIM. Pergamon Press. (Quarterly)
Sensor Review. IFS Publications. (Quarterly)

USA

IEEE Journal of Robotics and Automation. IEEE Robotics and Automation Council. (Quarterly)
Industrial Robots International. Technical Insights. (Twice monthly)
International Journal of Robotics Research. MIT Press. (Quarterly)
Journal of Robotic Systems. Wiley. (Quarterly)
Robotics Age. North American Technology. (Monthly)
Robotics Today. Society of Manufacturing Engineers. (Bimonthly)
Robotics World. Communication Channels. (Monthly)

The Netherlands

Robotics. North Holland. (Quarterly)

CONFERENCES

The latest editions of conference proceedings, like journals, provide up-to-date information on areas of application and advanced research. Due to the large number of these conferences, only five specifically concerned with robotics are listed here. (All are published by IFS Conferences Ltd.)

British Robot Association Annual Conference
International Conference on Advanced Robotics
International Conference on FMS
International Conference on Robot Vision and Sensory Controls
International Symposium on Industrial Robots

INDEX